T0319388

Software Reliability Techniques for Real-World Applications

Wiley Series in Quality & Reliability Engineering

Dr. Andre V. Kleyner
Series Editor

The Wiley Series in Quality & Reliability Engineering aims to provide a solid educational foundation for both practitioners and researchers in the Q&R field and to expand the reader's knowledge base to include the latest developments in this field. The series will provide a lasting and positive contribution to the teaching and practice of engineering. The series coverage will contain, but is not exclusive to,

- Statistical methods
- Physics of failure
- Reliability modeling
- Functional safety
- Six-sigma methods
- Lead-free electronics
- Warranty analysis/management
- Risk and safety analysis

Wiley Series in Quality & Reliability Engineering

Software Reliability Techniques for Real-World Applications
by Roger K. Youree
December 2022

System Reliability Assessment and Optimization: Methods and Applications
by Yan-Fu Li, Enrico Zio
April 2022

Design for Excellence in Electronics Manufacturing
Cheryl Tulkoff, Greg Caswell
April 2021

Design for Maintainability
by Louis J. Gullo (Editor), Jack Dixon (Editor)
March 2021

Reliability Culture: How Leaders can Create Organizations that Create Reliable Products
by Adam P. Bahret
February 2021

Lead-free Soldering Process Development and Reliability
by Jasbir Bath (Editor)
August 2020

Automotive System Safety: Critical Considerations for Engineering and Effective Management
Joseph D. Miller
February 2020

Prognostics and Health Management: A Practical Approach to Improving System Reliability Using Condition-Based Data
by Douglas Goodman, James P. Hofmeister, Ferenc Szidarovszky
April 2019

Improving Product Reliability and Software Quality: Strategies, Tools, Process and Implementation, 2nd Edition
Mark A. Levin, Ted T. Kalal, Jonathan Rodin
April 2019

Practical Applications of Bayesian Reliability
Yan Liu, Athula I. Abeyratne
April 2019

Dynamic System Reliability: Modeling and Analysis of Dynamic and Dependent Behaviors
Liudong Xing, Gregory Levitin, Chaonan Wang
March 2019

Reliability Engineering and Services
Tongdan Jin
March 2019

Design for Safety
by Louis J. Gullo, Jack Dixon
February 2018

Thermodynamic Degradation Science: Physics of Failure, Accelerated Testing, Fatigue and Reliability
by Alec Feinberg
October 2016

Next Generation HALT and HASS: Robust Design of Electronics and Systems
by Kirk A. Gray, John J. Paschkewitz
May 2016

Reliability and Risk Models: Setting Reliability Requirements, 2nd Edition
by Michael Todinov
November 2015

Applied Reliability Engineering and Risk Analysis: Probabilistic Models and Statistical Inference
by Ilia B. Frenkel (Editor), Alex Karagrigoriou (Editor), Anatoly Lisnianski (Editor), Andre V. Kleyner (Editor)
October 2013

Design for Reliability
by Dev G. Raheja (Editor), Louis J. Gullo (Editor)
July 2012

Effective FMEAs: Achieving Safe, Reliable, and Economical Products and Processes Using Failure Modes and Effects Analysis
by Carl Carlson
April 2012

Failure Analysis: A Practical Guide for Manufacturers of Electronic Components and Systems
by Marius Bazu, Titu Bajenescu
April 2011

Reliability Technology: Principles and Practice of Failure Prevention in Electronic Systems
by Norman Pascoe
April 2011

Improving Product Reliability: Strategies and Implementation
by Mark A. Levin, Ted T. Kalal
March 2003

Test Engineering: A Concise Guide to Cost-Effective Design, Development and Manufacture
by Patrick O'Connor
April 2001

Integrated Circuit Failure Analysis: A Guide to Preparation Techniques
by Friedrich Beck
January 1998

Measurement and Calibration Requirements for Quality Assurance to ISO 9000
by Alan S. Morris
October 1997

Electronic Component Reliability: Fundamentals, Modelling, Evaluation, and Assurance
by Finn Jensen
November 1995

Software Reliability Techniques for Real-World Applications

Roger K. Youree
Instrumental Sciences Incorporated
Huntsville, USA

Registered Offices
John Wiley & Sons, Inc., 111 River Street, Hoboken, NJ 07030, USA
John Wiley & Sons Ltd, The Atrium, Southern Gate, Chichester, West Sussex, PO19 8SQ, UK

Editorial Office
The Atrium, Southern Gate, Chichester, West Sussex, PO19 8SQ, UK

For details of our global editorial offices, customer services, and more information about Wiley products visit us at www.wiley.com.

Library of Congress Cataloging-in-Publication Data Applied for

Hardback ISBN: 9781119931829

Cover Design: Wiley
Cover Image: © Titima Ongkantong/Shutterstock

Set in 9.5/12.5pt STIXTwoText by Straive, Chennai, India
Printed and bound by CPI Group (UK) Ltd, Croydon, CR0 4YY

C9781119931829_231122

This book is dedicated to my wife, Susan.

Contents

Preface

Software reliability as a discipline started later than hardware reliability but has grown rapidly. Software reliability is an active area of research with new results, both theoretical and practical, being published regularly. As is often the case with a relatively young discipline, much of the material may seem unconnected, making it difficult to determine how to choose the right techniques for a given job and organization.

This book is a survey of techniques and approaches that can be used to produce reliable software in a cost- and schedule-efficient manner. It focuses on practical techniques and is tailored for practitioners and not academics. Software reliability is not in any one organization's domain, and this book takes a broad approach in that it considers all activities that affect the software, such as conceptual design and requirements development, even though they generally occur well before any coding takes place. Preventing or removing defects from these early activities will pay significant dividends.

The early chapters of this book are intended to provide an overall understanding of the nature of the problem, followed by more practical suggestions in later chapters. Chapter 2 covers some definitions and useful information about defects. Chapter 3 outlines an overall approach for developing reliable software, followed by Chapter 4, which describes different project phases or stages, how defects can enter the project in each phase, ways to mitigate these defects, and ways to monitor for defects applicable to the phase. Chapter 5 provides a summary and roadmap, along with some practical guidelines. The book concludes with Chapter 6 that gives more details on some of the techniques mentioned in Chapters 3 and 4.

Instrumental Sciences Incorporated
Huntsville, Alabama, USA

Roger K. Youree

Series Editor's Foreword by Dr. Andre Kleyner

The Wiley Series in Quality & Reliability Engineering aims to provide a solid educational foundation for researchers and practitioners in the field of quality and reliability engineering and to expand the knowledge base by including the latest developments in these disciplines.

The importance of quality and reliability to a system can hardly be disputed. Product failures in the field inevitably lead to losses in the form of repair cost, warranty claims, customer dissatisfaction, product recalls, loss of sale, and in extreme cases, loss of life.

Engineering systems are becoming increasingly complex with added functions and capabilities; however, the reliability requirements remain the same or are even growing more stringent. Increasing integration of hardware and software is making these systems even more complex and challenging to design. For example, in autonomous driving vehicles, software may play an even more important role than the hardware. All this brings ever-increasing attention to the topic of software quality and reliability.

The book you are about to read has been written by an expert and state-of-the-art practitioner in the field of software reliability. It covers a variety of topics critical to producing high-quality, malfunction-free software in a timely manner.

At present, despite its obvious importance, quality and reliability education is paradoxically lacking in today's engineering curriculum. Very few engineering schools offer degree programs or even a sufficient variety of courses in quality or reliability methods. The topics of reliability analysis, accelerated testing, reliability modeling and simulation, warranty data analysis, reliability growth programs, and other practical applications of reliability engineering receive very little coverage in today's engineering student curriculum. Therefore, the majority of the quality and reliability practitioners receive their professional training from colleagues, professional seminars, and professional publications. This book is intended to close some of these gaps and provide additional educational opportunities for a wide range of readers from graduate-level students to seasoned reliability professionals.

We are confident that this book as well as this entire book series will continue Wiley's tradition of excellence in technical publishing and provide a lasting and positive contribution to the teaching and practice of reliability and quality engineering.

Acronyms

There are many commonly used acronyms in software reliability, but some may have different meanings to different people. The following list of acronyms is used in this book:

AADL	Architecture Analysis and Design Language
ATAM	Architecture Tradeoff Analysis Method^{SM}
BDD	Behavior-Driven Development
BOM	Bill of Materials
CDP	Concept Development and Planning
CIL	Critical Items List
CONOPS	Concept of Operations
DC	Design and Coding
DOE	Design of Experiments
DRACAS	Defect Reporting, Analysis and Corrective Action System
DRB	Defect Review Board
DRE	Defect Removal Efficiency
EKSLOC	Effective thousand (Kilo) Source Lines of Code
FDSC	Failure Definition and Scoring Criteria
FEF	Fix Effectiveness Factor
FMEA	Failure Modes, Effects, and Analysis
FMECA	Failure Modes, Effects, and Criticality Analysis
FRACAS	Failure Reporting, Analysis and Corrective Action System
FRB	Failure Review Board
FTA	Fault Tree Analysis
IV&V	Integration Verification and Validation
LRU	Line Replaceable Unit
MOE	Measure of Effectiveness
MS	Management Strategy
MTBCF	Mean Time Between Critical Failures
MTBEFF	Mean Time Between Essential Function Failures
MTBF	Mean Time Between Failures
MTBSA	Mean Time Between System Aborts
MTSWR	Mean Time to SoftWare Restore
MTTF	Mean Time to Failure

NVA	Non-Value Added
ODC	Orthogonal Defect Classification
OM	Operation and Maintenance
OP	Operational Profile
PFMEA	Process Failure Modes Effects and Analysis
QA	Quality Assurance
QFD	Quality Function Deployment
RCA	Root Cause Analysis
RPN	Risk Priority Number
SDD	Software Design Document
SFMEA	Software Failure Modes Effects and Analysis
SFTA	Software Fault Tree Analysis
SLOC	Source Lines of Code
SPC	Statistical Process Control
SRE	Software Reliability Engineering
SRGP	Software Reliability Growth Plan
SRPP	Software Reliability Program Plan
SRS	Software Requirements Specification
SysML	Systems Modeling Language
TBA	To Be Added
TBD	To Be Determined
TBR	To Be Reviewed
TBS	To Be Supplied
TEMP	Test and Evaluation Master Plan
TDD	Test-Driven Development
UML	Unified Modeling Language
WBS	Work Breakdown Structure

Glossary

Defect: A defect is a problem that, if not corrected, could cause an application or product to either fail or to produce incorrect or unsatisfactory results.

Defect precursor: A defect precursor is an event that does not directly result in a defect being placed in the software but makes the introduction of a defect into the software more likely.

Error: An error is a human action that produces an incorrect result. Note that the word "error" is also a standard part of some software terms, such as "runtime errors" and "memory errors."

Essential function failure: An essential function failure is any incident or incorrect function that causes (or could cause) the loss of an essential function or the degradation of an essential function below a specified level. Essential functions are the minimum operational tasks that the system must perform to accomplish its mission or achieve acceptable customer satisfaction. A system abort (SA) is an essential function failure, but not all essential function failures are SAs.

Failure: There are two definitions of failure that are typically used:

1. A failure is the inability of a system or system component to perform a required function within specified limits.
2. A failure is the termination of the ability of a product to perform a required function or its inability to perform within previously specified limits.

Fault: Again, there are two definitions of fault that are typically used:

1. A fault is a defect in the software code that can be the cause of one or more failures.
2. A fault is a manifestation of an error in the software.

Nonessential function failure: A nonessential function failure is any incident or incorrect function that causes (or could cause) the loss of a nonessential function or the degradation of an essential function but not to an unacceptable level.

Operational profile: An operational profile (OP) is a set of relative frequencies (or probabilities) of occurrences of disjoint software operations during operational use.

Phase: A phase, or project phase, is a period in the life cycle of the project dedicated to a certain set of tasks and products. Other terms sometimes used are stages, increments, or sprints. Phases typically overlap.

Project: For this book, a project is defined as an organized undertaking to produce one or more products. We uses the term "project" rather than "program" to avoid confusion with the use of program to refer to a software program.

Project system: For purposes of this book, we define a project system to be the finished product, all of the intermediate products, tools, services, and documentation used to develop the finished product, and all of the processes used in the project. When the risk of confusion is small, "system" may be used in the place of "project system."

Root cause: The root cause of a defect, also called a primary cause, is the initial causal event or chain of events that results in a defect. The root cause is the fundamental reason for the defect and if corrected will prevent recurrence of these and similar defect occurrences.

System abort: A SA, sometimes known as a mission abort or operational mission failure, is an essential function failure that occurs during a mission or critical operations that prevents critical aspects of system performance. It usually results in terminating the mission or operations. A software crash is an example of a SA.

Software reliability: Software reliability is the probability that the software will not cause a system failure for a specified time period under specified conditions.

Software reliability engineering: Software reliability engineering (SRE) is defined in [1] as "the quantitative study of the operational behavior of software-based systems with respect to user requirements concerning reliability." It includes the following:

1. Software reliability prediction and estimation.
2. The use of attributes and metrics of the product design, development process, and operational environment to assess and improve software reliability.
3. The application of this knowledge to specify and guide design, development, testing, acquisition, use, and maintenance.

Software reliability estimation: There are two definitions of software reliability estimation in frequent use:

1. Reference [2] defines software reliability estimation as "The application of statistical techniques to observed failure data collected during system testing and operation to assess the reliability of the software."
2. Reference [1] defines software reliability estimation as the activity that "… determines *current* software reliability by applying statistical inference techniques to failure data obtained during system test or during system operation. This is a measure regarding the achieved reliability from the past until the current point."

Software reliability prediction: There are two definitions of software reliability prediction often used:

1) Reference [2] defines software reliability predictions as "A forecast or assessment of the reliability of the software based on parameters associated with the software product and its development environment."
2) Reference [1] defines software reliability predictions as the activity that "… determines *future* software reliability based on the available software metrics and measures."

Validation Validation of a product answers the question of whether the product meets the needs that prompted its creation.

Verification Verification of a product answers the question of whether the product satisfies its requirements.

References

1 Lyu M, editor. *Handbook of software reliability engineering*. Computer Society Press and McGraw-Hill Book Company, New York, 1996.

2 IEEE Standard 1633. *Recommended practice on software reliability*, 2017. Software Engineering Technical Committee of the IEEE Computer Society.

1

Introduction

Software is ubiquitous in today's world. It controls our home appliances, automobiles, phones, and many of our forms of entertainment. It increases our productivity at work, speeds our communications, and improves our medical care. It affects nearly every aspect of modern life. Software is also getting more complicated because of a number of reasons, such as an increase in the number and diversity of software applications, the more varied types of platforms for the software, and the increased reliance on other "third-party" software. Because of this, it is critical to produce reliable software. Software that fails often may mean that some entertainment application is not as entertaining as intended, or it could result in a life-or-death situation in a hospital or a mass transit system.

1.1 Description of the Problem

As mentioned above, software is everywhere and is becoming more and more complicated. It is largely "handmade" and subject to human errors. Also, most software contains, or at least interfaces with, software developed independently by other companies. As a result, software defects can be subtle and difficult to find, sometimes only manifesting themselves under very specific conditions. Unfortunately, when these conditions occur, the effects of a defect may be very serious, including loss of life. Even if lives do not depend on the software, litigations can seriously damage a company.

Software reliability tasks are often assigned to reliability engineering personnel. Many times, these people are more familiar with hardware reliability than they are with software reliability. Hardware reliability and software reliability are different, and hardware reliability engineers are frequently uncomfortable with software reliability.

There is more to the problem than just producing reliable software. There are budgets and schedules to meet. Whatever is done to produce reliable software must meet these constraints. Another consideration is the highly dynamic business environment typical of modern software products. Customer needs and wants are always changing, and if one company does not respond to them, another will.

Many times, software reliability is treated as an added task to be performed after the software has been developed and in the process of being tested. The importance of software reliability and the seriousness of the constraints that must be adhered to mean that there are often issues affecting software reliability that should be addressed early in the development

Software Reliability Techniques for Real-World Applications, First Edition. Roger K. Youree.
© 2023 John Wiley & Sons Ltd. Published 2023 by John Wiley & Sons Ltd.

process. Producing reliable software within the budget and schedule constraints requires embedding a software reliability mindset into the project from its start.

1.2   Implications for Software Reliability

Software reliability is ultimately about achieving customer satisfaction with a profitable product. This goal requires many things other than reliability, but it is unlikely to be achieved with a seriously defective product. The importance of software reliability, along with the complexity of the problem and the budget and schedule constraints inherent to the problem, means that a software reliability program should be planned and implemented early in the development effort and monitored and adjusted as needed. A company that does this successfully has a huge competitive advantage over a company that is unsuccessful at it.

Good software reliability practices are about doing things right the first time, and this effort starts at the beginning of the development effort. It is often said that doing a job right takes less time than doing it over, and this advice often holds. It is particularly applicable to software reliability given how difficult it can be to find and remove some types of software defects. Not all software defects are coding issues. Many are due to defects in products produced much earlier in the effort, and preventing or finding and removing them early before they become deeply embedded in downstream products can be very cost-effective and schedule-effective. Most people recognize the importance of software reliability for critical software, but many do not understand that good software reliability practices can reduce the cost of development and maintenance of the software. When properly planned and implemented, a software reliability program can significantly reduce the amount of rework required and rework costs money and can result in schedule impacts. One of the more obvious examples of reducing rework is with software testing. Software testing is expensive, and applying good software reliability techniques from early in the effort can mean significantly fewer faults found during software testing, resulting in less re-testing and shorter test cycles.

Choosing a good set of reliability techniques for a software project requires anticipating the types of defects and errors that are likely to occur in that project. However, our knowledge of the future is not perfect. It is said that in war, a general's plan for a battle never survives first contact with the enemy. Unfortunately, the same can often be said for plans for developing and supporting software. Things do not always go as planned, particularly in our highly dynamic and interdependent world. While starting the effort with a good set of software reliability techniques is important, monitoring results and then making appropriate changes are also a necessary part of the process. We live in a very dynamic world and need to get used to the fact that unexpected events will occur. We must continuously monitor and adapt while always trying to learn from events and see if we can do better next time. Managing for software reliability involves identifying and managing risks in an ever-changing environment.

There is no one set of techniques that is best for all software development efforts. The approach to software reliability should depend on the product, the software team, the company, and often the customer. This book therefore starts with a general understanding of defects that can affect software and what can be done about them and then progresses to more specific project areas. This book is also designed to be beneficial to a wide audience, such as software developers and software maintainers, producers and users of the software, and software for government and for commercial customers. More on the importance and implications of software reliability may be found in [1–3].

References

1 Lyu M. Software reliability engineering: A roadmap. *Future of software engineering*, pp. 153–170, IEEE Computer Society, 2007. https://www.researchgate.net/publication/4250863_Software_Reliability_Engineering _A_Roadmap. 22 Aug 2020.

2 Musa J. *Software reliability engineering: More reliability software faster and cheaper*. AutherHouse, 2004.

3 Neufelder A. *Ensuring software reliability*. Marcel Dekker, Inc., New York, 1993.

2

Understanding Defects

To prevent and control software defects, we need to understand them. This chapter explains the nature of software defects, including where they enter into the system, what effects they can have, how to detect them, and what causes them.

To reduce the number and impact of defects in our software, it is important to understand the nature of errors and defects. Almost any error on a project can affect the reliability of the software. Anything that makes it more difficult for project personnel to perform their tasks can negatively impact reliability, even if it does not directly result in placing a defect in the software code. A frustrated, angry, or confused programmer is more likely to make an error resulting in a software defect than a motivated, generally happy, and well-informed programmer. A poor work environment and a lack of good software development tools are examples of defect precursors. Defect precursors do not directly cause a software defect, but they make defects more likely and so are considerations for software reliability. Projects that produce high-quality software tend to be well-run projects. Not all errors or defect precursors result in defects, but reducing errors and precursors reduces the likelihood of defects. Similarly, not all defects produce software faults, and not all software faults result in software failures, but again, reducing them improves our chances of reliable software.

As we want to produce reliable software, our understanding of software defects needs to be tailored to that purpose. To this end, we consider the following:

1. Where defects enter the project system
2. Effects defects can have on the project system
3. How we can detect defects
4. What causes defects
5. How we can handle defects

The first four of these are addressed in Sections 2.1–2.4, while the fifth is covered in Chapter 3. Chapter 4 covers the material in more detail by addressing it for specific phases of a project.

2.1 Where Defects Enter the Project System

Knowing where defects can enter a project system is important because we can use this information to design mechanisms to prevent or detect them. When we think of software

Software Reliability Techniques for Real-World Applications, First Edition. Roger K. Youree.
© 2023 John Wiley & Sons Ltd. Published 2023 by John Wiley & Sons Ltd.

defects, we typically think of specific types of errors, such as typographical errors, logical errors, synchronization errors, resource errors, or interface errors, to name just a few, and the software defects that may result from them. These types of errors are obviously important, and we must be able to handle them; however, defects affecting the software can enter a system in almost any phase and through almost anything used to design or produce the software product. Processes and products in one phase are used by later phases to produce the final product, so defects in an early phase may propagate to the final product.

In Chapter 4, we describe six phases that are typical for a project. They are as follows:

1. Concept Development and Planning
2. Requirements and Interfaces
3. Design and Coding
4. Integration, Verification, and Validation
5. Product Production and Release
6. Operation and Maintenance

We also consider management impacts. All of these use processes and produce products that create opportunities to introduce defects. Examples of potential defect sources include a poor understanding of customer needs, imprecise requirements, and not following good configuration control processes. The first two examples are typically from the Concept Development and Planning phase and the Requirements and Interfaces phase, respectively, while the last example can be from any phase. It is also important to realize that defects can be introduced into software that has a low defect density, but these defects may have very serious consequences. Also, correcting a detected defect or adding a feature to mature software may introduce defects. Chapter 4 takes each of these phases and describes it, outlining what defects are typical for each phase and how they can enter the project system. It describes techniques and processes to mitigate these defects and lists some metrics to help monitor progress in each phase.

2.2   Effects of Defects

Software defects manifest themselves in many ways, and understanding this helps us produce more reliable software. Of course, a defect may never manifest itself. For example, if the defective part of the code is never executed, the defect never causes a fault or failure. As we generally try not to write unused code, we will assume that defects have some likelihood of being executed.

We commonly think of software defects as causing software crashes, infinite loops, or incorrect software results. Crashes and infinite loop tend to be readily visible. Incorrect results may be obvious or may be subtle. Other types of defects, such as memory leaks, may manifest themselves even more subtly. Software defects, or "bugs," are sometimes classified into two types:

1. Mandelbugs: A mandelbug is a software defect whose activation and subsequent behavior is complex and its behavior appears chaotic. An example of a mandelbug is a type of defect jovially referred to as a "heisenbug." Heisenbugs are altered by the

attempts to find them. They may be affected by the timing of the execution, by the memory addresses used, by having debugging tools connected to the system, or any of a large number of other factors. Once introduced into the software, heisenbugs, and mandelbugs in general, can be notoriously difficult to find.

2. Bohrbugs: A bohrbug is a software defect whose behavior is repeatable and predictable. Although the cause of the incorrect behavior may be unknown, they are repeatable if the right conditions are found and applied.

Knowing about these various types of defects helps us plan, carry out, and analyze software tests. However, the possible existence of these subtle and hard-to-find defects is one of the reasons why we should not rely solely on software testing to detect defects. It also adds emphasis to the fact that software testing can only show the existence of defects in software, not the absence of defects. Ultimately, it supports the idea that we need to put an emphasis of defect prevention.

If the only defects that we consider are defects in the software, we are missing opportunities to prevent defects from being introduced into the project system. As previously mentioned, almost any error or defect can increase the likelihood of software defects. For example, a poorly worded requirement may be interpreted differently by different software developers. If two developers are writing different software modules affected by this requirement, the different interpretations may mean that these modules do not work together correctly. Furthermore, the effects may be subtle and difficult to find, meaning that the most cost-effective and schedule-effective way to deal with the defect is by ensuring that the requirements are as clear and precise as possible.

Finally, not all defect effects are equally important. Defects that never manifest themselves are less important than defects that cause critical failures. Improving the reliability of software involves focusing on the defects that are most likely to occur and also on the defects that have the most serious consequences if they do occur.

2.3 Detection of Defects

An effective and efficient software reliability effort requires well-thought-out defect detection and monitoring. Good defect detection and monitoring should:

1. Find errors and defects early when it is most cost-effective and schedule-effective to correct them.
2. Be as complete as practical, finding a high percentage of the errors and defects, and finding them in all processes and products that can significantly affect the software product.
3. Be reliable by not missing too many errors and defects while also not creating too many false alarms and the ensuing unproductive effort.
4. Be cost-efficient and schedule-efficient to perform.

Good defect detection and monitoring should also add confidence in the software and related products. It should provide evidence that it is working, and project personnel should be able to trust the detection and monitoring processes and execution enough that the results can be used as a part of the final sign-off of the software.

Recognizing defect precursors is critical for preventing and removing defects efficiently. For example, knowing that a software defect may be due to a requirement defect informs us that we need to detect requirement issues and therefore institute appropriate processes for doing this. Process and product monitoring is important at each phase of the project, and Chapter 4 covers each in more detail.

There are many ways to identify an error, defect precursor, or defect. Some ways identify weaknesses or problems with the processes that produce a product and others identify issues with a product. Techniques to detect process defects and weakness include the following:

1. Use a process failure modes effects and analysis (FMEA)/failure modes, effects, and criticality analysis (FMECA).
2. Use process reviews, inspections, and independent assessors.
3. Use error brainstorming sessions. Those responsible for a task brainstorm on what errors could occur while performing the task. The list can be used to develop checklists for the errors, and the brainstorming process sensitizes the task performers to the errors.
4. Use a software reliability advocate to continually assess project processes for potential software reliability impacts.
5. Perform a premortem on the process to anticipate process defects.

Some techniques to detect defects in products are as follows:

1. Use product peer reviews, inspections, and independent assessors.
2. As with process defects, we can use error brainstorming and premortem sessions for the product.
3. Perform tests of code.
4. Use checklists of process steps to ensure that each step is followed when producing the product.
5. As with process defects, use a Software Reliability Advocate to continually assess project products for potential software reliability impacts.
6. Use a software reliability casebook to assess if all processes are correctly followed and if not to push for corrections and improvements.
7. Use requirements traceability analysis of a specification as a means of detecting potential requirement defects.
8. Also for requirements, let several people independently assess what would constitute verification of a specific requirement. Make the assessments specific enough that if certain criteria are met, the requirement passes, and if they are not met, it fails. Failure to agree on these criteria indicates the potential for confusion and for an inconsistent use of the requirement.

While detection of defects is important, we ultimately want to anticipate the chain of events that results in a defect and use this information to prevent the defect. Ideally, we prevent the first precursor, but realistically, we should also monitor for most if not all of the known precursors in the chain. We should also use "triggers." These are indicators that additional action is required for a monitored event. These triggers may at times be subjective, but early intervention increases the likelihood that a problem will be contained and will not spread damage to later phases of the project where it is increasing difficult to handle. Chapter 4 lists metrics and monitoring activities applicable to each phase.

Finally, the project should continuously assess how effective its defect detection processes are and always try to improve them. Avoid change for the sake of change as project changes can be disruptive. However, monitor the effectiveness of the detection and be willing to change a process if there is reason to believe that it will make a significant improvement.

The next section considers causes of defects. Knowing defect causes helps us prevent and remove defects. It also enables us to monitor events that trigger the creation of defects and therefore potentially detect defects earlier. For example, a defect may be caused by not following the processes used to create requirements, and not following a process may be caused by inadequate training. This information tells us that we should use skilled requirements developers or institute adequate training for requirements development and that we should also monitor training completions and adequacy.

2.4 Causes of Defects

To prevent or eliminate a defect, it is important to know the causes of the defect. Knowing defect causes helps us predict them and reduce their likelihood as well as to more efficiently manage resources. This strategy is analogous to the use of "Physics of Failure" techniques for hardware reliability. Defects usually have a causal chain, a sequence of events that ultimately results in the given defect. In this chain of causes, it may be that only a few of the causes are readily detectable. To choose the best place and approach to correct the problem, we need to understand this chain. It is also important to know that there may be more than one causal chain for a given defect, i.e. the confluence of two or more such chains results in the defect.

Consistent with the idea of causal chains, we distinguish between primary and secondary causes of defects. For purposes of this book, a primary cause of a defect is a root cause of the defect. Successfully addressing a primary cause not only addresses the specific instance of the defect in question but also prevents other similar defects from occurring and therefore improves the running of the entire project. Addressing a secondary cause may remove the current defect and may in some cases prevent other similar defects, but it does not address the more fundamental cause of the problem and therefore risks problem reoccurrence.

Examples of secondary causes include inadequate project objectives, unclear requirements, and excessively complex software code. Each of these causes provides useful information but is not the root cause of the defect. For each, we can constructively ask for additional information. For example, a requirement may be unclear causing unintended behavior from the resulting software code. The requirement can be clarified, and the code can be changed to address the clarified version of the requirement, thereby eliminating the defect. However, we need to ask if there is a way to prevent or reduce the likelihood of unclear requirements. We should ask what caused the unclear requirement and how we can improve the way that we produce requirements. Secondary causes are useful for helping us detect and analyze defects. They are covered for specific project phases in Chapter 4, but we need to understand defects at a deeper level to more effectively prevent or remove them.

Root cause analysis is the process of finding the primary cause of a defect or problem. At a high level, root cause analysis usually follows steps similar to the following taken from [1]:

1. Identify the problem.
2. Determine the significance of the problem.
3. Identify the causes (conditions or actions) directly preceding and surrounding the problem.
4. Identify the reasons why the causes in the previous step exist and work backward to the root cause.

A critical part of this analysis is to systematically work our way back to the root cause, and there are various techniques that can be used in this process. Several are listed below:

1. Five whys
2. Fault tree analysis (FTA)
3. Fishbone diagrams (cause/effect or Ishikawa diagrams)
4. Scatter plots and correlation analysis: These can be used to determine if two factors correlate with one another and aid in finding a causal relation.
5. FMEA/FMECA
6. Event and causal factor analysis
7. Barrier analysis
8. Change analysis
9. Human performance evaluation

See the topic on root cause analysis in Section 6.5 for more on these and other root cause analysis techniques.

In finding root causes, it can be useful knowing the categories applicable to most defects. Although the following list is not necessarily complete, most defects in software production or monitoring can be traced to these high-level issues:

1. Not producing or monitoring the right things: For example, we may have a software project with a significant number of interfaces, but we are not producing any interface documentation to specify them.
2. Poor processes for producing or monitoring a product or process: An example of this type of issue is having a product release process that does not ensure correct configuration control of the product, potentially resulting in the wrong product being released.
3. Not following the processes: This issue could occur if the process for creating software code is adequate, but because of schedule pressures and staffing issues, certain steps are not performed.
4. Following the process or monitoring poorly: With this issue, we use the procedure but do it poorly or intermittently. An example of this issue is having an adequate process for creating software code but using an inexperienced software developer who is unfamiliar with the process or is unable to follow the steps properly.
5. Non-human factors: The first four of these categories of failures are largely due to human errors, and humans make mistakes in spite of excellent processes, resources, ability, and training. However, some errors cannot reasonably be attributed to human error. For example, externally imposed constraints may make defects more likely. A sudden change in legal requirements may require a project change that negatively impacts software

reliability. Errors from this type of situation may appear at almost any time in any process or product.

As stated above, the first four of these categories of failures are largely due to human errors. Causes of human errors include the following:

1. Insufficient knowledge: This type of error is due to one or more task performers not knowing or not having access to relevant information.
2. Cognitive failure: A cognitive error occurs when a task performer is unable to correctly process the required task information.
3. Lack of needed skills: As the name indicates, this error is due to a task performer not having the correct skill set to perform the task.
4. Attention failure: Attention failures occur because of carelessness or loss of focus and a task that otherwise would be performed correctly is adversely affected.
5. Overload: Overload is due to too much work or too much multitasking.
6. Contradictory tasks: Sometimes, a task performer is assigned tasks or conditions that cannot all be satisfied, such as writing software code for contradictory requirements.
7. Lack of motivation: Lack of motivation is typically due to a lack of interest or a "bad attitude" and can be exacerbated by a poor work environment.
8. Misunderstanding: Poor communication between two or more employees can result in misunderstandings.

Using our example of a defect because of an unclear requirement, suppose that we have traced the original problem to an unclear requirement, giving us a secondary cause. With further analysis, we find that the requirement is unclear because there was a misunderstanding of who was responsible for the requirement and a "placeholder" was put into the specification until the issue was resolved. The issue was forgotten about because the requirement had no identification as being a placeholder. At this point, we need to know if this cause of confusion is an isolated incident or more systemic. If it is systemic, we may have other major issues to address. We also need to address how a "placeholder" mistakenly became a requirement and why it was forgotten. These mistakes could be due to the requirements processes not addressing placeholders, or someone not following the procedure, or perhaps other issues. With further analysis, we find that the process does address the placeholder situation, and the person who performed it incorrectly was temporarily assigned to the project to relieve a budget issue on a different project and had not been trained for the task. This information enables us to direct our efforts toward the root cause of the problem. For example, we could add training for temporarily assigned personnel, or if this is impractical because of schedule constraints, to add additional monitoring of products produced by these personnel.

Causes often suggest possible mitigations that may prevent or reduce the likelihood that such defects will occur in the future. For example, if a defect is caused by someone not having the right skill set, the person could be trained or supported by another employee with a stronger skill set or moved to tasks better suited to the employee's current skills. Non-human errors need to be addressed on a case-by-case basis. Chapter 3 covers mitigation of defects at a high level, while Chapter 4 details mitigation techniques and processes in more detail. Chapter 29 of [2] contains more information on human errors and reliability. See [3] and [4] for more details on the causes of software defects.

As a final note, knowing the causes of defects enables us to better predict defects and plan ways to avoid creating them. In Chapter 3, we cover planning the steps and processes needed to achieve our software reliability objectives within the given resources. Part of this plan is to create a list of what can go wrong and using this list to institute ways of preventing or detecting these problems. A sound knowledge of potential root causes can therefore help us prevent defects in a timely and cost-effective manner.

References

1 DOE. DOE-NE-STD-1004-92, root cause analysis, 1992. Available via http://everyspec .com/DOE/DOE-PUBS/DOE_NE_STD_1004_92_262/. Accessed 22 Aug 2020.
2 H. Pham, editor. *Handbook of reliability engineering.* Springer-Verlag, London, 2003.
3 Neufelder A. *Ensuring software reliability.* Marcel Dekker, Inc., New York, 1993.
4 Musa J. *Software reliability engineering: More reliability software faster and cheaper.* AutherHouse, 2004.

3

Handling Defects

To produce reliable software under cost and schedule constraints, we need to carefully plan our project activities and ensure that the plan is implementable by the team put together to do the tasks. This chapter outlines how to develop an overall strategy for software reliability. It then covers the nature of our software reliability objectives and provides details on how to plan the project to build reliability into the software with each project activity. We also discuss how to make the plan implementable. Finally, we discuss analogies between hardware reliability and software reliability engineering (SRE). As most practitioners are more familiar with hardware reliability, it is hoped that these analogies will help them better understand and more effectively implement software reliability practices.

3.1   Strategy for Handling Defects

In Chapter 2, we learn about errors, defect precursors, and defects, and in this chapter, we use this information to construct processes to handle these defects and to produce reliable software. To handle defects, we use four complementary approaches:

1. Prevent errors and defects by anticipating likely causes and providing mitigations.
2. Remove defects by monitoring and detecting defects and errors, preferably early when removal is more cost- and schedule-effective.
3. Design the system to be fault tolerant to reduce the impact of defects that are in the system.
4. Forecast defects and faults to manage project resources and to gain confidence in the reliability of the product.

Producing highly reliable software within project constraints requires clear goals, careful planning, and good execution. A standard overall process for achieving almost any goal is as shown in the accompanying Figure 3.1. This process is closely related to the "Plan–Do–Check–Act" Deming cycle. In more detail, this overall or high-level process consists of the following steps:

1. Determine objectives: Decide on reasonable objectives for software reliability consistent with the needs of the customer and the project resources.

Software Reliability Techniques for Real-World Applications, First Edition. Roger K. Youree.
© 2023 John Wiley & Sons Ltd. Published 2023 by John Wiley & Sons Ltd.

Figure 3.1 Overall Process.

2. Plan: Determine the steps and processes needed to achieve these objectives within the given resources.
3. Implementation and monitoring: Decide how to perform the plan and what the signs of success and of trouble are.
4. Feedback: Decide when feedback indicates that changes are needed, and if so indicated, determine what changes are appropriate and when to make them.

The next sections consider each of these steps in more detail.

3.2 Objectives

At a high level, a project wants to produce a profitable product with a high level of customer satisfaction. Both customer satisfaction and product profitability relate to software reliability. Not only is high software reliability important, we also need to have some level of assurance of its reliability. Expanding on this, we consider two main objectives for software reliability along with typical sub-objectives for each:

1. Objective 1: Create a highly reliability software product on schedule and within budget constraints. This objective can be further broken down into the following:
   (a) Sub-objective 1a: Prevent defects from entering into the product.
   (b) Sub-objective 1b: If a defect is in the product, design the product to perform adequately in spite of the defect.
   (c) Sub-objective 1c: If a defect is introduced into the product, find and remove it as soon and as economically as possible.
2. Objective 2: Know with a high level of assurance that the software is sufficiently reliable. The sub-objectives include the following:
   (a) Sub-objective 2a: Determine metrics and criteria for assessing the reliability of the software product.
   (b) Sub-objective 2b: Design methods to collect and analyze the information required to make the assessment.
   (c) Sub-objective 2c: Monitor the product, processes, and implementation of the processes to determine if the software is at risk of not being sufficiently reliable.

Ultimately, we want a satisfied customer; therefore, customer inputs and feedback throughout the project, but particularly with project objectives, can prove highly beneficial. Also, determining objectives is typically an iterative process. As the project progresses, objectives should be made more precise. Ideally, we have quantitative objectives that can be monitored and used to indicate when we are on track and when corrective actions are needed. However, we should consider our objectives carefully. Our objectives guide our plan and therefore our implementation of the plan. These can suffer if the objectives are not clear, motivated, and well accepted.

3.3   Plan

After determining our objectives, we need a plan to coordinate the efforts used to achieve them. This plan should provide guidance on the following:

1. How to prevent or reduce the impact of each type of anticipated error or defect affecting the software.
2. How to monitor for errors and defects, anticipated or not, in products, in process compliance, and in process effectiveness.
3. How to determine when the monitoring should trigger some form of action and what that action should be.
4. When and how to perform root cause analysis and how to use the analysis results and other monitoring information to make changes to the products, processes, and implementation of the processes.

The typical steps for designing the software reliability activities for a project are as follows:

1. Step 1: List steps that the project will perform to produce the software product.
2. Step 2: List what can go wrong in each of these steps.
3. Step 3: List how we can prevent these defects and errors or at least significantly reduce their likelihood and impact.
4. Step 4: List ways that we can quickly know if something goes wrong, i.e. list what monitoring is needed.
5. Step 5: List when the information from the monitoring indicates that we should do something different and what it should be.
6. Step 6: List how we will know if our processes and corrective actions are effective, and if they are not, list what we should do. We need to know how confident we can justifiably be in our product.

We elaborate on these steps below. It is important to note however that our plan and how we implement it are dependent on our objectives, our staff, and resources to implement the plan and the nature of the project. For example, we should consider the chosen software development process, such as spiral, incremental, or cleanroom software development when planning these activities. The results of following these steps can then form the basis of the Software Reliability Program Plan (SRPP).

Step 1: List steps to produce the product: To effectively reduce the impact of software defects, we need to identify where defects can be created, and this means understanding the processes and products used to produce and maintain the software. As a result, this first step requires that we create a list of project phases and the products that each produces. The list should also include the processes used to produce the products and where these processes are documented. If there are important processes that are not documented, the plan should note this and encourage the project to suitably document them.

Section 2.1 addresses this need at a high level. Chapter 4 provides more details, covering six typical phases of a project and the products each produces. It also covers the impacts that project management has on software reliability. These can all be used as guidance for producing the project-specific list required to complete this step.

For this step:

1. List the phases of the project.
2. For each phase, list the products produced.
3. For each product, list the processes used and where they are documented.

This part of the plan is complete when the project has a project-applicable and approved list of project phases and the products and processes applicable for each phase.

Step 2: List what can go wrong: The next step in producing our plan starts with the list from step 1, and for each project phase and its products and processes, we anticipate and list the types of errors, defect precursors, and defects that we are most likely to encounter so that we can plan for them. We also should understand the implications of these issues so that we can prioritize our efforts. Although not essential, it is advantageous to assign a likelihood and severity to each error or defect. By doing so, we can more easily allocate resources to potential issues. Assigning a likelihood and severity to each identifiable defect precursors is typically impractical, but a list of top candidates can be helpful. Assigned likelihoods and severities may be subjective and qualitative, especially early in the project, but they need to be solid enough that we can trust using them for resource allocations.

A useful process for predicting defects is to review historical failure reports and their root cause analyses. Unless significant differences have been made since these reports were written, they should provide a representative view of the types of defects to expect. In addition, Section 2.2 gives a high-level overview of the effects that defects can have. Chapter 4 details some of the things that can go wrong in each of the project phases and with its associated products. These sections can help produce the list of errors and defects for the project, but ultimately the list is project specific.

As the project progresses, we learn more about what can go wrong. Additionally, aspects of the project may change during this progression. It is therefore important to update the list of what can go wrong throughout the life of the project.
For this step:

1. Start with the list produced in step 1 of the phases of the project and the processes and products for each phase.
2. For each phase and each process and product of that phase, list potential errors and defects and notable defect precursors.
3. For each error, defect precursor, and defects listed, assign a priority.

Be sure to review and update this list regularly and as needed. This part of the plan is complete when the project has a project-applicable and approved list of prioritized issues that cover the reasonably foreseen defects and errors affecting the software product. The list should address each product and process in each phase found in the list created in step 1.

Step 3: List recommended techniques for preventing and tolerating defects: Once we have our list of prioritized errors, defect precursors, and defects, we need to choose mitigation techniques appropriate to both these issues and the project resources. We need to construct processes for preventing or reducing the likelihood of these issues and do so at as many issue entry locations as is useful and practical. Ideally, our plan will include mitigation techniques for each type of issue that has been identified. If we have prioritized defects by assigning likelihoods and severities, then we should at least include mitigation techniques for the more likely and most severe errors, defect precursors, and defects.

We listed three sub-objectives to objective 1. These defect mitigation approaches are "high-level strategies" (see, for example, [1]) often used with software:

1. Fault prevention
2. Fault tolerance
3. Fault removal

The first two of these, fault prevention and fault tolerance, are covered below and in Chapter 4. The third, fault removal, is addressed with step 5 below, as is fault forecasting. Fault forecasting also plays a major role in step 6, developing confidence in the software product.

Fault prevention: Fault prevention is the avoidance by construction of the occurrences of faults. For fault prevention, we consider all of the processes used in the production of the software and attempt to ensure that none of them result in a software defect. Fault prevention should be considered to be the primary way that software reliability is achieved. If a fault is never introduced into the project, then it does not need to be detected or removed. Successful fault prevention also prevents the "ripple effect" where the cause of one defect results in multiple additional defects, or the removal of a defect introduces one or more additional defects into the system. For example, a poorly worded requirement may be misunderstood, resulting in (for example) an incorrect software architecture design, resulting in an architecture that does not support an aspect of the executable software, and this lack of support may produce a fault while running the software under certain conditions. As the original defect may have gone undetected until the software is executed under possibly unusual conditions, correcting the problem requires changing multiple products. If the requirement defect had been prevented, multiple defects would have been prevented.

There are a variety of fault prevention techniques, such as formal methods and software reuse; however, we can prevent most faults by preventing typical errors and defects, and much of this prevention involves reducing human errors. In Section 2.4, we listed several causes of human errors. We repeat the list and add a number of suggestions to mitigate these errors:

1. Insufficient knowledge: Ensure that each employee is adequately trained and has a sufficient background for their assigned tasks. In addition, when someone is assigned a task, that person should be introduced to the types of defects someone performing the task is likely to encounter. This education should be performed early to make the performer sensitive to these defects. Performer feedback is also useful to both ensure that the concepts are understood and improve the process.
2. Cognitive failure: Design the work environment and processes with the employees in mind. Ensure that tasks are clearly defined and understood and that they do not require extensive multitasking. Make sure that employees are assigned tasks appropriate to their abilities and that lines of communication are clear.
3. Lack of needed skills: Hire highly skilled employees. Ensure that each employee is well trained for the tasks required. Rehearse tasks for future execution.

4. Attention failure: Stay organized. Keep a work environment that helps employees focus. Avoid multitasking. Allow for work breaks. Use checklists and other techniques to bring conscious attention to routine tasks. Keep the list of project-critical objectives short, consistent, and as stable as possible. Use task redundancies to catch errors, such as cross-checking data for data reliability.

5. Overload, fatigue: Match the workload and schedule to the work force. Try to keep tasks relatively short and manageable. Once a task is assigned, make a major effort to avoid changing it as "task churn" is a major cause of frustration and overload. Keep a constant focus on the few critical objectives.

6. Contradictory tasks: Have clear lines of communication. Make sure that everyone has an understanding of how their tasks affect everyone else's tasks. Make "internal contracts" that specify what a person or team will provide to another person or team, when it will be provided, and what its characteristics will be. Check the products of each phase for defects and contradictions. For example, check for contradictory requirements before passing the requirements to the next phase.

7. Lack of motivation: Understand what motivates (and de-motivates) the types of employees used on the project and make significant efforts to follow through and keep employees motivated. For example, adequate pay and a sense of value and appreciation are basic factors for employee morale and motivation. Also, make sure that each employee has a clear understanding of his or her role and why that role is important and valued.

8. Misunderstanding: Have clear communication and feedback to ensure that tasks and their objectives are understood. Use "internal contracts" so that people and teams know what to expect from other people on the project. Make sure that task descriptions are clear and complete.

It is difficult to successfully apply all of these remedies and the list is not complete. Much of its benefit can be gained by using the fact that to perform tasks well and largely error-free, employees need to be motivated and also:

1. Know what they are doing
2. Know why they are doing it, including understanding any constraints that may be required for the task
3. Know how to do it
4. Know when it needs to be done
5. Know what success looks like
6. Know who needs it and what they expect

This knowledge is dynamic, sometimes changing on a daily basis, but making sure that each employee knows this information is important for preventing errors and defects and therefore faults and failures. Having and following good processes helps, but errors can happen when a situation occurs that the writers of the process did not anticipate. Know the purpose of each step of the process and have a process for deviating from the process.

Fault tolerance: Fault tolerance is the ability to perform adequately in spite of the occurrence of faults. Fault tolerance usually involves either fault recovery or fault compensation. Fault recovery is where a fault-free state is substituted for an erroneous state. Fault compensation uses enough redundancy that the software continues to perform adequately in spite of the fault, such as the use of redundant software modules, each module being different in some way but each performing the same or similar tasks. Fault tolerance may allow fault-free performance in spite of a fault, or it may just allow for graceful performance degradation without a system crash or some other more serious response to the fault. A common fault tolerance technique is to check input conditions and follow preplanned steps if the input is expected to trigger a fault condition. Fault tolerance may use defensive programming to check inputs and outputs for illegal operations. It may use exception handling for unsuccessful operations to confine the defect. It may use checkpointing and rollback mechanisms to recover software operations from faulty conditions. Fault tolerance is a technique in Section 6.3 below. Also see [1, 2], and [3], as well as Chapter 12 of [4], Chapters 33 and 34 of [5], Chapter 11 of [6], and Chapter 14 of [7] for more on fault tolerance.

At this point, we have a list of defects that are important to us and have a better understanding of how fault prevention and fault tolerance can help us. However, no two projects are the same, and techniques that work for one may not be the most effective techniques for another. Chapter 6 lists many techniques that have been useful to many projects, and Chapter 4 helps determine which techniques apply in which project phases, but we are still left with the task of choosing which specific techniques to use for a specific project.

To make these choices, start with the list from step 2. This list contains the potential errors and defects that most concern us. Then,

1. Cross-match these defects and errors with techniques that can be expected to effectively prevent, tolerate, or find and remove them.
2. Identify any constraints and assumptions.
3. Create a list of criteria for selecting techniques. This list should emphasize the unique needs of the project, but as a reference, 4.1.5.1 of [8] lists the following criteria:
   (a) Improves the total quality of the product
   (b) Reduces the time to field the product
   (c) Is cost-effective
   (d) Reduces safety and environmental risks
   (e) Has the potential to achieve a "quantum improvement" in the product
   (f) Can be used within current resource constraints
   (g) Is proven
   (h) Is efficient and easily executed
   (i) Is directly focused on the product under development
   (j) Has management buy-in for use
   (k) Is perceived as "best-in-class"
   (l) Is planned for use based on previous experience
   (m) Is quantifiable
   (n) Can be easily introduced to those who must apply it
   (o) Is simple to apply
4. Choose and apply a decision process to determine the "best" technique or techniques for each potential error or defect of concern. Note that ideally, a single technique will cover multiple errors and defects. See the Decision Making Techniques topic in Section 6.5 for several potential processes.

5. Perform a sensitivity analysis to determine if the decision is repeatable. Make small changes to weightings and scores to see if the results change. Also consider using other decision making techniques to see if they give the same results.
6. Document the results and how the results are derived. This documentation should as a minimum include the requirements, assumptions, decision criteria with their priorities, and enough study details to reproduce the study results.

For this step,

1. Start with the list the phases of the project, processes, and products for each phase and potential errors and defects for each. This list is produced in step 2.
2. For each potential error or defect, list one or more processes or techniques that are likely to prevent or reduce the likelihood of the error or defect. Also, list approaches that make the process or product more tolerant of the error or defect.

This defect prevention and tolerance part of the plan is complete when each prioritized error, defect precursor, or defect from the list created in step 2 has been considered, and one or more prevention or tolerance techniques have been incorporated into the project addressing the item or an approved rationale for not addressing it has been provided.

Step 4: List what to monitor: Done correctly, implementing error, defect precursor, and defect mitigation techniques goes a long way toward producing reliable software, but if we stop there, we have an open-ended approach that does not adjust to changing situations. It also does not account for the possibility that we may not have correctly anticipated all of the most critical issues. The next step in developing our plan is to determine what to monitor and how to do the monitoring.

Recall that our objectives are to create highly reliable software and to know that it is reliable, all within our cost and schedule constraints. What we monitor and how we monitor it should be determined by these objectives. To develop the monitoring portion of our plan, we can use the following steps:

1. List what can go wrong (step 4.1): Use the list created from the third step to make a preliminary list of processes and products to monitor and the errors, defect precursors, and defects of each to monitor. This list also includes the prevention and tolerance techniques chosen for these issues as things can go wrong with these as well. With each error, defect precursor, and defect in this list (or at least the higher priority ones), list ways to monitor for the occurrence of the issue. We are interested in three high-level categories of items to monitor:
   (a) Errors and defects in the products
   (b) Process compliance
   (c) Process effectiveness
   Errors and defect precursors can create defects, and a defect in almost any product at any phase of the project can make a fault more likely. (If there is a task to produce some intermediate product and defects in this product have no impact on the quality of the final product, the project should consider whether this intermediate product is needed.) One common way that errors are made is by a failure to comply with the established processes. Another common way is to use incorrect or inadequate processes. A project

should monitor for all of these possibilities and also be mindful that what was a good process yesterday may be inadequate today. Also, particular emphasis should be placed on collecting enough information to find root causes of defects and errors. After completing this step (step 4.1), we have added potential items to monitor and associated monitoring techniques to the list from step 3.

2. List what we can practically monitor (step 4.2): Some things are very difficult to measure, and attempts to do so result in subjective or misleading information. After developing a list of ways to monitor the errors and defects, we need to determine if any of these monitoring procedures are impractical. If the monitoring for a particular defect is impractical but the information sought is particularly useful, look for multiple sources or monitoring techniques for it and if possible, design feedback mechanisms to check the results. After completing step 4.2, we have modified our list from step 4.1 to only include practical monitoring.

3. List what we can do with the resulting information (step 4.3): Even if we have good information about something that affects the reliability of the software product, we need to do something with that information. What we monitor is partially determined by what information will be most beneficial for improving the product, and this determination requires knowledge of how we will use the information. For example, we should determine if we are monitoring enough of the right things to have a reasonable likelihood of determining the root cause or causes of a defect. In this step, we consider the information collected from the monitoring listed in step 4.2 and list how we will use it. If the benefits are not large relative to the cost, remove it from the list and document the reason for removing it. Step 5 below covers how we can use the monitored information in more detail and consideration should be given to performing steps 4 and 5 together.

4. Check for good monitoring characteristics (step 4.4): After creating a list of products and processes to monitor and how we plan to monitor them, we should go through a checklist of the characteristics of good monitoring from Section 2.3. In the abbreviated form, we should ensure that our monitoring:
   (a) Finds errors and defects early.
   (b) Is complete, finding a high percentage of the errors and defects and doing so from all products and processes that can significantly affect the software product.
   (c) Has a low likelihood of missing too many defects or of creating too many false alarms.
   (d) Is cost and schedule efficient to perform.

Some monitoring applies to each phase and some are phase dependent. Examples of monitoring applicable to each phase include the following:

1. Feedback from the task performers.
2. Feedback from internal customers. Internal customers are project personnel who use the work of other project personnel.
3. Feedback from the product user or customer when appropriate.
4. Reviews and inspections of the products and processes.
5. Monitor that the tasks of each phase are on schedule with expected quality. Delays and poorer-than-expected quality may be signs of project trouble.

Some suggestions for applying these to the causes of human errors listed in Section 2.4 are as follows:

1. Insufficient knowledge: Know what qualifications each person should have based on their assigned tasks and monitor that each person is adequately trained or has shown sufficient competency for their tasks. Ensure that each person has been introduced to the types of defects someone performing these tasks is likely to encounter, and as this information is dynamic, monitor that this education is updated accordingly. Use surveys and employee feedback to ensure that the objectives and tasks are understood.
2. Cognitive failure: Use reviews and inspections of the processes and work environment to look for clear lines of communication and clarity of tasks and objectives. Also, use surveys and feedback from the task performers.
3. Lack of needed skills: Monitor hiring to ensure that those hired for tasks are those best qualified. Monitor training for effectiveness. Use employee surveys and feedback. Track employee turnover for loss of skilled employees.
4. Attention failure: Monitor the number of tasks assigned to each parson and if possible the complexity of each. Monitor task churn. Monitor the number of objectives and the amount of objectives churn. Compare these with schedule constraints applicable to those assigned the tasks as a high density of tasks per time period and particularly a high density of task churn per time period can increase attention failures. Use reviews and inspections of processes to determine if ways to enhance employee attention to the tasks are needed or if task redundancies or checklists need to be added.
5. Overload: Monitor assessed workloads and update the assessments based on the actual performance. Use employee surveys and feedback. Monitor task churn.
6. Contradictory tasks: Use reviews and inspections of the processes to ensure that process tasks are unlikely to conflict. Use employee surveys and feedback. Review assigned tasks for conflicts. Review objectives for conflicts as conflicting objectives will likely result in conflicting tasks. Monitor task descriptions to make sure that they are described clearly and completely to avoid misunderstandings that might result in task contradictions.
7. Lack of motivation: Use employee surveys and feedback. Monitor the employee benefits and work environments offered by competitors.
8. Misunderstanding: Monitor for clear communication and feedback. Use employee surveys and feedback.

Chapter 4 lists potential things to monitor for each phase of the project.

We tend to improve the things that we measure. If practical, monitor quantitatively, but make sure that differences in values are meaningful. Monitoring usually has "noise" in it, and so robust monitoring and feedback is needed. Finally, there are costs associated with monitoring, so it is important to have a purpose for each such item. Know the advantages and potential issues of each to avoid unnecessary effort. Understanding why something is monitored can also be important when there are changes to the project and to external situations that affect it. These changes may alter the meaning or significance of some of the monitoring results.

For this step:

1. Start with the list produced in step 3.
2. Based on this list, make a preliminary list of processes and products to monitor with an initial list of what to monitor for each. This step is step 4.1 above.

3. Modify this list based on what is practical to monitor (step 4.2 above).
4. Modify this list based on what we can reasonably do with the information obtained from monitoring (step 4.3 above).
5. Modify this list again to ensure that we are adequately considering good monitoring characteristics (step 4.4 above).

This part of the plan is complete when we have a list of products and processes to be monitored and how and when each is to be monitored. The list should include rationale for the decisions.

Step 5: List how to use the monitoring results: Monitoring is important, but we also need to use the information as feedback to positively affect the overall project. In this step, we consider two categories of uses. The first is fault removal, or more specifically fault, defect, defect precursor, and error removal. The second is forecasting. It should be noted that the results of the monitoring may also be used to indicate potential locations for process improvement. A process may not be defective but can still be improved. See the process improvement techniques entry in Section 6.5 for more information on this application.

Fault removal: Fault removal is the process of detecting and eliminating faults. This part of the plan requires that we take the results of the monitoring and find the root cause or causes of faults, defects, defect precursor, or errors and use this information to determine if we should make changes to the product or to the project processes. If changes are warranted, we decide what changes to make and when and how to make them. In the context of SRE, this process usually refers to testing the software, finding faults, isolating the causes of the faults, and removing them. Given the importance that other types of defects have to software reliability, this section deals with removal of defects in general.

Typically, the first step in removing a defect is detecting it. Faults in software code are usually found by testing. Traditionally, testing software occurs during the Design and Coding phase when designers perform unit tests on their code before release to the independent test group and in the Integration, Verification, and Validation phase when the code is gradually integrated, tested, further integrated, tested more, and so on to provide assurance that the product is ready for release. Sections 4.4 and 4.5 and Chapter 6 have more on tests to find software faults, including different types of tests, test plans, and test procedures. Detection of defects is covered more generally in Section 2.3 and above in step 4 of this section. Sections on the other phases discuss defect detection specific to the phase covered.

Defect and error source removal needs to be performed through a systematic and well-controlled process. Traditionally, fault removal for hardware and software faults employs a failure reporting, analysis and corrective action system (FRACAS) controlled by a Failure Review Board (FRB). The same type of systematic process should be used for the removal of the different types of defects and error sources that are created across the different phases of the project, so we refer to the system as a defect reporting, analysis and corrective action system (DRACAS) controlled by a Defect Review Board (DRB).

A project may use a single defect removal system and board or employee different systems and boards for different types of defects. The same basic process applies for defect removal in the various products and processes of the project as it does for the software itself. For example, requirement defects may go to a dedicated board responsible for seeing to it that requirement defects are identified, analyzed to root cause, assessed for corrective actions, assigned an approved corrective action if applicable, and tracked to closure. The board should also ensure that the information is recorded for statistics to use on this and other projects. These boards may have overlapping personnel and should be coordinated. For example, if a software fault is found to be due to a requirement defect, this defect is sent through the requirement defect process to find the root cause, ensure closure, and to add to the requirement defect statistics. If different systems are used, they need to share data, and an oversight board should be used to handle disagreements between the boards.

Finally, a few notes about defect removal in software are as follows:

1. The decision of when to remove a detected defect in code is usually dependent on the maturity of the code. If the code is still the responsibility of the programmer, the defect may be removed immediately. If the code is in the hands of a software test group, defects are typically removed as a block of changes after the current testing is complete. Unless the defect is critical, defects in released code are usually removed when there is a new code release.
2. Be sure to plan defect removals. If defect removal is delayed for too long, a defect build-up occurs, and there may be too many defects to remove and verify in a very short period of time. The result is an increased likelihood of defects being released into the finished product.
3. Decide when regression testing is needed and allow time and resources for it. Like any type of software test, regression testing requires people, time, and facilities.
4. Retain access to the developers of the code. Removal of defects can also introduce new defects, and if there are aspects of the code that are not fully understood, the likelihood of such an introduction increases.

Fault Forecasting: Fault forecasting is the process of estimating the existence of faults and the occurrences and consequences of the associated failures. For SRE, fault forecasting typically means estimating faults in the software product so that they can be removed or their effects mitigated. It is usually the most high-profile part of SRE and results in values for software reliability, availability, mean time between failures (MTBFs), and other metrics. Forecasting also helps a project plan and manage its resources.

Fault forecasting may be performed either before the development of software code or after the code has been produced. Predictions before having the code require information specific to the reliability model used but typically use information such as the estimated number of lines of code, type of code, and how the software will be used. The predictions are usually based on historical data. Results from this forecasting are used to allocate requirements and to plan and allocate resources, among other things. After executable code has been produced, fault estimations can be developed based on statistical analysis of repeated runs of the code. The results of these runs can be used to estimate the number of faults in the software and to predict reliability growth. Both pre-code and post-code forecasts are important. Pre-code predictions improve planning and resource use, possibly preventing serious disruptions to the project. Estimations using test runs are usually more accurate but require realistic test runs as the results can be heavily dependent on the operational profile. Reliability testing can also be time-consuming and expensive and so needs to be well planned.

The concept of forecasting can be extended to error and defect forecasting and therefore be instituted for products and processes in each phase of the project. Forecasting defects in a product enables us to predict the current and future "goodness" of the product. This information helps us allocate resources and determine when the product is finished. Defect forecasting for processes enables us to design better processes and continuously improve them. As a general rule, try to measure and forecast everything that is important to making the software reliable, keeping in mind that some things are difficult to measure accurately and even more difficult to create useful forecasts for.

As with forecasting software faults, forecasting defects can be performed before creating the product or by using data collected from defects found from the product or process. One strategy that can be used before creating the product is to perform a failure modes, effects, and criticality analysis (FMECA)-like analysis to identify and prioritize potential product defects. This strategy does not directly quantify the expected number of defects, but it does provide guidance for resource allocation. It also provides information that can be used for developing a defect forecast of the product. A second strategy is to determine the quantifiable aspects of the product, such as the expected number of requirements, and use these values as inputs to a prediction model. Such models usually rely on historical data.

For post-product forecasting, we can measure aspects of the product and use these quantities to predict the number of defects. Steps for developing forecasts may include the following:

1. Determine relevant quantifiable characteristics of the product to measure, what units to measure these characteristics in, and develop a database of these measurements.
2. Identify the current coverage of these quantities in appropriate units. For example, for requirements assessments, we might have a coverage of 120 out of 500 requirements.
3. Apply weights to the units when appropriate.
4. Assess the defects per unit found, weighted appropriately.
5. Based on the current coverage, weights, and defects found, project to the total expected defects for the item. This projection may depend on time and an expected rate of coverage as time progresses.

Forecasts should have the potential to positively affect the reliability of the software. Even forecasts that are not highly accurate can potentially provide an early warning of potential problems by indicating a trend in the wrong direction for a given quantity. One way to use forecasts is to develop expected curves, such as reliability growth curves, that are designed so that if the measured quantity follows the curve, the future value of the quantity is likely to adequately support the software reliability goals of the project. We can then set thresholds, or "triggers," that tell us when action is required based on the forecasted results. This approach enables an earlier and therefore more cost- and schedule-efficient response to potential issues. However, if it is improperly applied, it may result in too many false alarms.

Summarizing Step 5:

1. Start with the list of monitored items from step 4. If step 5 is done in conjunction with step 4, then start with step 4.3.
2. Decide if the project will use a single DRACAS and DRB or several. Note that the number may change during the life of the project.
3. If several are used, determine which monitored items go to which DRACAS.

4. Schedule when each system starts operating and when it ends. Some DRACASs and DRBs may be in place and be active for the duration of the project, while others may form and be active only for the duration of specific activities. However, there should be some DRACAS and DRB at some level of formality throughout the project.

5. Provide a high-level plan for each DRACAS and DRB. This plan should identify and describe the inputs to the DRACAS, what computer system implements it (or at least its characteristics), and the basic steps to be used in implementing DRACAS. The plan should also list the members of the DRB, their roles, and responsibilities and provide a schedule (such as weekly versus daily meetings) for the board meetings. This schedule may change as the project progresses.

6. For each monitored item, decide how to forecast the results. Determine if forecasting will be performed before development of the product or implementation of the process, after, or both. Provide a rationale for any monitored item that is not forecast.

7. Determine how to use the forecasted results, i.e. how the forecasted results will be used to impact the project.

This part of the plan is complete when each monitored item from the list from step 4 has been assigned to a DRACAS and DRB and each DRACAS and DRB has been planned. One or more forecasting methods should also be associated with each monitored item, along with how the forecasts will be used.

Step 6: Determine if we can have justifiable confidence in our software product: The last major piece of the plan is to provide means to determine if we can justifiably have confidence in the reliability of our software product.

The most basic means of confidence justification is a statistical estimation of one or more quantities measured from runs of the software, especially if the software is run on representative hardware and under conditions representative of the ways that the end users will use the software. Quantitative values often used to address this step are the reliability, availability, MTBF, mean time to failure (MTTF), and failure rate of the software. Software testing provides the most representative replication of the real-world application of the software, especially if representatives of the user community form one of the groups used to run the software. Software tests also have the advantage of providing a quantitative score for the complete software package, including the hand-generated, autogenerated legacy, and third-party code and how well these types of code are integrated.

Testing is rarely exhaustive, so it can be expected to miss defects. It can also be expensive and time-consuming, further limiting how inclusive the testing is. Finally, some defects are difficult to detect by testing but are easier to detect by other means. As a result, testing should be supplemented with other confidence-building techniques. Examples of other such confidence-building techniques to consider are listed below:

1. One important way to increase the confidence in the software reliability is to demonstrate that the software reliability plan has been followed, as is shown by implementing a software reliability casebook. More on software reliability casebooks may be found in Section 6.4. Related considerations include the following:

   (a) Show that one or more checklists have been completed. Such checklists may indicate that processes have been followed or that certain standards are likely met.

(b) Documentation of completion of various peer reviews and other validations of work products adds to our confidence in the software product.

(c) Checks for version control are recommended.

2. Formal methods can be used on the product, on parts of the product, or on intermediate products. Formal methods can be used to check software designs through model verification and software code through code verification. These methods can catch defects missed by peer reviews and software tests. More on formal methods may be found in Section 6.4.

3. Software quality control charts can be used to show adherence to predetermined threshold values and that there is no excessive product or process variability. More on software quality control charts may be found in Section 6.5.

4. Qualitative or pass/fail tests and checks may be used to supplement the measures of confidence in the product. These can, for example, be used on processes or intermediate products.

The following summarizes step 6:

1. Determine what measures of software reliability are to be used for justifying software reliability confidence. This determination should be based on the software reliability objectives and take advantage of customer feedback.

2. Determine how confidence in the software reliability measures is to be supported.

(a) Decide what testing, analysis, and other techniques will be used to justify the confidence in the final software product. Ideally, these techniques will be complementary, with strengths of one technique compensating for the weaknesses of another.

(b) Determine what software testing is required to demonstrate software reliability to the desired confidence level. Ensure that the testing is properly planned, including what constitutes adequate resources, staffing, and schedule.

3. Plan for a reliability casebook to show that the SRPP is followed and to document the results. If a casebook is not used, determine how compliance with the SRPP will be monitored and controlled.

4. If formal methods are to be used, plan for them. Determine what tools are required, what training is needed, what parts of the software will take advantage of these methods, and when and where the methods will be implemented.

This part of the plan is complete when measures to adequately justify confidence in the reliability of the software have been determined and planned for.

As with the final product, this plan is subject to monitoring and changes based on the results of this monitoring. For example, early in the project, we may suspect that a particular software reliability measurement will be the primary source of confidence in the product, but we may not have a required value for it until requirements are developed. It may also be the case that as the project progresses, we need to add another measurement, such as software availability, to ensure confidence. Neither of these examples represent ideal situations but projects rarely have ideal situations.

Figure 3.2 is a high-level diagram of designing and running a project. The first box, "Understand Project Needs," is covered in step 1 of our list of six steps. In step 1, we list the

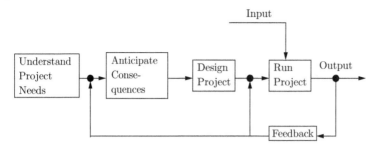

Figure 3.2 Designing and Running a Project.

steps that the project will need to perform to produce the product. This step is essentially making a model of the project that we can use to anticipate project events and plan for them. Steps 2 through 5, listing what can go wrong, how we can prevent these undesired events, ways that we can know that something is (or may soon be) wrong, and what to do with this information, are all about anticipating consequences based on the model from step 1. This anticipation, along with step 6, adding techniques to increase confidence, are used to design the project.

Next, we run the project. There are numerous inputs to the running of the project, some expected and some not. There are also various outputs from each phase of the project. Some of these outputs are metrics used to monitor the project and provide feedback. Sometimes, the feedback prompts us to alter inputs to a process, but the process is still adequate for the task. Other times, the feedback requires that we change parts of the project and its processes that are not working correctly. The same applies for the results of step 6, where we consider techniques to see if we can justifiably have confidence in the reliability of the product. If these techniques indicate that we should not have as much confidence as we would like, this information can be used as feedback to the current processes or potentially to change the processes.

Figure 3.2 can be applied at various levels. Not only does it apply to running the entire project, but it also applies to individual processes in a project. We first seek to understand the process by listing the basic steps required, then proceed with steps 2 through 5 to anticipate consequences and make plans for them. We then add step 6 to validate the results of the process. Anticipation is not perfect, so while using the process, we monitor for indications that additional actions may be needed to obtain the results that we want.

3.4 Implementation, Monitoring, and Feedback

Having a plan is not enough. The plan must be implementable by the responsible organizations, and poor implementation is just as likely to cause project failure as a poor plan. A good plan for one organization may not work for a different one.

One aspect of implementing a plan is setting up the processes that the project will use. Good processes are an important part of an efficient and smooth running project. They ensure a certain amount of consistency of performance and of product and therefore reduce uncertainty in the product. Ideally, they incorporate various lessons learned, enabling the project to continuously do better. Good processes should be:

1. Purpose directed: Each project process should have a definite purpose that is well identified and is needed. If a process does not have a needed purpose, it is probably superfluous.
2. Clear and understandable: A good process makes sense and people can understand it.
3. Something that people can follow: Not only does a good process make sense, but it is something that people can actually do.
4. Accepted: While it is important that people can follow a process, it is also important that people will accept the process so that they will follow it correctly. If people do not accept the process, it is more likely that they will cut corners with it and just "go through the motions."
5. Controllable and consistent: We want each process to produce the same results each time. To do this, the tasks involved must be within our control.
6. Effective: The process should do what it is designed to do and do it well. This characteristic may seem obvious, but it is critical.
7. Efficient: Not only should the process be effective and therefore "do the right things," it should do them with a minimum of wasted time and other resources.
8. Simple: All other things being equal, the more simple the process, the better.
9. Monitorable and measurable: A good process can be monitored and aspects of it measured so that we know if it is working correctly, and if it is not, then we can quickly determine that we have a problem. We can also use this information to fix problems and to continuously improve the process.
10. Adaptable and robust: The project environment is constantly changing, and a good process is able to handle these changes, either because the process already considers these situations or because the process can be readily adapted to fit them.
11. Sustainable: Some processes are stressful and can be performed for short bursts but not over an extended time period. If a process is to be used for an extended period, it must be designed so that people can do it regularly and without excessive stress. For example, if overtime is required each time the process is used, then it is probably not a sustainable process.
12. Documented: Each process should be well documented so that it is clearly defined and means the same thing to everyone. It should be clear who is involved and what each individual does and when each task is done. Any special cases should be identified, and it needs to be made clear when each special case is invoked. The documentation should clearly answer all reasonable questions about the process.

Processes are needed, but they are not our product. We should control our processes and not let them control us. If the number of project personnel whose main job is managing a process becomes too large relative to the number of personnel devoted to producing a product, it may be time to reconsider the project's processes.

Over-reliance and "blind" following of processes can result in inflexibility, preventing people from making good decisions. Processes obviously have a purpose and are needed, but they can over-control people. This is particularly the case when unusual situations occur that were not considered by the developers of the process. These unusual situations may be potential problems or potential opportunities. The project should allow for these types of situations by allowing controlled process deviations. Good employees are required to make this work. Hire good employees for the tasks required and give them enough freedom to do their jobs in a dynamic environment. Never think that a simple process only needs "simple" people. Processes should not be an excuse for not hiring quality people.

Another potential problem to consider when choosing or developing project processes is "process fads." Just because a famous company uses a particular process does not mean that it is right for every project. Processes are tools. It is sometimes said that if the only tool you have is a hammer, then everything looks like a nail. This caution applies to processes as well as to hammers. A process should fit a need and adapt to it rather than choosing a "highly recommended" process and trying to adapt the project to the process.

A final note on processes is that processes should be thought of as a collection and not in isolation. The collection of processes should work together supporting each other and ultimately supporting the project goals and objectives.

Implementation is more than just setting up processes and making assignments and due dates. The people, work environment, and implementation approaches must be coordinated. Implementation should be planned with the long-term goals and objectives in mind. Key implementation considerations include the following:

1. Balance planning and acting. A lot of time and effort can be wasted without a good plan, but ultimately, it is the actions taken that matter. Avoid "paralysis by analysis." The plan will never be perfect, and the longer you wait to implement a good (but imperfect) plan, the more opportunities will pass you by.

2. The right people should be tasked. For each critical task, and ideally for all tasks, determine the qualifications needed to properly perform the task. If the "right" person for a task is not available, make the best choice possible and consider added support and training. Ultimately, getting the right people and putting them in the right place is more important than the processes or organization. The right people, given the opportunity, will find out what processes work.

3. Determine the critical milestones and their schedules. Try to maintain flexibility, but critical milestones often drive much of the implementation approach.

4. Ensure that sufficient resources are allocated to do the tasks. As with choosing the right people, determine what resources are needed for tasks and when the resources will be needed.

5. Set up project lines of communication. Initially, the plan and the assigned tasks need to be communicated in an understandable manner. As the project progresses, status and changes need to be communicated in a timely manner to all affected personnel. Everyone on the project should have a clear and current understanding of all matters pertinent to their work and have enough of an understanding of the whole project to feel an ownership of the overall success of it.

6. The people tasked should be given a feeling of ownership of their tasks. They should know their roles and why these roles are important. They should also have the authority required to successfully perform their tasks and have accountability for their results.

7. Implement software reliability processes as soon as feasible. Even a few good software reliability techniques, properly applied, can make a big difference. Early implementation also helps to instill a "quality mindset" in project personnel.

8. Some decisions are dependent on the data available. For example, if the project does not have the historical data to use a particular model, a different technique must be chosen. However, the project can collect the required data for later use, possibly by a different project.

9. Implementation needs to be monitored and changes made if warranted. Changes may be required if the current approach is not working well, if there is evidence that a significant improvement can be made, or if there is a change in the project that affects the SRPP or its assumptions. Expect changes. Plans are imperfect and situations are dynamic. Let people challenge the processes and assumptions.

10. While changes to a project are often necessary, try to avoid making too many changes at once. Changes can be disruptive and a project can only absorb so much before performance deteriorates. It is a balancing act.

11. Attitude is important. Project leaders should exhibit enthusiasm and optimism and encourage these attitudes in others. It should not be a fake attitude. Sometimes, there are serious situations that require serious concern, but there should always be the assumption that the project can find a solution and be successful.

12. As much as possible, be prepared for the unexpected. It is rare when everything goes as originally planned for long, so keep implementation as flexible and "robust" as possible. Try to make the project adaptable to change with enough margins to handle unexpected events.

Throughout implementation, we need to monitor not just the product and processes used but also adherence to the plan and processes, the effectiveness of the plan and processes, and sometimes the pertinence of the objectives. We also need to monitor the state of the project personnel so as to build and keep a good work environment. Keep in mind that monitoring is required but is of minimal use without feedback. Feedback is a critical part of SRE. Feedback may affect the product, the implementation of the plan, or the plan itself. In rarer instances, it may also affect the objectives. Note that monitoring and feedback require that SRE personnel wear two hats. First, through monitoring, SRE personnel serve as unbiased assessors of the product, processes, and plan. The second hat is to serve as a resource and subject matter expert for using the results of the monitoring to ultimately improve the reliability of the software.

3.5 Analogies Between Hardware and Software Reliability Engineering

Most reliability engineers are more familiar with hardware reliability behavior and techniques than those for software reliability. Software has different failure modes than hardware and is produced using different processes. Hardware reliability engineers are usually concerned primarily with infant mortality failures and wear out failures. Software reliability engineers are mostly concerned with defects present in the delivered software, although software ages and software and its environment can change over time, potentially increasing the likelihood of failures in a manner analogous to wear out failures in hardware.

In spite of their differences, hardware reliability engineering and software reliability engineering have many similarities. Earlier, we listed four approaches used by SRE for handling defects. These are to prevent the defects and errors, remove them, design the software to be fault tolerant, and forecast defects and faults. Hardware reliability engineers use the same approaches for hardware. The following list gives hardware reliability examples of each:

1. Prevent: Based on reliability allocations, hardware design may choose highly reliable parts and line replaceable units (LRUs). Also hardware design chooses parts and LRUs that are designed for the environments that the system is expected to operate in. Physics of failure techniques are another hardware example of this approach.
2. Remove: FMEA and FMECA are used to find places to improve the design. Sneak circuit analysis (SCA) may be used to find design flaws in circuits that can cause unwanted behavior. Accelerated life testing (ALT) and highly accelerated life testing (HALT) are used to find parts of the design that are most likely to fail in the field. A FRACAS is used to find the root causes of hardware faults that are found as the project is designing, integrating, and testing the system.
3. Tolerate: Hardware designs often use redundant hardware parts and systems so that multiple failures of these aspects of the design are required to cause a reliability failure.
4. Forecast: Just as with software reliability, hardware reliability allocations and predictions are made and are checked against reliability estimations made from data collected during reliability testing.

Similarly, SRE uses many of the same approaches that hardware reliability engineers use. We mentioned physics of failure above. Physics of failure is a popular technique used by hardware reliability engineers. The technique begins by understanding the failure mechanisms of the materials and processes used by the hardware of the system and uses this understanding to address failure root causes early in the design process. From a software perspective, this process of understanding the failure mechanisms and using this understanding to address failure root causes early in the design process is covered in detail in Chapters 2 and 3.

There are also similarities between hardware maintainability and testability and software versions of these characteristics. Hardware designs try to make hardware LRUs easy to access and easy to remove and replace, particularly for the LRUs expected to fail most often. Making software maintainable involves, among other things, modularizing the software so that the parts of the software that are most likely to fail can be removed and replaced easily, ideally without affecting the rest of the software. Hardware designs also try to add self-tests to hardware LRUs and produce system-level tests to check for hardware failures and ideally to isolate to the failed LRU. For the same reason, software designs should design the software to be easy to test. It is often a good idea to log certain aspects of the performance of the software and look for specific anomalies, such as creeping memory requirements or slowing performance.

There are many other examples. Some of these examples are direct applications where the same method and approach is used for both hardware and for software, and some are analogies. Some examples of direct applications are as follows:

1. Reliability allocation
2. Reliability modeling

3. Reliability prediction
4. Reliability estimation
5. Reliability growth
6. Vendor control
7. FMECA
8. FRACAS, DRACAS

Other examples are less obvious. In the following list, we consider some techniques and approaches used for hardware reliability and then note how a similar, or at least analogous, approach can be used for software:

1. Choose reliable hardware parts: Prevent defects from being introduced into the deliverable software by using techniques such as pair programming, behavior-driven development, cleanroom methodology, or any of a number of other techniques.
2. Hardware redundancy: Software fault tolerance techniques, and for processes, peer reviews, and inspections.
3. SCA: Use code inspections and reviews to find unwanted software behaviors.
4. Rapid maintenance of hardware: Design for software maintenance, starting with the conceptual design and requirements.
5. Hardware testability: Software logging, software monitoring.
6. Hardware preventive maintenance: Software rejuvenation.
7. Reliability and maintainability design guidelines: Software best practices.
8. HALT, ALT: Robustness testing, bug bounties, and exploratory testing.

Many of the hardware and software approaches and techniques are similar or analogous because both hardware and software reliability activities aim to prevent field failures of products and to maintain the products so that they remain reliable. While many of these similarities are more analogous than direct, they show that strategies used in one discipline may be adaptable to the other, potentially resulting in a better overall reliability program.

References

1 Lyu M, editor. *Handbook of software reliability engineering*. Computer Society Press and McGraw-Hill Book Company, New York, 1996.
2 Pullum L. *Software fault tolerance: Techniques and implementations*. Artech House, Boston, MA, 2001.
3 Torres-Pomales W. *Software fault tolerance: A tutorial*. Technical report, Langley research center, Virginia, 2000. Available via https://ntrs.nasa.gov/citations/20000120144. Accessed 22 Aug 2020.
4 Kapur P.K., Pham H, Gupta A, Jha P.C. *Software reliability assessment with OP applications*. Springer-Verlag, London, 2011.
5 Pham H, editor. *Handbook of reliability engineering*. Springer-Verlag, London, 2003.
6 Pham H, editor. *System software reliability*. Springer-Verlag, London, 2006.

7 Lyu M. Software reliability engineering: A roadmap. *Future of software engineering 2007*, pages 153–170, 2007. Available via https://www.researchgate.net/publication/ 4250863_Software_Reliability_Engineering _A_Roadmap. Accessed 22 Aug 2020.

8 SAE International. *Reliability program standard implementation guide*, 1999. SAE JA-1000-1.

4

Project Phases

We have been considering a project as a whole, but now we consider specific tasks by the phase of the project in which they are typically performed. By doing this, it is hoped that it will be easier to recognize how the material covered in the previous chapters can be applied to a given project.

4.1 Introduction to Project Phases

Projects are typically segmented into sequences of activities that we will call phases, although others use terms such as stages, increments, or sprints. There are a variety of ways to divide a project into phases. For example, Chapter 14 of [1] defines the software life cycle as consisting of five consecutive phases: the analysis, design, coding, testing, and operation phases, whereas Chapter 32 of the same source lists the following: defining the user requirements, defining the software requirements, defining the architectural design, detailed design and production of code, transfer of the software to operations, and finally operations and maintenance. For our purposes, we use the following phase divisions:

1. Concept Development and Planning
2. Requirements and Interfaces
3. Design and Coding
4. Integration, Verification, and Validation
5. Product Production and Release
6. Operation and Maintenance

We also include a section on project management. This list of phases is not meant to imply any particular software development process or project management approach, and it is recognized that there is usually a considerable overlap in these phases. However, dividing the project into phases provides a way to organize the tasks required to produce and support the software. These phases may be sequential, overlapping, or simultaneous.

It is important to address each phase when preventing and removing defects. Table 4.1 comes from combining data from [2] and [3]. It lists typical defect introduction percentages per activity, although different projects will have different percentages.

The products of one project phase are often "early versions" of products of a later phase. They are the best representations of the later products that are available at the time.

Software Reliability Techniques for Real-World Applications, First Edition. Roger K. Youree.
© 2023 John Wiley & Sons Ltd. Published 2023 by John Wiley & Sons Ltd.

Table 4.1 Defects per Activity (%).

Software Development Activity	Percent of Defects
Requirements	10%–20%
Design	15%–30%
Coding	10%–35%
User Manual	12%
Error Correcting	8%

For example, even though the software requirements do not dictate how the software performs its functions, they do model the software from the perspective of what the software is expected do. If this "what" aspect is incorrect, the "how" aspect is also likely to be wrong. By considering the phases and the products produced in them, defects are more likely to be prevented at or near their source and if a defect is introduced, it can be detected and removed early before it affects too many other aspects of the project. Ideally, defects are detected and removed in the phase that they are introduced.

For each of the above phases, we include the following:

1. Description: We start with a description of the phase. This description includes a list of tasks that are typical for the phase as well as the potential products for the phase. These lists are representative and are not intended to limit potential products and approaches. For example, agile techniques tend to minimize documentation. This minimization can be good, provided that the right information is permanently available in the right form and is available to everyone as needed. These needs are project-specific. Always keep your focus on the few main objectives and build the project around them, with each task, plan, product, and process having a well-conceived purpose.
2. Defects: This section covers the nature of defects in the subject phase and how these defects can enter the project. General root causes for software defects are listed in Section 2.4. In these sections, we consider more detailed causes by phase.
3. Techniques: Next we cover techniques often applied in the given phase. Section 3.3 provides an overall approach for handling software defects. For the sections listed below, we list "General Techniques" that are often used in a given project phase to produce a good product, as well as "Anticipation Techniques" that can be used to anticipate defects and guard against them, and "Feedback Techniques" for finding and removing defects in a product.
4. Metrics: Finally, we discuss some metrics that can provide insights into a given phase. Section 2.3 considers the detection of defects in general. In the material below, we look at metrics that may be used to indicate defects or trends for specific phases. Some general guidelines when considering metrics include the following:
   (a) Choose metrics that are consistent with your objectives. Know why the project uses a particular metric. A good way to do this is to create a metrics plan for the project with the following information:
      i. Have a list of all of the metrics used on the project.

     ii. Provide a definition for each metric. Have a clear, unambiguous definition for each metric so that everyone understands it.

    iii. Have an explanation of why each metric is collected and what information is to be obtained from it.

    iv. Include a clear definition of how each metric is calculated, including how the input data are collected, who collects it, when it is collected, what level of precision the information is collected to, and other relevant information. Section 6.5 has a specific topic on data collection for metrics.

     v. Provide an explanation of what is expected from each metric and when a metric is indicating that there may be a problem. Know how to investigate changes in the metric to understand why the value of the metric changed, what is significant about it, and what should be done about it. Often, trends in the metrics are more important than absolute numbers. Section 6.5 has a specific topic on trend tests.

(b) Whether using a metrics plan or not, make sure that metric-related terms are defined clearly and consistently. Confusion can arise if, for example, two metrics use the word "throughput," but throughput means something different for the two metrics.

(c) Metrics that are predictive for one project may not be predictive for another, although they are more likely to be if the projects have similar products and development processes.

(d) Choose suitable metric measurement periods. Measuring too often or too infrequently can be counterproductive.

(e) Be aware that improving one metric may be at the expense of another metric. For example, we may be able to reduce the complexity of the code in a software module by increasing the number of lines of code.

(f) Consider if stakeholder buy-in is needed.

(g) Be rigorous in collecting the chosen metrics. Poor data collection can negate the value of the resulting metric.

(h) These metrics should not be used to appraise or threaten individuals or teams. Not only will that harm the *esprit de corps* of the project, but it will also encourage people to "adjust" metrics to avoid negative consequences.

(i) Automate metrics collection as much as possible. Automation typically reduces the workload, increases the consistency, and reduces human errors.

(j) Periodically reassess the metrics used for a project. If a metric is not providing useful information and its use is unlikely to result in project changes, then consider removing the metric.

As noted above, choose metrics that are suitable for the project at hand. However, most projects can benefit from three general types of metrics: project metrics, process metrics, and product metrics.

Project Metrics Project metrics consider how well the project is running. These metrics do not directly measure software reliability but rather provide an indication of how much "stress" the project is under. A "stressed" project is less likely to produce highly reliable software. Examples of such project metrics are as follows:

1. Staffing metrics: Staffing metrics can indicate project turmoil or inadequacies. Some examples of these metrics include the following:
   (a) Staffing numbers: A stable and expected value of this metric indicates adherence to the planned staffing profile, and large variances of it over time can show potential turmoil.
   (b) Employee mix: Not only is it important to have an adequate number of employees on the project, it is also important to have the right employees. A troubled project will sometimes move more experienced but more expensive employees to different projects to reduce project costs. This action is acceptable if the more experienced employee's expertise is no longer needed; however, if the expertise is still needed, the quality of the project's products may suffer.
   (c) Employee turnover rate: If there is a significant project employee turnover rate, it could indicate a troubled project. Project employees are often the first to notice that a project is troubled and they may attempt to leave while they can.
2. Metrics for adherence to budget and schedule constraints: There are a number of ways to assess whether a project is adhering to its budget and schedule constraints. A popular approach is to use the earned value (EV), actual cost (AC), and planned value (PV) metrics as defined below:
   (a) Earned value is the sum of the budgeted costs for all completed activities.
   (b) Actual cost is the sum of the ACs for the work completed to date.
   (c) Planned value is the sum of all of the budgeted cost estimates for work that has been scheduled to be completed to date.

   Combinations of these three metrics give the following:
   (a) Schedule performance index (SPI): SPI is defined as EV/PV and is a measure of the rate of performance of the project. A value greater than 1.0 indicates that the project is getting work done faster than was planned.
   (b) Cost performance index (CPI): CPI is determined by EV/AC and measures how well the project funds are spent. A value for CPI that is less than 1.0 indicates that less work is getting done per dollar than was planned.
   (c) Cost variance (CV): CV is defined as EV–AC and is a measure of budget performance. If the CV value is negative, the project has spent more than planned.
   (d) Schedule variance (SV): SV is defined to be EV–PV and is a measure of schedule performance. A negative value of SV indicates that the project is behind schedule.
3. Project "churn": Project "churn" is when there is excessive project change, particularly unexpected and uncontrolled change. Metrics for this potential issue include the following:
   (a) Number of project plan changes per time period
   (b) Average advanced notice of project plan change
   (c) Percentage of times that the project plan changes are due to current project plan deficiencies
   (d) Percentage of times that the project plan changes are due to the current project plan not being followed

(e) Percentage of times that the project plan changes for reasons other than voluntary plan enhancements.

(f) Number of times that the same aspect of the plan is changed.

Each of these metrics provides a different aspect of project "churn."

4. Customer sentiment: For some projects, the customer of the final product plays a part in the development of the product. Even if this is not the situation, each person in the project has an internal customer. This internal customer is the person or group that uses the results of the work that this person is doing. One way to measure the effectiveness of the running of a project is to determine who its customer or customers are, mutually determine what they need from the project, and let them assess the project based on the fulfillment of these needs. For example, a scale of 1–5 (with comments) could be used.

Process Metrics Process metrics consider whether appropriate practices and processes are defined, documented, and tracked and then assess how well a process is being followed and how well it is working. Some examples include the following:

1. Process time throughput: This metric is defined as how many output units are produced per unit time, such as how many requirements are produced per day.

2. Process cost throughput: This measures how many output units are produced per dollar spent on the activity, such as how many requirements are produced per thousand dollars.

3. Process efficiency: Here, we measure how many output units are produced per input unit. Process time throughput and process cost throughput can be considered to be special cases of this metric, with their inputs being time and cost, respectively. Other examples are the number of output units produced per number of steps in the process and the number of output units produced per number of people required to perform the process.

4. Process error rate: This is the ratio of errors per designated unit. An example is the number of software defects per line of code that this process lets through that are then caught in a later phase or by the customer.

5. Process cycle time: This measures how long it takes to complete the process.

6. Process complexity: This metric considers how many steps are required to complete the process. It can also consider how many feedback loops are required, how many times the process stops to wait on approvals, how many decision points there are in the process, and how many "special cases" there are. All of these intuitively imply greater process complexity.

7. Defect detection rate monitoring: We can use a prediction of the number of defects that the software will have to predict the rate at which defects should be detected as the software is developed. This defect detection rate prediction is based on the observation that throughout the development of the software defect detection should follow a Rayleigh distribution. By comparing the predicted defect detection rate with the actual detection rate, the project may have an early indication that defect detection processes are not working correctly. More on this technique may be found in Section 6.4.

Note that these metrics should indicate how well the process is working, both in terms of how well the process is performing its intended function and how expensive and time-consuming it is to the project. They are not specifically designed to assess the quality of the product itself.

Each project phase has its own often unique processes, but some typical processes and potential effectiveness metrics applicable to each of them are as follows:

1. Reviews and inspections: Reviews and inspections are used to find defects in a product, so the effectiveness of a review is dependent on the fraction of the total number of defects found. The metrics for this fraction include the following:

    (a) Process error rate: The process error rate is the number of errors or defects caused by (or not detected by) the process divided by the size of the item going through the process. For the review of a document, it could be defined as the number of defects that the review misses divided by the number of pages reviewed. The missed defects are those defects found later after the review. Over time, the process error rate gives an estimate of how well the reviews or inspections are detecting defects. It provides its results in later phases and possibly for use by later projects.

    (b) Capture–recapture: Capture–recapture allows for a current assessment of how many defects remain in the reviewed or inspected product. However, it is recommended that there be at least four reviewers. See [4] for more details.

2. Perform an activity, such as produce a document or perform an analysis or test: There are a variety of quantities to measure the performance of a task. The process metrics are listed above, such as process time throughput and process cost throughput measure efficiency of the performance. We also want to assess how effective or how "good" the process is. Candidate metrics for this type of metric include the following:

    (a) Customer satisfaction: From one viewpoint, the customer for a process or activity is the person or a group of people who use the process. One way to measure the effectiveness of a process is to have the people who use it rate it. For example, a questionnaire with rating criteria using a scale of 1–5 could be used. The questionnaire should also allow for comments.

    From another point of view, the customer is the person or a group that needs the output of the process. This person or group could rate the quality of the product that the process produces, providing an indication of the quality of the process.

    (b) Error rate: As the project progresses, defects may be found in products that use a particular process. During root cause analysis, determine if a given defect was caused by the subject process or could have been prevented or its likelihood significantly reduced by the process or by a reasonable modification to the process. The number of such defects, divided by the total number of defects for that product, when subtracted from 1 is a measure of the "goodness" of the process.

    (c) Rework: Another metric for process quality is the ratio of the process rework time divided by the total time spent on the process to produce a product. Rework time is the time spent doing some or all of the process again, or time spent doing an off-nominal part of the process, due to a potential product defect.

    (d) Clarity: Processes should be documented and the documentation should make the process clear and understandable. Failing this, the process may not be followed correctly resulting in more defects and cost and schedule impacts. An indirect way to measure process clarity is through process complexity. Another metric for process clarity is to use process user feedback. The questionnaire used to measure user satisfaction can have one or more questions related to the clarity and understandability of the process. Metrics such as the Flesch reading ease test score, the Flesch–Kincaid grade level, or the Fog index can be used to measure the readability of the documented process. Finally, a metric can be developed based on the ratio of defects in the process products due to not following the process as intended to the total defects in the same products.

3. Validate a product: There should be one or more identified needs for any product. Validation of the product answers the question of whether the product meets the needs that prompted its creation. A metric for that validation should indicate if the product validation process is effective and efficient. Potential metrics for a product validation process include the following:

   (a) Number of validated products that are later found to be insufficient and should not have passed validation. These failures could be due to not following the validation process correctly or due to a validation process defect. Each of these can serve as a process metric.

   (b) Cost of validation: The amount of time or money per validated item

   (c) Customer assessment of the validation process: The person or a group in need of the product can review the product validation process and assess how well they consider it to meet the validation needs.

4. Defect reporting, analysis and corrective action system (DRACAS) and defect review board (DRB): There are numerous metrics for a DRACAS and DRB. Some examples of some commonly used metrics for this purpose include the following:

   (a) Time open: Time open is the total time from the occurrence of the defect to verification that the root cause has been successfully addressed and the defect report has been closed.

   (b) Reporting lag: This metric measures the time from occurrence of the defect to when it is reported to the DRACAS.

   (c) Root cause rate: This is the fraction of reported defects that have identified and verified root causes.

   (d) Defect reduction: This metric measures the rate of occurrences of a specific defect before corrective action versus the rate after corrective action.

   (e) DRACAS expense: This is the total cost in dollars or in time per defect.

   (f) DRACAS complexity: This is the average of the total number of steps required to complete and close a defect.

   (g) Repeat defects: Another metric is the average number of times that a defect is reported. Multiple reports of the same defect incident are not included in this metric.

   (h) Defect removal efficiency (DRE): There are two popular versions of this metric. The first applies to the delivery of the product. For it, DRE is defined as $DRE = E/(E + D)$, where E is the number of defects found before delivery and D is the number of defects found after delivery. Another version of DRE considers different phases of the project and uses $DRE = E_i/(E_i + F_i)$, where E_i is the number of defects found during phase i and F_i is the number of defects found after phase i but attributable to phase i. For both of these versions of a DRE metric, a value close to 1 is ideal.

5. Configuration control: Some examples of metrics for configuration control are as follows:
   (a) Percentage of configuration items successfully audited
   (b) Number of configuration items introduced per time period
   (c) Number of configuration items addressed to completion per time period
   (d) Number of configuration items with missing or incomplete configuration information
   (e) Number of configuration items with deviations between stated and actual configuration
   (f) Number of unauthorized configuration items
   (g) Average time to detect configuration item errors
   (h) Cost of configuration management per configuration item
6. Use of metrics: An important process on any project is the use of metrics. Collecting metrics takes time and effort and therefore should be performed with specific purposes in mind. Every metric should be collected for a documented purpose. Metrics for these purposes include the following:
   (a) Number of actionable trends per time period identified by a given metric
   (b) False alarm rate: The number of times per time period that a metric indicates that action is required and it is then found that it is not required.
   (c) Cost per metrics: The number of hours, or the amount of money, per time period required to collect, analyze, act on, and manage a given metric or metric system.

Product Metrics   Product metrics typically look at the quality of the product being produced. They vary depending on the product being produced, but examples include the following:

1. Stability of the requirements
2. Number of untraceable requirements
3. Number of untestable requirements
4. Complexity of the software
5. Number of lines of code of the software
6. Number of defects found in the product
7. Predictions (per a specified prediction method) of the reliability of the software
8. Estimates (per a specified estimation method) of the reliability of the software
9. Customer satisfaction: The customer for a product is the a person or group of people that the product is produced for. A project produces many products and most are for internal customers, such as a specification produced for a design team. One way to measure the effectiveness of a product is to identify its customer or customers, mutually determine what they need in the product, and let them assess the product based on these needs. For example, a scale of 1–5 (with comments) could be used.

Next we consider each of the project phases listed at the start of this chapter by presenting a description of the phase and its tasks and products, giving a list of typical defects for the phase, describing examples of techniques that can be applied in the phase, and listing metrics that may be useful in the phase. It is hoped that this approach of considering project phases will reduce confusion and help the reader determine what needs to be done in each phase and for each major task of the project.

4.2   Concept Development and Planning

The Concept Development and Planning (CDP) phase of a project provides early planning for the project, such as developing an understanding of the customer needs and turning these needs into project objectives. It is also where we choose processes and standards to be used on the project.

4.2.1   Description of the CDP Phase

A project needs a clear idea of what needs to be accomplished and the time frame and resources involved. It also needs to know and fully understand the rationale for these needs. The CDP phase is where these needs are initially addressed. Typical tasks in this phase of a project include the following:

1. Understand the customer's needs and wants: To provide a product that the customer will buy, it is critical to understand the customer's needs and wants. This understanding extends to how the customer defines software reliability and its criticality. For example, one customer may emphasize the length of failure free time periods while another may emphasize software availability. Note that this understanding of customer needs is often considered to be a part of requirements elicitation and as such could be a part of the next phase, requirements and interfaces.
2. Translate customer needs and wants into project objectives: Understanding the needs and wants of the customer is important, but they must be put into a form that the project can act on. This translation of customer needs and wants affects how the project is organized and run. It is also eventually flowed down to the requirements of the product. Again, customer needs and wants for software reliability need to be translated into software reliability objectives.
3. Develop a conceptual design: Based on the needs and wants of the customer, one or more conceptual designs are produced. These designs are usually at a high level but provide enough details to show the feasibility of achieving the needs of the customer. There should also be measures of effectiveness (MOEs) that can be used to compare the different design options. Other important conceptual design considerations are cost, schedule, and risk.
4. Translate project objectives into project activities: This task is to determine what activities are needed to take the objectives and conceptual design that satisfy the customer's needs and wants and produce a corresponding product. Some examples of sub-tasks for this task are as follows:
    (a) Determine what activities are required to accomplish each project objective.
    (b) Define the scope of each of the project activities, including a clear description of the activity and its objectives and expectations, and descriptions of constraints and resources. The scopes of software reliability activities are included in this task.
    (c) Evaluate technical alternatives and key technologies, including identifying previously developed subsystems for reuse in the current effort. Also consider how different technical alternatives impact software reliability.
    (d) Develop acquisition strategies. If software is acquired from third parties, the reliability of that software should be one of the decision criteria.

(e) Choose or develop software processes and standards. Section 3.4 lists characteristics of a good process and discusses limitations of project processes.

(f) Assess the feasibility and risks of the project.

5. Plan and implement the project: Project activities must be planned and staff and resources must be managed. This task includes making work plans, estimating the required staff, scheduling the activities of the project, and determining and planning for the resources and budgets required for the work.

Products for this phase often include the following:

1. Statement of needs: A needs analysis produces a statement of needs that documents the customer needs and wants. A statement of needs often includes preliminary feasibility analysis.

2. Operational requirements: Based on the customer needs and a feasibility analysis, a set of operational requirements is produced. Operational requirements are broad statements of the objectives of the system or product. Sometimes, a product is specifically designed to address a subset of the customer needs. If this is the case, this decision should be made consciously and its rationale documented. Other tasks related to this task are as follows:

(a) Relate customer needs to operational requirements.

(b) Development of a system description.

(c) Create a description of how the system will be used, *i.e.*, a system concept of operations (CONOPs).

(d) Produce a description of the environment that the system will be used in, such as what hardware will run the software, what else will be running on that hardware, what operating system will be used, and numerous other potential factors.

(e) Document the validation of the operational requirements by means of an updated feasibility analysis for the operational requirements.

3. Conceptual design: A documented conceptual design, along with the analysis results supporting the design, should be produced. This design indicates that satisfying the operational requirements is feasible and provides a path forward. The design is not necessarily optimal or final but rather serves as validation of the operational requirements and as a starting point for the design. Ideally, the conceptual design will provide early estimates of the size of the software and possibly a preliminary list of software line replaceable units (LRUs) and the relationships between them. Early software reliability predictions are based on this conceptual design.

4. Project strategy: Develop a strategy for achieving project objectives. Sub-tasks related to this task are as follows:

(a) Evaluate the scope of the project

(b) Evaluate technical alternatives

(c) Develop risk assessments

(d) Analyze risks and plan to mitigate them

(e) Apply lessons learned from previous projects

(f) Determine what processes and standards will be used

(g) List assumptions made

(h) Develop a design effectiveness model to use for comparing alternatives

(i) Update the feasibility assessment

(j) Lay out an overall strategy for meeting the operational requirements

Note that these sub-tasks overlap some of the tasks required for the previous three products (statement of needs, operational requirements, and conceptual design), but the previous three products are important enough to merit separate mention.

5. Project plan and budget: Develop a project plan and budget sufficient to make the project strategy happen. This product includes the following:

   (a) Major project milestones

   (b) Staffing estimates

   (c) Roles and responsibilities for activities and organizations (This product is often in the form of a Responsibility Assignment Matrix that includes roles, authorities, responsibilities, and required skills.)

   (d) Work breakdown structure (WBS)

   (e) Bill of materials (BOMs)

   (f) Training requirements

   (g) An assessment of what role software will play in the system

   (h) A preliminary determination of what type of development process will be used for the software (waterfall, spiral, incremental, agile, etc. see [5] for more details)

   (i) Production of initial development plan, maintenance plan, configuration management plan, test and evaluation master plan (TEMP), and plans for other critical activities

Other products more directly related to software reliability that should be developed in this phase (although possibly refined later) include the following:

1. A reliability program plan should be produced in this phase.
2. A reliability growth plan should be developed either as a part of the reliability program plan or as a separate plan.
3. Develop criteria for what constitutes a system failure or a software failure and how to assess the seriousness of a failure. This set of criteria is often referred to as the failure definition and scoring criteria (FDSC) and is explained in more detail in Section 6.4.
4. Develop an initial operational profile (OP). More on this technique may be found in Section 6.4.
5. Perform a risk assessment for the project. Consider cost, schedule, and performance risks, as well as reliability, safety, security, and other potential vulnerabilities.
6. Develop a data collection plan.
7. Early (conceptual) reliability models of the system should be developed.
8. Early software reliability predictions should be made. These are likely to be very approximate but can serve as guides for planning and resource allocations. More on this technique may be found in Section 6.4.
9. Determine the level and type of validation required for each product. Products with a significant "down stream" impact need to be validated carefully before being considered ready for use. For example, the specifications may need several steps of validation, such as results from automated requirements tools, traceability analysis, checklists, peer reviews, formal inspections, and customer reviews. The early reliability model

may be validated with less effort, such as the use of checklists and peer reviews. These validations should be documented in the reliability casebook.

10. Produce high-level reliability allocations. These are used to help assess concept feasibility and to plan resource allocations. More on this technique may be found in Section 6.4.
11. Begin assessing reliability metrics in this phase.
12. Plan and set up the DRACAS and DRB. Determine how the system and its board will operate, what tools will support it, what defects are applicable, determine roles and responsibilities, and when the system should begin operations. Different project areas may have their own DRACAS and DRB, but these separate systems should be coordinated by an oversight system. More on DRACAS may be found in Section 6.4.
13. Determine if any reliability trade studies are needed and if so, plan them.
14. Determine how software reliability activities will be integrated into the project. It is important to consider how the project will view and handle reliability while being an effective and motivated team.

It is important to understand the purpose of each product listed in this section and to determine how the project addresses each of these purposes. For example, sometimes a formal, controlled document is needed for a given product and other times, less formal products will suffice. Each project is unique, and tasks and products should be tailored to the needs of the specific project.

4.2.2 Defects Typical for the CDP Phase

Defects in this phase can cause serious problems if not remedied soon. For example, a poor understanding of customer needs can result in software that is unacceptable to the customer and correcting this problem too late may require an expensive and time-consuming new release or even a new product. Some examples of defects for this phase are as follows:

1. Lack of a solid understanding of customer needs: Not understanding customer needs means that it is unlikely that the software will meet those needs. While it is possible to produce reliable software that performs a different set of functions than those desired by the customer (a marketing problem), it is also likely that there will be a lot of project churn to correct this situation and project churn increases the likelihood of defects. A well-run project will make it a priority to understand the customer needs, document those needs so that project personnel are constantly aware of them, and address them throughout the project.
2. Lack of an adequate set of operational requirements: Needs are translated into requirements so that designers have something to design to. Operational requirements, what the system needs to do in order to satisfy the customer needs, are the first step. If these requirements are done poorly, then lower-level requirements are likely to be adversely affected and the released product is at a greater risk of being faulty.
3. Inadequate conceptual design: Although conceptual designs are typically at a high level, they should give enough details to provide confidence that the operational requirements can be satisfied in a cost- and schedule-efficient way at an acceptable risk. If the conceptual design does not consider the key design aspects, or if the analysis of the conceptual

design is inadequate to find these weaknesses, the project may unknowingly proceed with an expensive, high-risk effort.

4. Poor translation of project objectives into project activities: Even with a solid set of operational requirements, a project needs a strategy for achieving the requirements. A clear understanding of the activities required to go from the operational requirements to a released, maintainable, and supported product is essential. Not understanding the programmatic aspects of the project or the technical constraints, unrealistic expectations and risk assessments, a lack of or inadequate standards and processes, all stress the project and make software defects much more likely.

5. A poorly planned or budgeted project: Once the project activities are established, they must be planned and implemented. Poor planning or implementation will result in a poorly run project and probably result in a poor product.

4.2.3 Techniques and Processes for the CDP Phase

In this section, we start with general techniques typical of the CDP phase. Project management techniques are covered in Section 4.8.3. After general techniques, we consider anticipation techniques that anticipate defects typical of the CDP phase and aim to prevent them, and feedback techniques that are used to find the existing defects and remove them. This is a sampling of techniques and not a complete or mandatory list.

1. General techniques for the CDP phase:
   (a) Requirements elicitation: In this phase, requirements elicitation is used as a part of the development of the statement of needs and the operational requirements. It is used to determine the performance and features required to meet the customer needs and wants. Important goals of this task are to determine what problems need to be solved and to identify stakeholders and system boundaries. Techniques and advice include the following:
      i. Requirements elicitation plan: Determine ahead of time how the project will perform requirements elicitation. It is a good idea to start elicitation with a problem statement, describing the need or needs to be addressed by the product.
      ii. Stakeholders list: Identify and involve all of the stakeholders in the requirements elicitation process.
      iii. Traditional data gathering techniques: Traditional data gathering techniques include surveys, interviews, focus groups, and reviews of existing documentation.
      iv. Group techniques: An example of a group technique is the use of customer meetings to discuss user needs and operational requirements.
      v. Product tracking: With this technique, we track what is being said about our product, competing products, and related products.
      vi. User stories: User stories are short, uncomplicated descriptions of a functional requirement for a system. They are often written by the customer.
      vii. Use cases and user scenarios: These techniques are primarily used for functional requirements. Use cases provide a visual representation of interactions between actors. User scenarios provide textual descriptions of these interactions.

viii. Quality function deployment (QFD): QFD is a customer-driven process for determining customer "must haves," "wants," and "wows" and prioritizes these so that they can be converted to engineering requirements and assigned to the appropriate organizations.

ix. Prototyping: Prototyping for requirements elicitation improves user involvement and communication between users. A prototype is a mock-up or partial implementation of a system. It is a model that forms a pattern or a structure to base a more detailed design on. Examples include storyboarding, screen mock-ups, and pilot systems.

x. Model-driven techniques: State modeling techniques are used to make sure that all credible inputs and requests are considered under all credible conditions. Universal Modeling Language techniques can aid in this process.

xi. Checklists: Checklists can be used for many purposes, such as ensuring that all of the characteristics of a good requirement are addressed.

xii. Document the rationale: Document the rationale for each requirement. This documentation clarifies the requirement, helps reduce requirements misunderstandings, and aids verification. It also tends to reduce the number of requirements by helping to identify redundant or overlapping requirements and unnecessary requirements.

xiii. Actively elicit requirements: To develop your requirements, do not simply rely on customers to provide their needs. Get inputs from internal customers, such as software architecture team members and specialty engineers. Active elicitation can add requirements that experienced personnel realize are needed even though the customer may not realize it, but care must also be taken to ensure that the resulting system is not over designed with "nice-to-have" features that are not worth the cost to the customer and that potentially add complexity and therefore risk to the software reliability.

Requirements elicitation is also covered as a topic in Section 6.2, as are many of the techniques just covered for the topic.

(b) Operational profiles (OPs): Develop an OP for the software. More on OPs may be found as a separate topic in Section 6.4.

(c) Effectiveness models and simulations: Effectiveness models and simulations can be used to test alternative operational requirement choices and to assess conceptual designs. The models and simulations provide MOEs for the various alternatives, and these can be used to rate options.

(d) Conceptual designs: A conceptual design is used to show that there is a feasible way to satisfy the operational requirements and statement of needs and provides a path forward that leads to more detailed requirements and design. Techniques applicable for conceptual design work include the following:

i. Plan the development of conceptual designs: Decide what the conceptual design activities will be, when they will occur, and who will do them. Also decide how the project will know if these tasks have been completed successfully.

ii. Brainstorming: Brainstorming is an idea generation technique to produce a broad range of ideas on a topic.

   iii. Affinity diagrams: Like brainstorming, affinity diagrams can be an important idea generation technique and is often used in conjunction with brainstorming.

   iv. Measures of effectiveness (MOEs): Decide early what MOEs are most important for the system. Define them carefully to avoid confusion and include how to measure them and what constitutes acceptable and unacceptable values.

   v. Checklists: Checklists can support a variety of conceptual design activities. For example, a checklist of the main operational requirements that the conceptual design must meet can be made and used to modify the potential concepts.

   vi. Prototyping: Prototyping the conceptual design increases our understanding of it and enables us to better explain it. As an example, consider expanding the prototypes used for requirements elicitation to prototype the conceptual design.

   vii. Document the rationale: Document the rationale for the conceptual designs. This documentation clarifies the strengths and weakness of each potential design and after one has been chosen, aids in its validation.

(e) Set up project processes and procedures: A few example tasks for this activity are listed as follows.

   i. Set up a configuration control system: A configuration control system needs to be set up early in the life of the project. Levels of control may change as the project progresses, but these changes should be planned for specific purposes.

   ii. Determine what software processes, coding practices, best practices, and standards are to be used. Examples of software development processes include waterfall, spiral, incremental, agile, cleanroom, and eXtreme programming, to name a few. See [5] and [2] for more details.

   iii. Decide what metrics to use, how to collection them, and how to use them. For example, determine what to use as triggers (what is measured and what is acceptable versus unacceptable for each of these measured quantities) and what to do if a trigger is reached.

   iv. Define reliability terms for the project, such as software defect, defect precursor, error, fault, and failure. Consider developing a FDSC.

   v. Develop criticality definitions for both reliability and safety. Also consider including criticality for vulnerabilities (hacking, malware, etc.).

   vi. Make early assessments of how much new code, modified code, re-used code, auto-generated code, etc. is to be used.

   vii. Select tests for the project and begin planning for them.

   viii. Characterize the operational environment (with user inputs if possible).

   ix. Determine how software maintenance will be performed and by whom. Start planning what maintenance features the software will have and how to ensure that maintenance is considered in the design and development of the software.

   x. Make a reliability risk assessment.

   xi. Determine which decisions require a project-standard decision-making process and plan for it

   xii. Select the models and techniques to be used.

   xiii. Start recording lessons learned. Make them as measurable as possible.

   xiv. Choose process improvement techniques and plan when and how to use them.

xv. Database baseline: Consider using a single database as the project baseline, making it assessable across the project. Make sure that all phases and project users, processes, and products are considered when choosing and integrating such a database.

(f) Use a software reliability advocate: A software reliability advocate continuously assesses the project for potential software reliability issues and advises project personnel on the impacts and mitigations of these issues.

2. Anticipation techniques for the CDP phase:

(a) Project Premortem: This is a technique typically used at the start of a project. Project stakeholders meet and pretend that the project has failed. They then list reasons why. After developing and prioritizing this list of reasons, solutions are developed and assigned. This technique can also be used for various aspects of the project. For example, software reliability stakeholders can use it to determine "why the software has so many defects."

(b) Benchmarking: This can help determine possible goals and standards for the project. Like brainstorming and affinity diagrams, it can be used to generate new ideas on how to do better.

(c) Process failure modes effects and criticality analysis (PFMECA): A PFMECA can be used to anticipate potential process defects and determine mitigation techniques for these defects. Ideally, this analysis has already been performed on standard processes, but new or modified processes should be investigated for where defects can enter the system. Even if a process failure modes and effects analysis (PFMEA) has been performed on standard processes, check to ensure that the previous analysis still applies.

(d) Reliability program plan: The reliability program plan should detail the techniques and processes that are planned for each phase of the project along with their rationale. The plan should cover each phase of the project. There may be a separate plan for software reliability but if so, it should be coordinated with the overall project reliability plan.

(e) Reliability allocations: Develop high-level software reliability allocations or goals. Reliability allocations for the software aid in setting priorities and in assessing the feasibility of the design. See Section 6.4 and [6] for potential allocation techniques.

(f) Reliability predictions: Use reliability prediction techniques based on the information available in this phase of the project. This information typically includes estimated lines of code, type of code, and other considerations. What constitutes required information is dependent on the prediction model chosen. These predictions of the reliability of the software can be used as software metrics. A number of useful prediction techniques are found in Section 6.4 and [6].

(g) Software defect root cause analysis: A software defect root cause analysis is an analysis to identify the most likely causes of software defects for a project to enable the project to manage defect detection, analysis, and reduction more effectively. More on software defect root cause analysis may be found in Section 6.4.

3. Feedback techniques for the CDP phase:

(a) Daily stand-up meetings: These short meetings allow feedback from project personnel. They also give management an opportunity to provide project personnel with a quick project status update.

(b) The use of quality circles: Quality circles allow project personnel to identify, analyze, and correct project problems.

(c) Peer review and inspections: Peer reviews and inspections can be used to find defects in the statement of needs, plans, procedures, operational requirements, and other documents.

(d) Reliability casebook: A reliability casebook helps the project manage the reliability of the software and assess if processes are followed adequately. More on reliability casebooks can be found in Section 6.4.

(e) Orthogonal defect classification: Orthogonal defect classification is a technique for classifying defects to make them easier to analysis. See Section 6.4 for more on this topic.

(f) Pareto analysis: Set up a Pareto analysis of defects based on dollars or time using a Pareto distribution of defects ordered by money or time lost. The analysis is used to prioritize the effort to ensure that the top priorities are handled first. See Section 6.5 for more on this topic.

4.2.4 Metrics for the CDP Phase

Metrics in the CDP phase not only help set metrics baselines for later phases, but they also can provide early warnings of potential trouble with the project. Metrics applicable to the CDP phase can be categorized as project metrics, process metrics, and product metrics for products of this phase. Project metrics indicate how well the project is running and are covered at the start of Section 4.1.

Process Metrics Recall that the process metrics monitor the processes, such as monitoring the processes for applicability, if they are carried out correctly, and if they are effective. A number of applicable process metrics are listed in Chapter 4. These metrics can be applied to, among other things the following:

1. Planning the project
2. Performing peer reviews and inspections of products
3. Producing documents, such as operational requirements and a reliability program plan
4. Using requirements elicitation techniques and conceptual design techniques
5. Performing analyses, such as reliability allocations, reliability predictions, and process FMECA
6. Setting up and operating a DRACAS and DRB
7. Performing configuration control activities
8. Using metrics

Product Metrics Product metrics are typically used to assess if some aspect of a given product is adequate. Below we cover product metric examples for products developed in this phase as well as some metrics used to support later phases of the project:

1. Product metrics for the products of this phase:
   (a) Statement of needs: Examples of metrics applicable to a statement of needs are as follows:
      i. Use case size (for functional requirements)

    ii. Customer satisfaction surveys

    iii. Peer review and inspection results, such as defect density

    iv. Checklist completion metrics

(b) Operational requirements: Product metrics for operational requirements are similar to those for the requirements specifications found in Section 4.3. These metrics measure quantities such as the size of the requirement specifications, their volatility, completeness, and traceability. If a quantitative effectiveness model is used, MOEs can be used in the assessment of the operational requirements.

(c) Conceptual design: Product metrics for a conceptual design include the following:

    i. Traceability of conceptual design characteristics to operational requirements

    ii. Measure of effectiveness analysis results and associated cost and risk

    iii. Customer satisfaction metrics, where the customer referred to here is not just the external customer but also the internal customers. For example, customer assessment of a prototype used for conceptual design can provide valuable information.

    iv. Risk assessment results

    v. Number and severity of defects found in reviews and inspections of the conceptual design

(d) Requirements elicitation: Examples of metrics for requirements elicitation include the following:

    i. Fraction of requirements validated with the user

    ii. Comparison of elucidated requirements from the use of different elucidation techniques

    iii. Volatility of requirements

    iv. Elucidation efficiency (a measure of the amount of rework as a fraction of total work)

(e) Process FMECA, reliability program plan, reliability casebook, reliability allocations, reliability predictions: Peer review or inspection results, such as number of defects per item

2. Product metrics in support of later phases: Some metrics are initially developed in this phase to be used for planning for later phases. Examples include the following:

(a) Expected number of source lines of code (SLOC)

(b) Details of the types of code to be used

(c) Predictions of the software reliability (used in this and later phases)

Examples To better understand how to plan tasks in the CDP phase to produce reliable software, we present two examples. The first example is of a small company that produces a commercial software product for hospitals. The second example is of a large company that works on large government contracts. These examples are extended to the other project phases in the corresponding sections.

Example 4.2.1 Our first example provides an idea of how CDP techniques and metrics can benefit a small software company using an agile philosophy and associated processes to produce a commercial product. This small software company, Company A, develops a software product to help hospitals track their use of medicines. They have several different versions of the product, each tailored to a different customer. Because of the rapidly developing market and a concern for competitors entering the market, Company A puts an emphasis on understanding customer needs and quickly providing a quality product that satisfies those needs. They typically produce a different version of their product for each customer. One of the founders of the company worked in the hospital business and has insight into how the hospital business works.

Since Company A wants to understand and satisfy customer needs quickly, they have adopted agile approaches for software development and project management. They develop their software using behavior-driven development (BDD) while planning and implementing the work in a sequence of work sprints, each typically two weeks long. The sequence of sprints usually starts with a planning sprint and ends with a transitional sprint before release. The company is continually seeking customer needs and wants, and when there are a sufficient number of them, they assess if a new product version is warranted. If it is, a planning sprint is initiated as soon as people are available. After finishing the transitional sprint, the new product is expected to be ready for release. It should be noted that Company A does not have an identifiable CDP phase but rather has tasks appropriate to such a phase. For Company A, some of the important objectives associated with the CDP phase are as follows:

1. Understand the customer's needs and wants
2. Translate customer needs and wants into project objectives
3. Develop a conceptual design
4. Translate project objectives into project activities
5. Plan and implement the project

Understanding the customer's needs and wants is critical to Company A's business strategy. The company needs to know the needs and wants of not just its current customers but also of potential new customers. Understanding customer needs is an ongoing process with Company A and not typically assigned to any particular sequence of build sprints. They understand the need for documentation but want to keep their processes quick and inexpensive. They also understand that the ultimate users of their products are often not software savvy and so they need a simple intuitive interface and failure-free results. To these ends, Company A uses the following techniques:

1. Requirements elicitation plan: Company A keeps a list of all current customers and potential customers and has a separate plan for each. Each plan is fairly informal, but includes known needs of the hospital or hospital chain as it relates to Company A's products, potential issues that the hospital may be having, a list of contacts and their roles, and information about each of the contacts. They also keep information about which elicitation approaches have worked well in the past with contacts and which have not.
2. Group techniques: To understand a given customer's needs, Company A stays in regular contact with existing customers to understand their needs and how Company A's software can help satisfy them. They also meet with the end users of the software to better understand how their product is actually used and if these end users are satisfied. They try to understand what new features may be desirable and what existing features can be improved or removed, how to make the software more user friendly, and to determine if

the users have experienced events that they consider to be defects. Company A realizes that just because the user considers a certain software behavior to be a defect does not mean that the software designers consider it to be one; however, if a behavior annoys the user enough for the user to consider it as such, it is likely to be a behavior that needs to be changed. When possible, Company A also meets with potential new customers to describe the type of the product that they can provide and to ask the potential customer for features and behaviors that they would like to have in such a product. They try to make such a meeting as customer-driven as possible.

3. Product tracking: In addition to information from the customer, Company A follows information about what others are saying about their products and about any competing products.

4. User stories: Company A makes extensive use of user stories to try to understand and document their customer needs.

5. Five whys: Before a user story is finalized, it goes through a "five whys" session to make sure that it has a business purpose.

6. Checklists: To organize tasks in a simple, easy-to-follow manner, Company A frequently uses checklists.

7. Document the rationale: For each of their products, Company A keeps a database containing the rationale for each behavior that has been designed into the software.

8. Actively elicit requirements: Finally, Company A has periodic meetings and brainstorming sessions to determine new behaviors that may be of interest to various existing or potential customers.

For the task of translating customer needs and wants into project objectives, Company A does not create a series formal specifications but rather creates a list of software behaviors for each version of their product that can be informally approved by the customer of that software version. This approach is more in line with the rapid software development and product release that the company aims for.

Company A needs to keep conceptual designs flexible and simple as much of their effort goes to changing their existing products based on changes in customer needs. Most of Company A's products are continuously modified to add new behaviors, improve old behaviors, and remove unneeded behaviors and features. By doing this, a previous version serves as a good representation of most of the next version of the product. They hold meetings to discuss how the software can perform certain behaviors and record the results of these meetings. These meeting results include sketches and diagrams of software architectures, data flows, user interface behaviors, and other ways to describe and document the conceptual design. They also include a collection of user stories to describe the behaviors that the software will have. For more complicated behaviors, use cases may be used. Company A also uses the following techniques for conceptual designs:

1. Plan the development of conceptual designs: Updating the conceptual design for a product is a part of the process for determining whether to create a new version of a software product. The initial conceptual design work for a new product version is performed outside of the sequence of build sprints and is then refined during the planning sprint. A new conceptual design is created if a new product is being considered or if an existing product has been modified extensively.

2. Brainstorming: This is a favorite technique at Company A and is in keeping with the agile philosophy emphasis on collaboration. For conceptual designs, brainstorming is typically used when working on new products or significant changes to the existing products.

3. Checklists: Company A creates a checklist of the required software behaviors and for each asks if the conceptual design is adequate. They also consider the collection of behaviors as a whole and assess the adequacy, consistency, and various checklist items from the company's lesson learned files.

4. Document the rationale: Company A is a small, tight-knit company that makes many of its decisions in meetings, and this holds for conceptual designs. Notes from these meetings serve to document the rationale for the resulting conceptual design.

To translate project objectives into project activities and to plan and implement them, Company A relies on agile techniques. The company needs to continuously determine customer needs, both the changing needs of current customers and the needs of potential customers. It must take these needs and plan a software version that meets these needs and then design, code, test, and validate the new product. These new products are then made ready for product production so that they can be released on demand. Company A is repeatedly performing these tasks and so has designed its operations around them. Once they have a new set of customer needs, they determine if a new version of their product is warranted. If so, they create a planning sprint, followed by a sequence of build sprints to modify a version of their product, incrementally adding new behaviors or changing or removing old behaviors. The sequence of sprints ends with a transitional sprint to test and assess the new product, ensuring that it is a quality product ready for release. The following planning activities are some of the activities performed in the planning sprint:

1. Determine processes: Company A does not have an extensive set of procedures and processes. There are some flowcharts and lists of steps to follow for the main tasks, but the company values flexibility and makes extensive use of often informal meetings to determine how to proceed. However, the company does value recording the results of these meetings and following up to determine lessons learned.

2. Set up a configuration control system: Company A manages the configurations for each customer individually, and since the company regularly updates their products to new versions, they have a standard, mostly automated process for new product versions. Configuration control of the software uses version control software. Other items, such as conceptual designs and meeting minutes, are converted to a common electronic format and stored in a database with easy read access. The configuration control processes are documented in flowcharts with narratives to explain the required steps.

3. Make risk and improvement assessments: Company A has a standard checklist of risk areas and questions that it goes through when determining whether to make a new version of a product and at the start of a new software version sequence of build sprints. During the planning sprint, project personnel meet to discuss risks and their mitigations, technical alternatives, areas for improvement, and other related topics.

4. Determine metrics: Company A has a standard set of metrics that it uses. Meetings are held periodically to assess the metrics, and one of the standard topics of these meetings is whether new metrics are needed or the existing metrics need to be changed or removed.

5. Make reliability risk assessments: Company A does not specify a reliability value for any of its products but does value the length of failure-free performance time for its products. It collects as much failure information from its customers as possible and has used this information and information from internal testing to develop its own method of estimating the mean failure-free performance time, measured as MTBF, for its products. They apply this method to each version of each product. The resulting estimation is primarily used as a part of the release decision. The company also uses a DRACAS and DRB. They record each defect in a database along with the information about the defect. They use the information to find the root cause of the defect and to determine ways to mitigate the defect and prevent such defects from occurring in the future.

Other planning sprint activities specifically related to software reliability include the following:

1. Update the software reliability program plan (SRPP): Since Company A frequently updates its products and each of its products performs many of the same behaviors, updating the SRPP starts with the standard SRPP and during the planning sprint, it is assessed to see if modifications are needed for the current effort.
2. Characterize the new software: For Company A, a new product is typically an updated version of an existing product, so some software LRUs are likely to be unchanged. Based on the number of new behaviors and their nature, each LRU is assessed for potential changes and these assessments are used to estimate the number of new and modified lines of code required for each LRU.
3. Develop criteria for what constitutes a failure and assess its criticality: Company A again takes advantage of the frequent updates to its products, allowing this task to start with the criteria used for the previous version. During the planning sprint, these criteria may be modified based on the new behaviors to be added or changed.
4. Develop an OP: Again, the company's products are similar and are used in much the same ways. However, different customers have different procedures and needs, so OPs are tailored to the customer.
5. Reliability allocations and predictions: Company A does not advertise any quantitative reliability values for their products, but they still consider product reliability to be very important. They are particularly interested in the mean failure-free performance time of each of their products. In the planning sprint, an allocated MTBF value is developed for the product and allocated to the LRUs. Also, each LRU is assessed and an initial prediction of its MTBF is made.
6. Software failures modes and effects analysis (SFMEA): As with other aspects of Company A's products, it has a SFMEA for the previous version of the software and during the planning sprint, updates it for the next version.
7. Reliability casebook: Company A stores artifacts for each version of the software for each customer. In the planning sprint, the reliability casebook for the subject software version is created.

Knowing that a significant source of software defects is human errors, Company A uses automation where feasible, encourages working in groups, and performs plenty of reviews and inspections.

Company A also uses metrics to assess the progress and alert for potential problems. The company uses the following metrics for the tasks in the CDP phase:

1. Project metrics: Company A uses the following metrics to track the project
    (a) EV, AC, and PV: EV, AC, and PV are measured for each customer project. There may be multiple projects for a given customer, such as maintaining one version of a product and developing several new versions to handle different combinations of issues or desires of the customer. These three metrics can be used to develop other metrics, and metrics from customer projects can be rolled up to the customer level. They can also be rolled up to see the metrics across all customers and potential customers. These can be used to assess concerns such as company profitability and which customers are the most profitable and why.
    (b) Project churn: Company A also measures the number of project plan changes per month. Each such change is recorded along with the reasons why.
    (c) Customer sentiment: Since Company A relies on satisfied customers, external customer sentiment is an important metric. External customer sentiment is difficult to quantify, especially considering the different personalities and tendencies of different customers. Company A has developed its own customer sentiment metric on a scale of 1–10 based on factors such as the customer's willingness to recommend product changes and new behaviors, feedback from the customer on how satisfied they are with a product, the tone of their comments, and their reactions to suggestions from Company A for product changes. The metric is adjusted based on the individual customer's tendencies and company climate.
2. Process metrics: For process metrics, Company A emphasizes process errors through the use of the following metrics
    (a) Process errors: When defects in a product are found, root cause analysis is performed and part of this analysis is to determine if the root cause is a process defect or if a reasonable change to some process would have prevented the defect.
    (b) Peer review results: Periodically, Company A chooses a specific process and performs a peer review of it. The number and severity of findings is treated as a process metric.
    (c) Rework per software version: The number of rework hours per total work hours is measured and is used to assess whether a version is excessively difficult, complex, or confusing.
3. Product metrics: Examples of Company A's product metrics are as follows
    (a) Number of new behaviors per version of the software product
    (b) Number of modified behaviors per version of the software product
    (c) Number of removed behaviors per version of the software product
    (d) Number and severity of comments from peer reviews on the list of behaviors elicited from the customer for the next product version
    (e) Number and severity of comments from peer reviews on the conceptual design
    (f) Number and severity of comments from peer reviews on the MTBF predictions
    (g) Number of behaviors per product release that are changed from the end of the planning sprint to release

Example 4.2.2   Our second example is of Company B, a large company working on a large government contact. Company B must deliver a hardware and software system at the end of a multi-year contract. The contract also requires that they develop and deliver a collection of documents and meet a series of milestones, such as a system design review (SDR), preliminary design review (PDR), critical design review (CDR), test readiness review (TRR), and functional configuration audit (FCA). The government has given Company B a statement of work (SOW) and a high-level specification for the contracted system as well as other guidance documentation and has personnel on site to provide help and to monitor progress.

The contractor is required to produce lower-level specifications on down to hardware, software, and interface specifications. They also produce drawings and other design documentation. A variety of plans are required as deliverables, such as a systems engineering master plan (SEMP), a quality assurance (QA) plan, a configuration management plan, a TEMP, and a reliability, availability, and maintainability (RAM) Plan. Company B is using an incremental develop approach where tasks from the CDP and Requirements and Interfaces phases are performed early, followed by several DC and Integration, Verification, and Validation (IV&V) phases. Each of these iterations produces a version of the final product with more features and capabilities than the previous version or build. At the end of the contract, the product should meet the full set of requirements and be ready for limited production. Assuming that the customer is satisfied with the resulting product, it is expected that there will be a future contract for limited and then full-scale production of the product.

Company B won the contract after a competition with several other large companies. Company B responded to a request for proposals and turned in their winning proposal at the end of the proposal time period. Tasks performed during this proposal time period provide an early effort at the tasks of the CDP phase. Recall that the CDP phase tasks include the following:

1. Understand the customer's needs and wants
2. Translate customer needs and wants into project objectives
3. Develop a conceptual design
4. Translate project objectives into project activities
5. Plan and implement the project

Much of understanding the customer's needs and wants comes from the proposal activities; however, after contract award, the government can speak more openly with Company B and clarify what is wanted. Part of the reason for having on-site government representatives is to facilitate this understanding. Company B makes frequent use of this opportunity to ask for clarifications and then documents the results and obtains customer agreement with these clarifications. Needs and wants also develop and mature as requirements are refined and are associated with potential design solutions. Finally, the government customer is frequently under pressure from higher-ups to modify aspects of the specifications or contract based on the latest government needs and wants. The government customer understands that contract churn makes for inefficiently run contracts and increases the likelihood of errors, but some such changes are inevitable. Knowing this, there is an engineering change proposal (ECP) process where the contractor can identify the impacts of these changes and assess their cost and potential schedule impacts.

The customer has provided Company B with a system specification that goes a significant way toward translating needs and wants into project objectives. Another aspect of this translation is working with the customer to obtain and refine a system description, the CONOPs, and a detailed description of the operational, maintenance, transportation, and storage environments for the system.

Company B's winning proposal serves as a starting point for the conceptual design. This conceptual design is modified and matured in the early part of the contract based on an improved understanding of the customer's needs and to better satisfy the system specification. The more detailed conceptual design also allows for a more detailed assessment of what hardware and software will be required. At this time, there is an early identification of the software LRUs and estimates of their SLOC and types of code, and there are early reliability allocations and predictions for them.

Company B plans to use the government-provided system specification to produce lower-level specifications and then design and build the product to these requirements. Work is often performed in parallel. For example, the hardware design and software design organizations are producing preliminary designs based on the system specification and the conceptual design. These preliminary designs help solidify the conceptual design and also help to determine what lower-level requirements are likely to be feasible. The test and evaluation, production, and support organizations are also working with requirements and design personnel to ensure all aspects of the product are addressed and are feasible. Risk assessments are performed throughout. Processes and procedures are chosen or developed, as are standards and best practices. Various organizations and suborganizations such as configuration control, quality control, security, subcontract management, contract compliance, legal, and facilities are formally set up.

Planning and implementing the project involves a wide range of tasks. A few examples are as follows:

1. Staffing, budgeting, and resource estimates: The project must determine how many people are required to perform the tasks that the project requires. In addition to determining how many people, it must decide what skills these people should have, determine when and where they are needed, and what training is required. Other resources, such as facilities, computers, and various other supplies must also be planned and budgeted. An earned value management (EVM) system is set up to manage these items.
2. Project scheduling: Company B creates a WBS to decompose and organize the project deliverables. The project also creates an integrated master plan (IMP) to document and tie together important events and accomplishments. The IMP is then used to develop the integrated master schedule (IMS). This schedule keeps track of the critical path and is used to adjust tasks and resources as needed to keep the project on schedule and in budget.
3. Plans: Another important CDP phase task is the development of project plans. Even if a plan is for tasks and events that are to occur near the end of the contract, developing an initial plan for them is important. This initial plan makes sure that the requirements and constraints that these tasks impose are properly considered early in the project when they can be scheduled and budgeted most efficiently. Examples of such plans include the TEMP, SEMP, configuration management plan, RAM plan, and metrics plan.

4. Communication: Good project communications is an important part of any well-run project. Smooth communication paths from management to employees, from employees to management, and between employees are all important. Various means of communication such as meetings, virtual meetings, instant messaging, and e-mails are some of the approaches that can play a role in this effort. Company B has set up recurring project updates, procedures for unexpected occurrences, safe feedback mechanisms for employee feedback, and other communication paths.
5. Planning for formal reviews: Formal reviews, such as the SDR, PDR, CDR, TRR, and FCA, play very important roles in the project. In addition to being contractually required, they also serve as views of the project to the outside world. This outside world includes the people that the government customer must keep satisfied. Poor performance at one of these reviews can result in negative publicity, significant additional work, reduced fees for the contractor, or even contract cancelation. To address these reviews, Company B sets up teams devoted to planning for each review and works with the customer to make sure that review entrance and exit criteria are well understood and are manageable.

Company B plans to use a variety of techniques to produce reliable products. It uses a single "master" DRACAS and DRB but with several specialized subservient DRACASs, such as for requirements issues or for issues in software design, with each having its own board. These subservient DRACASs are used to focus on key areas. Reviews and inspections are used for all major document products, such as specifications, software code, and hardware drawings. Defect detection rate monitoring using a Rayleigh distribution is used to assess if defect detection appears to be adequate. Use cases and use scenarios are used to clarify requirements. Prototypes are developed to reduce risks. Process failure modes and effects analysis (PFMEA) is performed for select critical processes.

The project has an eight-hour reliability requirement for the hardware and software system. This requirement is to be verified through a reliability demonstration test. Company B activities traditionally associated with reliability engineering include working with the customer to develop and refine the OP, FDSC, and operating environments, define reliability terms, perform reliability risk assessments, produce and maintain the RAM plan and the reliability casebook. The project produces a software defect root cause analysis, a software FMEA, and reliability block diagrams to model the hardware and software reliability. These models use the early software LRU list and their assessed SLOC and code types to allocate reliability to the software LRUs and to predict the software reliability. The list of hardware LRUs is used in the hardware reliability allocations and predictions. The project also develops parameters for the reliability demonstration tests for each build, and the results of these tests support the reliability growth tests. Finally, formal methods are used on some of the safety critical software.

A variety of metrics are used to assess progress and alert for potential problems. The company uses the following metrics for tasks in the CDP phase:

1. Project metrics: Some of the metrics that Company B uses to track the project are as follows:
   (a) Staffing: Company B uses staffing numbers, employee mix, and employee turnover rate to monitor for potential staffing issues and as an early indication of employee discontent.

(b) EV, AC, and PV: As with the previous example, Company B uses EV, AC, and PV as a measure of cost and schedule performance. These three metrics can be used to develop other metrics, such as SPI and CPI.

(c) Project churn: Company B also measures the number of project plan changes per month, the average advanced notice of a project plan change, the reasons for the changes and their percentages, and the number of times that the same aspect of the plan is changed. These metrics provide an insight into why churn is occurring.

(d) Customer sentiment: As a contractor having prime responsibility for a large government project, Company B's contractual performance is periodically rated by the customer and a poor rating may have financial implications. Company B tries to obtain high ratings and works with the customer to determine where issues may be and what actions are likely to relieve the customer's concerns.

2. Process metrics: For process metrics, Company B is primarily assessing how well a process is working and some of the metrics used to do this include the following:

(a) Process time throughput: Company B is working on a large contract with many processes. Poorly working processes can severely affect schedules, so detecting such issues early is important. Recall that this metric is defined as how many output units are produced per unit time, such as how many requirements are produced per day.

(b) Process cost throughput: Just as issues with processes can negatively impact schedules, they can also negatively impact budgets. Process cost throughput, which measures how many output units are produced per dollar spent on the activity, can detect such issues.

(c) Process error rate: Processes are not perfect. Even a process that has performed well over the years may start introducing more errors because of a change in some aspect of the project. The process error rate is the ratio of errors per designated unit, such as requirements errors that the requirements inspections did not catch.

(d) Peer review results: Periodically, Company B chooses a specific process and performs a peer review of it. The number and severity of findings are treated as process metrics.

(e) Rework per process: This metric looks at how much rework there is due to a process error or defect for an item that has been through the defective process.

(f) Defect detection rate monitoring: It is important to detect defects early, and indications that the project is not detecting enough of them early enough can be important. Defect detection rate monitoring using a Rayleigh distribution provides an early indication that defect detection is not progressing as expected.

3. Product metrics: Examples of Company B's product metrics applicable for this phase are as follows:

(a) Traceability of conceptual design to system specification

(b) Number and severity of comments from peer reviews on the plans

(c) Number and severity of comments from peer reviews on the conceptual design

(d) Number and severity of comments from peer reviews on the reliability allocations

(e) Number and severity of comments from peer reviews on the reliability predictions

(f) Customer feedback on products

(g) Comparison of reliability allocations and predictions

4.3   Requirements and Interfaces

As the name implies, the Requirements and Interfaces phase takes the conceptual design and produces requirement specifications and interface documentation used to create the design. These products are important bridges between customer needs and product design.

4.3.1   Description of the Requirements and Interfaces Phase

In this phase of a project, operational requirements are refined into design requirements that are sufficient to design the product. System and software requirements and interfaces are developed. Functions are assigned to hardware and software and the criticalities of these functions are determined. These requirements include the reliability requirements for the software and how they are to be measured. This phase also produces interface documentation for both internal and external interfaces. These interfaces may be project-developed interfaces or developed external to the project. Throughout Section 4.3, when we refer to requirements, we also include interface requirements unless otherwise noted. Tasks for the Requirements and Interfaces phase of a project typically include the following:

1. Requirements elicitation: Requirements elicitation is the process of discovering requirements for a system. In requirements elicitation we determine the performance, features, and behaviors required in the functional, non-functional, and interface requirements to ultimately meet the customer needs and wants. Although typically a requirements process, this book includes elicitation of high-level requirements in the CDP phase as that phase develops the statement of needs and operational requirements.
2. Requirements analysis: Requirements analysis is used to take user needs and decompose them into clear, doable requirements and to check the requirements for necessity, consistency, completeness, feasibility, unambiguousness, affordability, and verifiable. Ideally, the requirements are refined until there is agreement between all of the stakeholders.
3. Requirements documentation: We need to document the requirements in a complete, well-organized manner and make sure that the entire team has access to the requirements.
4. Requirements verification: Ultimately, requirements verification is performed in the IV&V phase, but a determination of how to verify them is made in this phase. Another aspect of requirements verification for this phase is to make sure that each requirement meets quality standards.
5. Requirements validation: Validate the requirements by performing an examination of the requirements as a whole to ensure that the collection meets the operational requirements and therefore the customer needs.
6. Requirements management: Developing and maintaining a good set of requirements is a complex task and requires well-planned management. Requirements management should consider the elicitation, analysis, documentation, verification, validation, and changes in requirements. It considers priorities, budgets, and schedules to ensure that quality requirements are produced within budget and schedule constraints.
7. Update products from the previous phase: As the project matures, new information allows us to produce better versions of products created in previous project phases.

Products from previous phases, such as the BOM, staffing estimates, and project milestones, should be reviewed and updated as needed.

Other products more directly related to software reliability that should be developed or refined in this phase (and possibly refined again later) are included in the following tasks for this phase:

1. Refine the FDSC.
2. Refine the OP.
3. Update the risk assessment of the project.
4. Review the reliability program plan and update as needed.
5. Update the reliability models of the system.
6. Update the software reliability predictions.
7. Update the assessment of the level and type of validation required for each product and document them in the reliability casebook.
8. Refine the reliability allocations. These allocations play an important role in determining the reliability requirements for the specifications.
9. Continue assessing reliability metrics, adding new metrics, and removing unused metrics as appropriate.
10. Continue and expand the DRACAS and DRB activities as necessary.
11. Performing reliability trade studies are needed.

The main products of this phase are specifications that can be used by the design team to design the product. For complex products, there is usually a hierarchy of specifications, such as a system specification, specifications for subsystems, and then a hardware specification and a software specification for each subsystem. A software specification is often called a software requirements specification or SRS. One or more interface specifications are also produced to cover interfaces. A number of recommendations for writing good requirements can be found in [7]. Other material on requirements may be found in [8].

There are various software development processes that may be used to develop software, such as spiral, incremental, waterfall, agile, or cleanroom approaches, and the choice of development process may affect the form of the requirements and interfaces documentation. For example, with agile processes, "executable specifications" are often used. These specifications are in the form of a collection of tests. While the techniques and processes used for a project may vary depending on the chosen software development process, requirements and interfaces remain critical aspects of producing reliable software.

4.3.2 Defects Typical for the Requirements and Interfaces Phase

Inconsistent or poor requirements can easily result in software defects. For example, different programmers may be responsible for the development of code to satisfy the inconsistent requirements. Each programmer may produce good code satisfying their requirements but when combined, software failures occur. Similarly for interface errors, different programmers may be responsible for different sides of the interface, and while each programmer produces good software with respect to their side of the interface, the overall software is defective. Incorrect or inconsistent requirements or interfaces can be difficult to find and may result in difficult to find software defects, increasing the likelihood that the defects find their way into the released product. The following examples of defect types apply to specifications and interface requirements:

1. Missing requirements: For this defect, a requirement is needed but is not in place. For example, not properly covering the complete operational environment in the requirements can result in the occurrence of an unexpected situation that, because of a missing requirement, the software is not designed for. The requirements need to consider all of the operational environments and modes of the system. Likewise, all interfaces need to be specified, including interfaces between project-developed software, project-developed and third-party software, database interfaces, and user interfaces.

2. Unnecessary requirements: Just as not having requirements that are needed results in defects, having unnecessary requirements causes excess work and possibly reduces design options or causes requirement inconsistencies.

3. Incorrect or unrealistic requirements: Although basic, all requirements should be correct and realistic. Each requirement statement should accurately and completely state a system need. Also, each requirement, and the requirements as a whole, should be realistic.

4. Unnecessarily restrictive requirements: Requirements should ensure that the specified product meets the needs of the customer, but they should also be written so as to not unnecessarily restrict the designers when they take the requirements and design the product. A common problem with requirements is over specifying by putting design details in the specification.

5. Conflicting requirements: Conflicting or contradictory requirements may be subtle but are a common cause of software defects. Even though the individual requirements may be correct for the situations considered, as a collection the requirements, in certain environments or modes of operation, may be in conflict.

6. Unverifiable requirements: All requirements should be stated in a manner that allows for a clear and unambiguous "pass/fail" assessment.

7. Requirements churn: Requirements sometimes need to be changed but doing so too frequently (requirements creep or requirements instability) can cause turmoil in a project. Even what appears to be a small change may have significant impacts to later phases of the project. Make sure that everyone knows the cost of requirements changes. Changes to requirements also need to undergo the same rigor and thoroughness as the initial requirements and changes need to be communicated throughout the project and undergo proper configuration management.

8. Misunderstood requirements: Different people may interpret a requirement differently and use the requirement per their interpretation. Requirements may be misunderstood because of ambiguous or poor wording, hidden assumptions, poor communication, excessive complexity, or a variety of other reasons.

   Requirements use imperatives such as "shall," "will," "is required to," "are applicable," and "is responsible for." These may have different meanings. For example, sometimes a requirement with the imperative "shall" is a "must meet" requirement, whereas a requirement with a "will" is a "nice to have" requirement. Any such distinctions should be clearly stated in the specification.

Sometimes the true meaning of a requirement is not determined until a "pass–fail" verification statement is determined. If this occurs late in the project, the various (possibly inconsistent) interpretations may have become embedded into the design and software.

9. Traceability and allocation issues: A parent requirement in a high-level specification may not allocate to a lower-level, more detailed requirement in a lower-level specification. Also, a lower-level specification may have a requirement that does not trace to a requirement in a higher-level specification. Either of these cases indicates a potential requirements defect that could cause a software defect.

10. Poor decomposition: Systems are decomposed in a specification. Often system functions are decomposed into subfunctions. For a complex system, a hierarchy of specifications is typically used. How the complex system is decomposed into a hierarchy can significantly impact downstream activities. For example, a well-thought out decomposition makes interfaces much more simple and verifiable than one that is poorly thought out.

11. Inadequate validation of requirements: Requirements should be validated to ensure that each meets a need. If requirements are not validated or are poorly validated, the project runs the risk of having unnecessary requirements and not having requirements that are needed to make a product that supports the customer.

4.3.3 Techniques and Processes for the Requirements and Interfaces Phase

This section starts with general techniques for the Requirements and Interfaces phase and then covers some techniques that anticipate requirement defects and aim to prevent them and techniques that are used to find existing defects and remove them. Note that a project does not have to do all of these techniques to produce reliable software; however, a judicious choice of some of these techniques, and possibly others not included here, should be made for this purpose. This is a sampling of techniques and not a complete or mandatory list.

1. General techniques for the Requirements and Interfaces phase:
   (a) Requirements management: Develop a requirements plan with processes to manage the requirements elicitation, modeling and analysis, documentation, validation, and changes. Requirements changes will happen and are not necessarily a bad thing, but they must be managed. Know when to baseline requirements. Use proper configuration control methods with change impacts to ensure that requirements changes are properly assessed, impacted, budgeted, and scheduled. Expect that there will be rework of some of the requirements and design the schedule accordingly.
   (b) Requirements format: There are several requirements specification formats. These formats are designed to ensure that critical information is included and when multiple specifications are used, that they are consistent in format.
   (c) Requirements traceability and allocation analysis: Every requirement should be traceable to a source, and every requirement should be in a path that allocates to design elements. Ultimately, there should be a justifiable path from customer need to design implementation. Requirements should also be allocated to verification and validation elements. This analysis applies to non-functional as well as functional requirements. Use tools and databases to help automate and enforce this process.

Finally, traceability and allocation of requirements should start early in the project and not be something done later to "tidy up loose ends."

(d) Modeling to analyze the requirements and interfaces: A model is an abstraction that helps us represent complex information and situations in ways that are easier to work with. They help us organize, analyze, and communicate information. Modeling could be of the organization in which the system will operate, modeling of the data and data flows, modeling of system behavior, or modeling other aspects of the system and its environment. Use cases are often used for requirements modeling. Use case modeling defines functional requirements in terms of use cases and actors, often using Unified Modeling Language (UML) or Systems Modeling Language (SysML). These languages are used to represent software concepts and software-intensive systems.

We can also use effectiveness models and simulations. As the requirements are refined from the operational requirements and more detailed requirements are introduced, the effectiveness models and simulations used with the operational requirements can be refined and used to test alternative requirement choices.

(e) BDD for requirements development: BDD creates user stories with short narratives and acceptance criteria or scenarios. Two critical parts of this approach to specification generation are the early development of acceptance criteria and an emphasis on stakeholder participation.

(f) Automated requirements support tools: Requirements management is almost always a major effort, and automated tools such as requirements databases can reduce the workload and reduce the risk of human errors. For example, such tools can support traceability analysis and configuration control. Also consider automated tools for requirements modeling, such as requirement diagrams. Ideally, these modeling tools will be easily integrated into the requirements management tool or tool set so that the suite of tools is easy to use and therefore reduces human errors. There is typically upfront cost to the automation, but afterward, the automation should save time and money. However, consider all project users, processes, and products before choosing the tools to make sure that they are consistent with the needs of the project.

(g) Use a software reliability advocate: A software reliability advocate continuously assesses the project for potential software reliability issues and advises project personnel on the impacts and mitigations of these issues.

2. Anticipation techniques for the requirements phase:

(a) Checklists: Checklists can be used to ensure that all of the characteristics of good requirements are addressed. They can also be used to ensure that all types of relevant requirements are included in the specifications. Checklists can also be used to know and emphasize where schedule and budget impacts typically occur for requirements activities and ensure that critical tasks are performed.

(b) Functional FMEA or FMECA: By performing a functional FMEA or FMECA, we take high-level functions, systematically look for ways that they can fail, assess the impacts of these failures, and address potential mitigations. By doing this, we can determine if additional or modified requirements are needed to reduce the likelihood of these failures.

(c) Training: Train requirements writers, peer reviewers, and other stakeholders in the characteristics of good requirements.

(d) Requirements validation: For requirements validation, we ask if the requirement specifies the right product. Each requirement should ultimately address one or more customer needs or wants. (In addition, some requirements may address "business non-value-added activities" that do not add real value from the standpoint of the customer, but are still required. An example is a regulatory requirement.) Requirements validation may use any of a number of different techniques, including the following:

   i. Prototyping
   ii. Modeling
   iii. Reviews and inspections
   iv. Validation checklists

These typically compare the requirements against earlier artifacts such as the operational requirements, use cases, or user scenarios.

(e) Customer feedback: Make sure that requirements writers have access to internal and external customer inputs and feedback. Involve internal customers (for example, members of the software architecture team and of the test team) in the development of the requirements and in changes to them.

(f) Restricted languages for specifying requirements: Requirements written in natural language can be vague and ambiguous. Because of this, restricted languages for requirements writing have been developed. A restricted language forces the requirements writer to be more precise and in some cases may make the transition from requirement to code more defect free. However, restricted languages reduce expressiveness and often require training.

(g) Fault tree analysis (FTA) on software requirements: One way that FTA can be useful in preventing requirements defects is to use a fault tree to determine ways that the performance of a high-level requirement may go wrong and to modify the requirement or to create lower-level requirements to protect against these defects.

(h) Reliability predictions: In this phase, more is known about the software than in the CDP phase. Use this updated information to update the reliability predictions from the previous phase. These predictions can then be used to assess if the software is likely to meet its reliability requirements and goals. A number of useful reliability prediction techniques are found in the Software Reliability and Availability Predictions topic in Section 6.4 and in [6].

3. Feedback techniques for the requirements phase:

(a) Peer reviews of requirements: Peer reviews are a way of "paralleling" a task so that if one person errs, a second person may provide the "redundancy" to catch the error.

(b) Statistical process control and trend tests: Consider the use of statistical process control and trend tests to assess peer review results. These approaches may enable a more precise assessment of the implications of the peer review results.

(c) Formal inspections of requirements: Formal inspections usually catch more defects than the less formal peer reviews but are more "expensive." As a result, both have their place. One of the more noted formal inspection techniques is the Fagan inspection (see [9]), named after its creator Michael Fagan.

(d) Static requirements tools: There are a variety of static requirements tools. One use of such tools is to look for vague wording in requirements and therefore reduce ambiguity.

(e) FTA on identified software requirements defects: FTA is listed in the anticipation techniques as a means of looking for what may go wrong. We can also use FTA as a feedback technique for finding the root cause of a detected and identified requirements defect.

4.3.4 Metrics for the Requirements and Interfaces Phase

As with the CDP phase, metrics are also important during the Requirements and Interfaces phase. Project metrics are covered at the start of Section 4.1. Examples of process and product metrics applicable to this phase are covered below.

Process Metrics Recall that the process metrics monitor the processes for, as a minimum, applicability, correctness, and effectiveness. A number of applicable process metrics are listed earlier in this chapter. In the Requirements and Interfaces phase, these metrics can be applied to, among other things, the following processes:

1. Planning and managing the requirements and interfaces processes
2. Performing peer reviews and inspections of the products
3. Producing and documenting specifications and interface documentation
4. Using requirements elicitation techniques
5. Performing analysis, such as analysis of requirements, FTA, functional FMECA, and reliability predictions
6. Developing and using prototypes
7. Using DRACAS and DRB
8. Performing configuration control activities
9. Using metrics

Product Metrics Examples of product metrics for specifications and interface documents include the following:

1. Number of lines of text: This number is a metric of the size of the specification.
2. Number of times that to be determined (TBD), to be reviewed (TBR), to be supplied (TBS), or to be added (TBA) is used: These phrases imply a lack of completeness.
3. Number of imperatives/requirements: An imperative implies that an action is required. This number is used to measure the total number of requirements. Example imperatives are "shall," "will," "must," "is required to," "are applicable," and "is responsible for." Note that there may be different imperative counts because of different implications. The word "shall" is sometimes used to state a "must meet" requirement, whereas "will" may be used for a requirement that is "nice to have." If there is a difference in the meaning of imperatives, the specification should make these meanings clear.
4. Number of directives: A directive is a reference to a figure, table, or note.
5. Number of continuances: A continuance is a phrase that follows an imperative and that introduces lower-level requirements. This metric is also a part of the requirements

count but is for lower-level requirements. Continuances include phrases such as "following," "as follows," "as listed," "in particular," and "support."

6. Number of options: Options are words or phrases that seem to give the software developer alternatives in the design although they may also introduce ambiguity. Examples of words that imply options are "can," "may," and "optionally."

7. Specification reading ease: Metrics such as the Flesch reading ease test score, the Flesch–Kincaid grade level, or the Fog index can be used as a metric for reading ease. Note that a higher score (meaning that the specification is on a lower reading level) is not necessarily better. The idea is to be on the reading level of the target audience.

8. Weak phrases: Weak phrases are phrases or clauses that are likely to introduce ambiguity and uncertainty. Weak phrases are often indicated by phrases such as "as applicable," "adequate," "as appropriate," "as a minimum," "effective," "be capable of," or "be able to,"

9. Number of requirements without a verification approach: The final meaning of a requirement is determined by what is considered to be a successful verification of the requirement. It is particularly important for potentially confusing requirements to have a "pass/fail" verification approach to minimize confusion about the meaning of the requirement. In addition, determining a verification approach is a part of ensuring that the requirement is verifiable.

10. Number of defects: When performing reviews or inspections, defects are likely to be found. The quantity per type of review is a useful metric. For example, a review of the requirements decomposition may find significantly more defects than reviewing the requirements for conflicting requirements or unrealistic requirements. A more long-term metric for requirements defects is to count the number of defects that through root cause analysis are traced back to the given specification. These metrics can be subdivided by phase that the defect is found in.

11. Traceability and allocation metrics: These metrics include items such as the number of requirements that do not have a parent, or the number of requirements without a child or an implementation (such as to be implemented in a particular module of code). Ideally each requirement should be traceable to a user need and allocated to some aspect of the design. (Note that some requirements are necessary because of regulations or other constraints.)

12. Requirements volatility: This metric measures the rate of change in the requirements. One way of quantifying it is as the ratio of requirements changes (additions, removals, and modifications) to the total number of requirements in a given period of time. Different versions of the metric occur when we start the counting at different points in the project. Different versions can also be created by considering the volatility of requirements of a particular priority.

Another version of this type of metric is to count the number of requests for changes. The ratio made from this quantity adds information about potential turmoil in requirements understanding or potential implementation issues.

Examples We end this section by continuing the pair of examples introduced in the CDP section. Example 4.3.1 continues Example 4.2.1 from the CDP section and is of a small company that produces a commercial software product for hospitals. Example 4.3.2 continues Example 4.2.2, which is of a large company that works on large government contracts.

Example 4.3.1   This example continues Example 4.2.1 from the CDP phase through the Requirements and Interfaces phase. Recall that Company A is a small software company that produces software for hospitals. It follows an agile philosophy and uses agile processes such as BDD to produce its products. Company A does not create a series of formal specifications but rather creates a list of software behaviors for each version of their product.

The company elicits a new list of desired behaviors for its software before the start of a sequence of build sprints. After elicitation, it decides when to start a sequence of sprints to produce a new version of their product, with the first sprint being a planning sprint and the last being a transitional sprint. Recall that the main tasks for the Requirements and Interfaces phase include the following:

1. Requirements elicitation
2. Requirements analysis
3. Requirements documentation
4. Requirements verification
5. Requirements validation
6. Requirements management
7. Update products from the previous phase

Requirements elicitation for Company A is described in Example 4.2.1. The company makes extensive use of group and other techniques to continuously update their lists of customer needs and wants.

Most of the requirements analysis for a new sequence of build sprints is performed during the planning sprint, although typically each build sprint has some as well. Requirements analysis for a sequence of build sprints consists of a collaboration between a software developer, someone with the business perspective, and someone with a quality perspective. The requirements start as user stories in an "As a – I want – so that" format with "Given – When – Then" acceptance criteria. The collaboration team first ensures that user story desired behavior is clear, so the "Given – When – Then" format is often used to create example verification tests for the behavior. Once the requirement is clear, the team determines its size in story points. The team also uses a checklist of questions to assess the necessity, consistency, completeness, feasibility, unambiguousness, affordability, and verifiable of the requirement.

Requirements are initially documented as user stories expressed in an "As a – I want – so that" format with "Given – When – Then" acceptance criteria. As BDD progresses, the behaviors are expressed by executable tests. These artifacts are stored in a database for easy project access.

Requirements verification is performed as the behaviors are coded. As Company A uses BDD for their software products, each behavior is added by first creating a test that the software is expected to fail (because the new behavior has not been added to the code yet), then adding the new behavior, and retesting for the behavior. The behavior is verified by passing the new test and all previously passed tests.

Recall that requirements analysis is performed by a collaborative team consisting of a software developer, someone with the business perspective, and someone with a quality perspective. Requirements validation is primarily performed by meetings with this team and a collection of members of the team that elicited the original behaviors from the customer, and if possible customer representatives. These meetings review and assess whether the tested behaviors properly represent the targeted customer needs.

Company A manages requirements by the above-mentioned processes of eliciting needed behaviors, assessing when to create a new software version and what behaviors to include in it, and using a planning sprint, a sequence of build sprints, and a transitional sprint to plan, create, and assess the new software version for release.

Products from the previous phase, such as characterizing the software, OPs, and MTBF predictions are assessed throughout the sequence of sprints and the resulting products are updated as needed.

As with the CDP phase, Company A also uses metrics in this phase to assess progress and alert for potential problems. As product metrics for the CDP phase are often metrics of behaviors, and behaviors become the requirements of the Requirements and Interfaces phase, metrics of these two phases overlap. Examples of other product metrics used in the Requirements and Interfaces phase include the following:

1. Progress toward incorporating the behaviors assigned to the build
2. Progress toward incorporating behaviors from the customer's "wish list" (some of which are to be incorporated in the subject software version)
3. Defects found before software release that can be partially or fully traced back to a requirements defect
4. Defects found after software release that can be partially or fully traced back to a requirements defect
5. Requirements volatility

Example 4.3.2 This example continues Example 4.2.2 from the CDP phase through the Requirements and Interfaces phase and is of a large company, Company B, working on a large government contact. Company B must deliver a hardware and software system at the end of the multi-year contract. The contract also requires that they develop and deliver a collection of documents and meet a series of milestones, such as a PDR and a CDR. The government has given Company B a SOW and a high-level specification for the contracted system as well as other guidance documentation and has personnel on site to provide help and to monitor progress. It also requires that Company B produce lower-level specifications on down to hardware specifications, software specifications, and interface specifications. Company B is using an incremental develop approach where tasks from the CDP and Requirements and Interfaces phases are performed early, followed by several Design and Coding (DC) and IV&V phases. Each of these iterations produces a version of the final product with more features and capabilities than the previous version, or build. At the end of the contract, the product should meet the full set of requirements and be ready for limited production. Recall that the main tasks for the Requirements and Interfaces phase include the following:

1. Requirements elicitation
2. Requirements analysis

3. Requirements documentation
4. Requirements verification
5. Requirements validation
6. Requirements management
7. Update products from the previous phase

Company B is given a high-level set of requirements by the customer, so requirements elicitation is mostly a matter of developing a deeper understanding of how the customer interprets the requirements. Also, new and modified requirements are needed because of the information learned from the conceptual designs and other sources. Much of the requirements elicitation is through meetings with the customer, however prototyping and use cases and user scenarios also drive requirements elicitation.

Company B puts the project requirements into a large database and uses software that is designed to analyze and manage requirements. This analysis software helps traceability analysis, requirements allocation, and requirements readability analysis, to name a few uses. Company B also uses use cases and user scenarios to analyze requirements. Reviews are performed on requirements from various different viewpoints, such as readability and clarity, verifiability, format, and consistency.

The requirements are documented in a deliverable hierarchy of specifications. There are hard copies of these specifications, but most specification users use the requirements database, which contains the latest approved copy of each specification. The entire project and the customer have access to the database.

One of the viewpoints used for requirements reviews is requirements verifiability. Each requirement has an associated verification method, which is either verification by test, analysis, demonstration, or inspection. Requirements verification is a major effort resulting in plans and procedures for how to verify each requirement, describing the expected results, and producing an unambiguous pass/fail criterion.

Requirements validation is the process of ensuring that the requirements define the system that the customer actually wants. In the Requirements and Interfaces phase, Company B uses customer reviews as the major means of performing this task.

How requirements are managed is documented in a requirements plan, along with a number of processes and procedures for specific aspects of requirements engineering. Company B uses a database to store and manage many aspects of the tasks and its use allows for automation of some of the more tedious tasks. A related part of managing requirements is change management and change processing. Requirements change not just when going from one build to the next but also within a build, and careful control of these changes is critical.

A number of products from the previous phase are updated when applicable, such as the FDSC, OP, reliability models, reliability allocations and predictions, risk assessments, the reliability plan, characterizing the software, and running the DRACAS and DRB.

As with the CDP phase, metrics used in the Requirements and Interfaces phase are used to assess progress and alert for potential problems. Company B continues to use the project and process metrics through the life of the project. Their metrics for requirements are measurements of the size of the requirements tasks to help with resource management, status metrics to determine if alternate actions are called for, and metrics for requirements quality.

Examples of their product metrics more specific to the Requirements and Interfaces phase include the following:

1. Requirements size
   (a) Number of requirements
   (b) Number of directives
   (c) Number of continuances
2. Requirements status
   (a) State of each requirement (such as proposed, approved for incorporation into a specification, approved for incorporation into a specification but deferred, incorporated into a specification, reviewed, incorporation into a specification has been approved, to be changed, and rejected)
   (b) Number of times that TBD, TBR, TBS, or TBA is used
   (c) Number of requirements without a verification method
   (d) Number of requirements without an approved verification plan and procedure
   (e) Percent of requirements that have been validated
   (f) Requirements volatility
3. Requirements quality
   (a) Total number of requirements defects found from the requirements inspection
   (b) Defect density for requirements found from the requirements inspection
   (c) Requirements inspection results, by category (such as the number of requirements that are missing, unnecessary, incorrect or unrealistic, unnecessarily restrictive, conflicting, unverifiable, or ambiguous)
   (d) Traceability and allocation metrics
   (e) Defect detection rate monitoring results to provide an early indication if requirements defect detection is progressing as expected.
   (f) Defects found before software release that can be partially or fully traced back to a requirements defect
   (g) Defects found after software release that can be partially or fully trace back to a requirements defect

4.4 Design and Coding

In this phase, we produce the software architecture, choose or develop our algorithms, and design and code software. This phase continues through to the successful unit testing of the code.

4.4.1 Description of the DC Phase

Once design-level requirements exist, the project proceeds with designing the software (although there is often considerable overlap in these two phases). Typically a software requirements specification (SRS) is used to guide the software architecture and software algorithm decisions, and these are used to design each of the software modules. The SRS and the software architecture, algorithm, and software module designs are used to produce a software design document (SDD) to describe the software architecture and design. Next the SRS, SDD, and coding standards, guidelines, and best practices are used to develop software code. Various software development processes may be used, such as spiral, incremental, waterfall, agile, or cleanroom approaches and the choice of development process may affect how the SRS and SDD are developed and what form they take.

Software design is often divided into "high-level design" and "low-level design." High-level design is sometimes referred to as software architecture design. The low-level design, also called "detailed design," details and refines the high-level design before the actual coding. Software architecture provides an overall top-level organization of the software. It breaks high-level functions into sub-functions (functional decomposition) and establishes relations and interconnections between them. It establishes software controls, hierarchies, data structures, interface characterizations, procedural details, and precedences. It determines what software language to use for the various parts of the software. The software architecture plays a large role in the performance, survivability, availability, modifiability, and reliability of the software. A well-chosen highly-cohesive software architecture provides a foundation for fault prevention and fault tolerance as well as for reducing operational and maintenance costs. Software architecture decisions tend to be costly to change once implemented.

The low-level design expands on the high-level design by refining the high-level design and by designing components of the software. Roughly speaking, high-level design goes down to the module level and low-level design covers the modules. The low-level design provides details such as the workings of the internal logic of the software, how the classes relate, and descriptions of the software modules. A good low-level design makes the coding easier, quicker, and easier to maintain.

After low-level design, design is translated into code. This code may be new "hand-generated," auto-generated, third-party, modified, or reused code. This part of the design process also creates any interfaces the software requires to use third-party, modified, or reused code. Coding typically includes unit tests, which are limited tests of the code before code integration and higher-level testing. Unit testing focuses on small units of the software. Part of the software design process should consider a software decomposition that makes unit testing productive. More on unit testing may be found in [10]. DC activities often include the following:

1. Software architecture design:
   (a) Using the requirements from the Requirements and Interfaces phase, create a software architecture specification. This specification defines the intended functions of the software, its interfaces, interoperability requirements, target platforms, distributability, maintainability, and other requirements needed to specify the architecture. Also, software systems change with time, so be sure to specify an architecture that is robust enough to handle these changes.
   (b) Choose the software language or languages along with rationale supporting these decisions. These may be documented in a design rationale document.
   (c) Determine programming support tools, such as editors, debuggers, compilers, linkers, and loaders.
   (d) Using the software architecture specification, design the software architecture.

(e) Verify the software architecture design. Verifying the software architecture design through the use of peer reviews, inspections, or some other approach to determine that the software architecture design meets it requirements.

(f) Validate the software architecture. This validation may be through peer reviews or inspections of the software architecture specification and documentation or based on techniques such as Architecture Analysis and Design Language (AADL), architecture tradeoff analysis Method[SM] (ATAM), other analysis methods, or combinations of methods. The validation is to determine that the software architecture meets the needs of the rest of the system and therefore of the customer.

(g) Produce a software architecture decision document describing the software architecture choices and the rationales for these choices. This activity usually includes developing architecture diagrams.

2. Detailed design:

(a) Develop a list of required algorithms and their interconnections necessary to satisfy the requirements with the specified architecture.

(b) Produce a specification for the algorithms.

(c) Choose the algorithms and provide rationale for the choices.

(d) Produce an algorithm description document.

(e) Develop class diagrams.

(f) Develop a software LRU list and the relationships between the LRUs.

(g) Update the size estimates for the software, including an estimate for each software LRU.

(h) Update the BOM.

(i) Verify and validate the detailed design.

3. Software code:

(a) Determine the type of code to use for each module, such as new "hand-generated," auto-generated, modified, reused, or third-party code. Document each decision and its rationale. These decisions may be documented in a design rationale document.

(b) Produce software implementing the algorithms. It is advantageous to produce test cases and to test the code incrementally as the code is developed. For arguments in favor of producing the test case before the code, see [2].

(c) Perform code inspections, unit testing, and validation techniques on the code.

(d) Document the code inspections, unit testing, and validation techniques performed on the code.

Potential products for this phase include the following:

1. Software architecture design:

(a) A software architecture specification

(b) A software architecture decision document describing the software architecture choices and the rationales for these choices

(c) A software architecture design

(d) Documentation of the software architecture validation

2. Detailed design:

(a) A list of algorithms

(b) Algorithm specifications

(c) A software detailed design

(d) Documentation of the analysis that is used to choose the algorithms

(e) Design documentation that documents the detailed design and trade-offs made between performance, reliability, maintenance, security, safety, and other key aspects of the software design

3. Software code:
   (a) Test cases
   (b) Code
   (c) Code documentation
   (d) Unit test results
   (e) Documentation of the results of the code validation

4. Update products from previous phases: As the project matures, new information allows us to produce better and more current versions of products created in previous project phases. Products from previous phases, such as the BOM, project schedules, and the risk assessments for the project, should be reviewed and updated as needed.

Other products more directly related to software reliability that should be developed or updated in this phase (and possibly refined again later) include the following:

1. Update the reliability models and predictions for the system. Updates should be made as the design progresses to help assess design choices. The models and predictions at the end of this phase should be for the full design.

2. Update the reliability program plan to keep it current with the latest project developments.

3. Update the OP as needed based on the latest understanding of how the software will be used.

4. Update the FDSC as needed based on the latest understanding of how the software will work.

5. Continue assessing reliability metrics, adding new metrics as appropriate and removing metrics that are not useful.

6. Continue and expand the DRACAS and DRB activities as necessary.

7. Perform reliability trade studies as needed.

Designers should work with testers to ensure that the software is testable and work with maintainers to make sure that the software is maintainable. There may also be software prototyping to reduce risk by constructing and testing immature and higher-risk subsystems or components.

4.4.2   Defects Typical for the DC Phase

The DC phase is the phase that most people think about when considering software defects. It is the phase that significant code is first created. Examples of defects for this phase include the following:

1. Software architecture design defects: An architectural defect can result in software architecture behavior that is inconsistent with the system requirements and a poorly designed

architecture can make software defects much more likely. Defect examples include the following:

(a) Inadequate software architecture specification: Defects can enter the software because of the software architecture specification not adequately accounting for the requirements.

(b) Poorly designed software architecture: A software architecture is poorly designed if it is unable to perform an action that is required of it, if it performs the wrong action, or if it performs an action inadequately. An inadequately performed action may result in missing actions, incorrect or missing connections, or incorrect events. It may be due to performing an action at the wrong time, too slowly, in the wrong sequence, leaving the system in the wrong state after the action, or other incorrect or unwanted results. These unwanted results may be errors or mismatches in data types, missing or unintended data flows, or other issues.

(c) Poor documentation: A poorly documented software architecture may result in defects because it may result in project personnel misunderstanding the architecture and incorporating that misunderstanding into their products. For example, designers may use the documentation to design the software or testers to develop tests. Incorrect or ambiguous documentation may result in a mismatch between the architecture and the software that is designed at least partially based on the documentation. It may also be used as a part of the software architecture validation process and defects in the documentation may negatively affect the validation. Documentation of the analysis that is used to choose the algorithms, documenting trade-offs made in performance, reliability, maintenance, security, safety, and other key aspects of the software design are all important as well.

(d) Poor choice of software languages: Different software languages have different strengths and weaknesses and a poor choice of language can make development of the software more difficult and make the software more likely to have defects.

(e) Poor choices of tools: Compilers, interpreters, debuggers, and a host of other third-party tools may have defects or may be poorly documented and poorly understood. Even if they are reliable and well understood, they may not be suitable for the tasks that require them, making defect introduction more likely. These defects may be difficult to find and result in defects in the final product.

(f) Validation errors and defects: Errors and defects in validation may allow software architecture defects and inadequacies to go unnoticed and therefore uncorrected.

2. Detailed design defects: Examples of detailed design defects include the following:

(a) Algorithm errors: Incorrect algorithm choices or designs may result in poorly performing software. Even if "good" algorithms are chosen and each is designed well, they must be compatible relative to the needs of the project. In other words, they must work well together. Algorithm choices, specifications, and documentation all play a significant role in the number and severity of defects in the software.

(b) Poor choices in third-party software: Typically a software product is a combination of "company-developed" software and purchased third-party software, ideally working together to produce a quality product. A poor choice in third-party software, either poorly designed, poorly understood, or inadequate for the required tasks, can result in a defective overall product.

(c) Interface issues: Interface issues include interface incompatibilities, receiving or sending data incorrectly, at the wrong time, at incorrect rates, to or from the wrong locations, or various other issues.

3. Coding errors and defects: When we think of "software problems," we typically think of coding errors and defects. Examples of these include the following:

(a) Inadequate code: This type of defect occurs when we take good requirements and use them to produce code that does not conform to these requirements. Three sources of code inadequacies are as follows:

   i. Poor coding: One source of software defects is poor coding practices and techniques. Structuring the code, organizing it, consistent naming conventions, and other coding practices are important aspects of writing good code. Poor coding can result in complex code, code that is difficult to maintain, debug, or test, poorly commented code, or code that is incomplete because of aspects of the design not coded.

   ii. Human errors during coding: Even with good coding practices, code can be defective. Coding errors due to human errors are a common cause of software defects. These types of errors include errors in typography, syntax, logic, data typing, error handling, and so forth.

   iii. Misunderstanding the design: Even if the documentation is good, the design can still be misunderstood because of human error, lack of training, or other reasons.

(b) Testing errors: A good programmer tests code as it is being developed. Failure to test, or errors in this testing, can result in defects propagating to the next phase and possibly beyond.

(c) Poor installation scripts: Software installation is often performed through a software script and an error in this script can result in uninstalled software or defects in the installed software. Installation scripts should be written to cover all target environments. If this is not feasible, instructions for manual installation and when manual installation is required should be provided.

(d) Documentation errors: Poor documentation can easily result in incorrect decisions from those who use the documentation to understand and use the system. This documentation includes documentation for the users, documentation of the design and its implementation, and documentation of the validation of the design and code.

4.4.3 Techniques and Processes for the DC Phase

No single approach can be relied on as the sole technique for preventing or finding and removing software defects from the software and its design. Some types of defects are more likely to be found or prevented by one technique, while others are better handled by a different technique. A combination of techniques should be employed to produce high-quality software. The following is a sampling of techniques and not a complete or mandatory list.

1. General techniques for the DC phase:

(a) Use checklists based on best coding practices: A few examples for such a list are as follows:

   i. Simple, modular designs: Simple modular code is easier to understand, code, test, maintain, modify, and debug.

    ii. Reuse of certified code: Reuse of quality code can save time and expense and also improve software reliability.

    iii. Refactor: Refactoring code is the process of restructuring or changing code to improve its performance, readability, or quality without adding new features or external behavior. Be sure to refactor often.

    iv. Several desirable characteristics of a software design are listed in [2], including minimal complexity, ease of maintenance, high fan-in, low-to-medium fan-out, loose coupling, extensibility, and the use of standard coding techniques. These and other checklists in [2] can be used as a part of one or more coding checklists.

More on the use of checklists and on program best practices are covered in Section 6.5.

(b) Use a consistent development method and approach: Consistency reduces confusion and as a result reduces errors.

(c) Pair programming: Consider the use of pair programming to improve code quality.

(d) Test-driven development (TDD): TDD develops code incrementally making small changes to the software and with each increment, it first creates a test case for a requirement that the code should meet. It then implements the requirement by changing the code to pass the specific test case.

(e) BDD: Like TDD, BDD produces code incrementally, developing tests first, then producing code to pass the test. However, TDD generally focuses on producing code one feature at a time, BDD emphasizes developing code one behavior at a time.

(f) Models: Using models in the DC phase can take many forms. Examples of models for designing software using a structured analysis approach include functional flow block diagrams, data flow diagrams, entity relationship diagrams, and state transition diagrams. Models applicable for object oriented programming often use UML, robustness diagrams, and function class decomposition. Software architectures can be modeled using AADL or other architecture design languages to describe and analyze a software architecture.

(g) Simulations: Simulations can be used to aid in choosing algorithms. They can also be used to test parts of the code.

(h) Architecture Tradeoff Analysis Method^{SM} (ATAM): This method of analyzing a software architecture gathers stakeholders to analyze system goals and requirements, resulting in a collection of scenarios that are used to create software architecture tradeoffs. Risk analysis is performed on these tradeoffs resulting in architecture decisions. See Section 6.3 and [11] for more on this method.

(i) Algorithm description document: An algorithm description document documents descriptions of the algorithms and serves as an intermediate step between requirements and software design.

(j) Software architecture decision document: A software architecture decision document captures the decisions that go into the software architecture design effort. These decisions should be made with inputs from all relevant stakeholders, and the decisions should be captured and formally documented to be used and adhered to throughout the program. Relevant stakeholders include the following:

    i. End users

    ii. Designers

    iii. Testers

    iv. Component vendors

    v. Reliability engineering

    vi. Safety engineering

    vii. Security and vulnerability engineering

    viii. Software administrators

    ix. Maintainers

(k) Continuous integration: Continuous integration is the practice of frequently committing code to a version control repository and building and testing the code each time the code is committed. Typically each person commits code at least daily. This process reduces integration problems by forcing associated efforts to be more coordinated. There are various tools to help manage the continuous integration process. More on continuous integration may be found in the topic on Continuous Delivery and Continuous Delivery Pipeline in Section 6.3.

(l) Requirements allocation and tracing: Each documented design item should be traceable to a requirement, and each requirement should be allocated to a point of implementation. This process is an important part of verifying the design.

(m) Good documentation practices: Although often considered more as an after-thought, good documentation practices from the start of the design process can save the project considerable pain. Poorly documented decisions can lead to confusion when implementing or verifying these decisions. Poorly documented code often results in unnecessarily complex code and misunderstandings.

(n) Configuration control: Maintain good configuration control throughout the design phase (and all other phases).

(o) Use a software reliability advocate: A software reliability advocate continuously assesses the project for potential software reliability issues and advises project personnel on the impacts and mitigations of these issues.

2. Anticipation techniques for the DC phase:

(a) Quality objectives: Set quality objectives for the programmers. Quality objectives should be specific, measurable, achievable, relevant, time-based, and agreed upon with both the person assigning it and the person responsible for it. However, balance these objectives with a goal of not "micro-managing" employees.

(b) Style guides: The use of coding style guides can improve the uniformity of the product.

(c) Best practices: Using software DC best practices can increase the consistency of the code and reduce errors.

(d) Lessons learned checklists: Like the use of best practices, a lessons learned checklist can improve consistency and reduce errors.

(e) Prototyping: Developing prototype code for some of the higher risk software functionality is an example of a risk-reduction technique. It demonstrates software functions and capabilities to assess if the proposed design meets the user needs. Prototypes are typically "throw-away" code and are not developed for usability, supportability, or other aspects that final code should have.

(f) Logging: Good logging can be critical to finding software defects.

(g) Refined reliability allocations: Allocate reliability-related requirements to the software modules. Reliability-related requirements, such as MTBF or reliability, are largely meaningless to the person writing the code. (How does the coder know that the code has a specific reliability value or that it has all of the defects that it is allowed to have?) The purpose of such values is to give an idea of priority to the coder and give "goals" for the test processes to achieve. To aid the software developer, provide them with reliability-related goals expressed in terms that are at least somewhat meaningful to them, such as the maximum allowable number of defects per thousand lines of executable code. Even non-quantitative (ordinal) assessments can be useful to prioritize the reliability efforts for different modules of code.

(h) Fault tolerance techniques: During the high-level design, determine what fault tolerance techniques to use and how to apply them.

(i) Design for non-functional requirements: During high-level design, consider reliability, maintenance, and testability. Choices made for other non-functional requirements, such as software safety and software vulnerabilities, can also affect software reliability. Early decisions, such as choices of modules and their interfaces, can have a significant impact on these requirements. For example, testable software is less prone to defects and code that is easily tested is easier to release and cheaper to maintain.

(j) Choice of operating system and compiler or interpreter: The operating system and compiler or interpreter can introduce defects into the system. More established operating systems and compilers or interpreters tend to be more reliable than the newer ones.

(k) Update reliability predictions: In this phase, more precise values for lines of code, types of code, and other factors are available. Use these to update the reliability predictions. A number of useful techniques are found in the Software Reliability and Availability Predictions topic in Section 6.4 and in [6].

(l) Software failure modes effects and criticality analysis (FMECA) or failure modes effects and analysis (FMEA): A software FMEA or FMECA is a bottom-up approach to analyzing the software for potential failure modes, their effects, and potential mitigations.

(m) Continuous improvement: Throughout the design phase, analyze each product and each process for potential improvements.

(n) Software metrics: Use software metrics to find routines and modules that are more likely to have defects. For example, more checking can be performed on the most complex routines. There are several commonly used complexity metrics with the most well-known being the McCabe complexity metric.

3. Feedback techniques for the DC phase:

   (a) Author reviews: Code authors should review their own code before peer reviews or inspections. Checklists can be helpful in ensuring that all code authors cover the same checks before peer reviews or inspections.

   (b) Peer reviews: Peer reviews can be performed with varying levels of formality. Both design products and code products should be considered candidates for these reviews.

(c) Inspections: Formal inspections can be a cost-effective technique for removing a significant percentage of the defects in a software product. Per [2], formal inspections of the design and of the code can remove 70%–85% of the defects in the products.

(d) Unit testing: Although unit testing is a type of software test, it is usually performed by the code developers and therefore is included here. Unit testing involves testing small units of code during code development to ensure that defects are caught and corrected early. Frequent unit testing is recommended throughout code development. More on unit testing can be found in [10].

(e) Statistical process control and trend tests: Consider the use of statistical process control and trend tests to assess peer review and unit test results. These approaches may enable a more precise assessment of the implications of these feedback results.

(f) DRACAS with root cause analysis: When a defect is indicated, be sure to perform a root cause analysis and use the results to prevent similar occurrences in the future. It is important to continually grow the reliability of the product and of future products.

(g) Demonstrations: Making frequent use of demonstrations, often with prototype software, can provide a sequence of progressive performance milestones for the project and for the customer.

(h) Static and dynamic analysis tools: Static analysis tools do not execute the code whereas dynamic analysis tools do. Each has its strengths and weaknesses. Static and dynamic analysis tools can be used to detect most defects in the use of the software language.

4.4.4 Metrics for the DC Phase

There are many metrics to choose from for this phase of a project. We again consider process metrics and product metrics. As with the other phases, project metrics are covered at the start of Section 4.1. Examples of process and product metrics applicable to this phase are listed below.

Process Metrics A number of applicable process metrics are listed in the first section of this chapter. In the DC phase, these metrics can be applied to, among other things, the following:

1. Performing analysis for high-level and low-level designs
2. Documenting the analysis and design
3. Producing code
4. Planning, performing, and documenting unit tests
5. Developing and using prototypes
6. Conducting peer reviews and inspections of products
7. Performing configuration control
8. Using DRACAS and DRB
9. Using metrics

Product Metrics Example product metrics for the DC phase include the following:

1. Software architecture design: Architecture design and detailed design metrics for this phase typically do not rely on the software code but rather are often computed from

the requirements specifications and design documents. They tend to be based on code functionality rather than code implementation. For example, semantic metrics are often applicable, as are morphology metrics. With call and return architectures, the software architecture is represented in terms of module calls and returns. A directed graph, or digraph (where the terms "graph" and "digraph" are from the mathematical topic of graph theory), is constructed where each node of the graph is a module and each arc is a call (from calling module to providing module) or a return (from providing module to calling module). Examples of metrics for these characteristics are as follows:

(a) Number of nodes
(b) Number of arcs
(c) The sum of the number of nodes and arcs
(d) Degree, in-degree, out-degree of a module: The in-degree of a module is the number of arcs terminating at the node. The out-degree of a module is the number of arcs initiating from the node, and the degree of a module is the sum of its in-degree and its out-degree.
(e) Average path length: The (directed) path length from node n_i to node n_j in a digraph is the number of arcs needed to go from node n_i to n_j, requiring that we travel in the direction of the arcs. (If there is no such path, the path length is said to be infinite.) The geodesic distance from node n_i to node n_j is the shortest such path length. The average path length of a digraph is the average of the geodesic distances from all pairs of nodes in the digraph.
(f) Eccentricity of a module: The eccentricity of a node n in a digraph is the maximum geodesic distance from node n to any other node in the digraph.
(g) Radius: The radius of a digraph is the smallest value of eccentricity over all nodes in the digraph.
(h) Diameter: The diameter of a digraph is the largest value of eccentricity over all nodes in the digraph.

Other metrics covering aspects of software architectures include the following:

(a) Results of peer reviews and inspections of the software architecture specification
(b) Measures of completion of software architecture checklists
(c) Number, severity, and estimated likelihood of software FMECA software architecture defects identified
(d) Estimated size and complexity of the software code based on the software architecture
(e) Metrics for the implied "costs" of the software system due to the software architecture choices. Here "costs" can mean any of a variety of things, such as the size and requirements of any databases that the architecture imposes on the system, the number of interfaces, how much third-party software is required, and the level of confidence we can have in those products, and numerous other implications of the architecture.

2. Software detailed design: There are many metrics for software design and each project should decide which are most useful for that project. Examples of software design metrics include the following:
   (a) Number of defects in algorithms: This metric consists of the total number of defects found through peer reviews, inspections, or both.

(b) Defect density for algorithms: Algorithms can be reviewed or inspected with the resulting defects being used to calculate a defect density with respect to the number of algorithm steps.

(c) Required memory, throughput, and other "costs" of the algorithms

(d) Potential sources of coded algorithms. For example, if the project has access to a trusted software package that has a given algorithm, this algorithm may rate higher than a new, yet-to-be programmed algorithm.

(e) Estimated size of the coded algorithms

(f) Number of parameters

(g) Number of modules

(h) Function point metrics

(i) Fan-in, fan-out: Fan-in is the number of other functions that call a given function in a module. Fan-out is the number of other functions that are called from a given function in a module.

(j) Complexity: Metrics for structure, data, and system complexity can be used.

    i) Structural complexity: The structural complexity for a given module is defined to be the square of the fan-out of that module.

    ii) Data complexity: The data complexity of a module is the number of inputs and outputs to the module divided by one plus the module fan-out.

    iii) System complexity: The system complexity for a module is the sum of its structural complexity and its data complexity.

Note that different sources define software architecture and software design differently, so per some definitions, some of these software design metrics may be software architecture metrics.

3. Software code: There are many metrics for software code. A few are as follows:

(a) Size metrics: The main software size metric is SLOC. The more lines of code, the more possibility of defects. Several variations include the following:

    i. Total lines of code

    ii. Non-comment, non-blank lines of code

    iii. Executable lines of code

    iv. Logical lines of code

    v. Commented lines of code

    vi. Effective thousand source lines of code (EKSLOC): Not all code is expected to have the same defect density (defects per thousand lines of code). For example, auto-generated code typically has a much lower defect density than newly developed "hand-generated" code. EKSLOC, is a weighted average of new, modified, reused, and auto-generated code. See [6] for more details.

    vii. Function points: A function point is a functional size metric, providing a measure of the amount of functionality that the code has.

Size metrics are typically applied to the overall code and also to modules of code.

(b) Halstead's metrics: Halstead's metrics, or Halstead's complexity metrics, use four primitive quantities,

    i. η_1: The number of distinct operators in the software program

    ii. η_2: The number of distinct operands in the software program

    iii. N_1: The total number of operators in the software program

    iv. N_2: The total number of operands in the software program

These primitive metrics are used to calculate several non-primitive metrics. A few of these are as follows:

   i. Software program length N: $N = N_1 + N_2$

   ii. Software program volume V: $V = N\log_2(\eta_1 + \eta_2)$

   iii. Estimated effort to write the software program E: $E = V(\eta_1 N_2 / 2\eta_2)$

Other non-primitive metrics are potential volume, level, difficulty of the computer program, time needed to write the program, and number of delivered defects. See [12] for more on these metrics.

(c) Complexity metrics: Excessively complex code is more likely to be defective. Several metrics for complexity are as follows:

   i. Cyclometric complexity: Cyclometric complexity is a measure of code complexity through the use of control statements. Fewer control statements generally means more simple and less defect-prone code. Cyclometric complexity is a measure of the number of linearly independent paths through a program unit. Empirically, McCabe found that a cyclometric complexity of 10 or less is desirable. See [13] for more on this metric.

   ii. Class coupling: Class coupling is a measure of the coupling of one class to other classes.

   iii. Code coverage metrics: These metrics can be useful for developing and implementing unit tests.

   iv. Fan-in, fan-out: Large values indicate more complexity.

(d) Software reuse metrics: These metrics consider the amount of reused code. Examples of these types of metrics include the following:

   i. Number of unmodified reused software modules

   ii. Fraction of software modules that are unmodified reused code

   iii. Fraction of lines of code that are unmodified reused code

   iv. Number of slightly modified reused software modules

   v. Fraction of software modules that are slightly modified reused code

   vi. Fraction of lines of code that are slightly modified reused code

   vii. Number of moderately or heavily modified reused software modules

   viii. Fraction of software modules that are moderately or heavily modified reused code

   ix. Fraction of lines of code that are moderately or heavily modified reused code

(e) Modularity metrics: Software modularity implies low coupling and high cohesion. Examples include the following:

   i. Number of modules

   ii. Module size

   iii. Object-oriented metrics: Several of the object-oriented metrics listed below measure modularity.

   iv. Semantic metrics: As with the object-oriented metrics, several of the semantic metrics listed below measure modularity.

(f) Object-oriented metrics: Examples include the following:

   i. Methods per class

   ii. Coupling between object classes

    iii. Lack of cohesion in methods

    iv. Inheritance dependencies

    v. Weighted methods per class

    vi. Degree of coupling per object

    vii. Depth of inheritance tree

    viii. Degree of reuse of inheritance method

    ix. Class hierarchy nesting level

    x. Number of abstract classes

    xi. Number of children per class

(g) Semantic metrics: The metrics listed above are generally classified as syntactic metrics because they use the source code syntax to determine the metric value. These metrics reflect how the software is represented but not what the software does. Semantic metrics are another type of metric. This type of metric is based on the function of the software within a given software environment. Semantic software metrics are less well developed than syntactic software metrics. A few examples for object-oriented software (found in [14]) include the following:

    i. Class cohesion

    ii. Class domain complexity

    iii. Relative class complexity

    iv. Class interface complexity

    v. Class overlap

    vi. Documentation quality of a class

More on semantics software metrics can be found in the Semantic Analysis of Code topic in Section 6.3 and in [14].

(h) Code churn: Code churn is a measure of the amount of change that a module has undergone. There are different ways to measure this phenomenon, including the following:

    i. Churned lines of code: The number of added lines of code plus the number of changed lines of code relative to the baselined version of the module.

    ii. Deleted lines of code: The number of deleted lines of code relative to the baselined version of the module.

    iii. Weeks of churn: The number of weeks that the code is open for editing from the baselined version of the module.

(i) Defect containment: A defect introduced in a particular program phase is ideally found in that phase. The further from that phase that the defect is found, the poorer the defect containment score.

(j) Agile metrics: These metrics are potentially useful in improving the software development processes.

(k) Software runtime defect metrics: These metrics use runtime data from software runs to measure defects. Some examples include the following:

    i. Defect density: Defect density is the number of confirmed detected defects per software size during a defined development or operation time period. Thousand lines of code or effective thousand lines of code are often used for software size.

    ii. Mean time between failures (MTBFs): This metric measures the average time between software failures. Time is typically operation or CPU time. It is closely related to the reliability of the software. This metric typically depends on the operational mode of the system.

   iii. Mean time to recover/repair (MTTR): This metric is the average time to restore the software after a failure.

   iv. DRE: This metric is calculated as the number of defects found before product delivery (E) divided by the sum of E and the number of defects found by the customer after product delivery (D). The result is a measure of how effective the project's defect handling is.

   v. Number of defects found before release

   vi. Number of defects found after release

   vii. Number of defects found before beta testing

   viii. Number of defects found after beta testing

   ix. Number of defects found that are traced to requirements issues

   x. Number of defects found that are traced to design issues

   xi. Number of defects found that are traced to coding issues

   xii. Number of defects found that are traced to testing issues

(l) Defect metrics for installation code: The above metrics are usually applied to the main product code but are generally applicable for installation code as well.

Product metrics for software code are also classified as either static or dynamic. A static software metric is a metric that can be collected without running the software, whereas a dynamic software metric requires that it be run.

One important application of these metrics is determining if the software is ready for IV&V. IV&V is expensive and efforts to reduce the cost and schedule impacts of it are important.

Examples We end this section by continuing the pair of examples introduced in the CDP section and continued in the Requirements and Interfaces section. Example 4.4.1 below continues Example 4.3.1 to the DC phase and is of a small company that produces a commercial software product for hospitals. Example 4.4.2 continues Example 4.3.2, which is of a large company that works on large government contracts.

Example 4.4.1 This example continues Example 4.3.1 from the Requirements and Interfaces phase through the DC phase. Recall that Company A is a small software company that produces software for hospitals. It follows an agile philosophy and uses agile processes such as BDD to produce its products.

The company elicits a new list of desired behaviors for its software prior to the start of a sequence of build sprints. It then decides when to start a sequence of sprints to produce a new version of their product, with the first sprint being a planning sprint and the last being a transitional sprint. Recall that the main tasks for the DC phase include the following:

1. Design the software architecture design
2. Perform a detailed design
3. Code the software

Company A routinely modifies its software to create new versions, and therefore they intentionally use a software architecture that allows for frequent behavioral updates.

However, the nature of these updates is not always predictable and so assessing if the software architecture is adequate for a proposed new version of a software product is part of the process of deciding when to create a new version of the product. The assessment team must consider whether the new version can be adequately supported on the software architecture of the existing product. If not and the team decides that an update is still a good business decision, then there may be a sequence of sprints that creates a new or modified software architecture in addition to incorporating the new behaviors. Most of the architectural design is in the planning sprint and earlier sprints although each build sprint may also have some architectural design.

New behaviors are added during build sprints and part of this process involves software design. Changes required to implement behaviors are kept small and incremental so that any design changes are likely to be simple and manageable. As with the architectural design, most of the detailed design is in the planning sprint and earlier sprints although each build sprint may have some as well.

Software is coded using the agile BDD technique. Using BDD, a test for the new behavior is developed and is used to test the existing code. The existing code is expected to fail. The code is then changed to add the new behavior and the new code is tested and is now expected to pass the new test and all previous tests. After passing the tests, there is typically some code refactoring and the process repeats with the next behavior on the list.

As noted in Example 4.3.1, products such as characterizing the software, OPs, and MTBF predictions are assessed throughout the sequence of sprints, and the resulting products are updated as needed.

Other techniques related to the DC phase include the following:

1. Company A often uses pair programming for code development.
2. There is a major emphasis on keeping the code modular. Modularity aids not only design but also test, production and release, and maintenance.
3. The design is documented through meeting minutes plus sketches and diagrams that often cover software architectures, data flows, user interface behaviors, and other ways to describe and document the DC decisions. Once the behaviors have been coded, Company A uses the code and automated tools to generate technical and end user documentation.
4. Earlier we stated that each documented design item should be traceable to a requirement, and each requirement should be allocated to a point of implementation. Company A keeps the path from user stories through meeting minutes with diagrams and sketches through to end documentation clearly documented and accessible. This way, the path from each customer need through company decisions to implementation in the code and on through tests that the behavior had to pass and ultimately to where it is described for the end user is available and can be used for analysis of defects and lessons learned. The company also makes sure that there is a path in the opposite direction, with each behavioral description in the user documentation being traceable through the testing and design implementation to a user story and customer need.
5. Exploratory testing is used multiple times during each build sprint.
6. In the transitional sprint, there are peer reviews covering the software architecture, design, and code. The peer reviews are a part of the verification and validation processes.

7. Company A uses a coding style guide, a list of best practices, and lessons learned checklists.
8. Code is committed at least daily to a version control repository.

As with the previous phases, Company A also uses metrics to assess progress and alert for potential problems. Examples of product metrics that the company uses in this phase are as follows:

1. Number of defects traceable to the software architecture, both before product release and after release
2. Number of defects traceable to the detailed design, both before product release and after release
3. Number of defects traceable to the coding, both before product release and after release
4. Code size
5. Code churn
6. Defect containment
7. Software runtime defect metrics, such as defect density, mean failure-free performance time, and DRE

Example 4.4.2 This example continues Example 4.3.2 from the Requirements and Interfaces phase through the DC phase. Recall that Company B is a large company working on a large government contact and must deliver a hardware and software system at the end of the multi-year contract. They are also required to develop and deliver a collection of documents and meet a series of milestones, such as a PDR and a CDR. Company B uses an incremental develop approach where tasks from the CDP and Requirements and Interfaces phases are performed early, followed by several DC and IV&V phases. Each of these iterations produces a version of the final product with more features and capabilities than the previous version, or build. At the end of the contract, the product should meet the full set of requirements and be ready for limited production. Recall that the main tasks for the DC phase include the following:

1. Design the software architecture design
2. Perform a detailed design
3. Code the software

Company B uses the software specifications to guide the development of the architecture design. They use Systems Modeling Language (SysML) to express and document the design. Key drivers are identified and options are listed. Fault tolerance, reliability, safety, and security are considered at the architecture level to provide as much flexibility as possible. There is then a risk analysis for each acceptable option. Using these results, the major stakeholders then meet to prioritize the options, modify them if needed, and then decide which to choose. A software architecture decision document is used to describe and document architectural decisions.

Software design is performed by the software organization. As with the software architecture, SysML is used to express and document the design. An algorithm description document is created to provide an intermediate step between the requirements and the detailed design, resulting in a hierarchical approach for organizing the code and recording algorithm requirements and decisions. An overall software design strategy is developed and briefed to all software designers. Pseudo-code, process diagrams, data flow diagrams, and detailed logic diagrams are all used in the design process. These are captured and used to document the detailed design and trade-offs made between performance, reliability, maintenance, security, safety, and other key aspects of the software design.

Software coding is the responsibility of the software organization. They use a coding style guide, best practices, and lessons learned checklists. Code is committed at least daily to a version control repository. There are also automated analyses and metrics taken of the code with each commitment. These are used to assess if these aspects of the code are as expected and within acceptable ranges. Author reviews and unit tests are performed and passed before considering the code ready for peer reviews. The code undergoes internal validation by tracing the code and design back to the requirements. In this phase, external validation is through demonstrations and customer reviews of the products.

In the Requirements and Interfaces phase, members of the software team work with the requirements team to develop the specifications. The software experts help the requirements team write doable software requirements and by working with the requirements experts, the software team members have a better understanding of the requirements that they must design to. Good communication between different disciplines is emphasized to make sure that the project is coordinated and fewer mistakes are made. Reviews of the architecture design, algorithm choices, detailed design, and software all help to find defects early. When found, defects go through the DRACAS and DRB. Also during this phase, a software FMECA is produced at the design level and is updated based on DRACAS results. Additionally, software reliability models, predictions, and potentially allocations are updated based on the latest product information.

Company B uses a variety of metrics for the DC phase tasks. Examples of product metrics that the company uses in this phase are as follows:

1. Graphical metrics, such as number of nodes, number of arcs, degree, in-degree, out-degree, and related metrics
2. System "costs" (such as number of interfaces, size of databases, quantity of third-party software) due to software architecture choices
3. Number and density of defects found in the reviews of the algorithms
4. Required memory, throughput, and other "costs" based on the algorithms
5. Structural, data, and system complexity
6. Code size
7. Code churn
8. Code complexity
9. Coupling metrics, such as efferent coupling, afferent coupling, and instability
10. Lack of cohesion in methods metrics
11. Number of defects traceable to the software architecture, both before product release and after release
12. Number of defects traceable to the detailed design, both before product release and after release
13. Number of defects traceable to the coding, both before product release and after release
14. Defect containment

4.5   Integration, Verification, and Validation

The next phase takes the software developed in the DC phase and integrates it into the overall system, verifies it, and then validates it.

4.5.1   Description of the IV&V Phase

IV&V activities take software from the DC phase that the design team has released for testing and applies a preplanned set of tests and analyses to it. Once the software modules, or "parts," have been produced, they must be integrated into the target hardware and tested as a whole. Integration testing ensures that the modules are compatible and have been integrated properly. Verification is used to determine if software satisfies its requirements. Validation is used to determine if the integrated system meets the needs of the customer. Validation determines if the right product is being built and verification determines if the product is being built right. Testing is typically the main method of performing both verification and validation. Although the emphasis of this phase is on the deliverable product software, installation software and documentation should also undergo verification and validation.

Many verification and validation activities are performed in parallel with the development of the software, not after the development. Not only are IV&V personnel learning about the requirements and design during earlier project phases, they are also providing inputs to increase the likelihood of IV&V success. In addition, each phase has its own verification and validation activities. Feasibility and risks are assessed in the CDP phase. Requirements are analyzed and validated as a part of the Requirements and Interfaces phase. The software architecture is peer reviewed and inspected. The code from the detailed design undergoes unit testing. All of these activities are important for successful IV&V.

To add independence to the process, integration testing, verification, and validation are often performed by an independent organization. Validation may be performed by non-project personnel to add even more independence. As a minimum, these IV&V activities should determine if the software is integrated into the system successfully and meets the software requirements and customer needs. Other tests are specifically designed for other purposes, such as to find software defects, to estimate the reliability of the software, or to determine if the software is ready for delivery to the customer. (See Section 4.6 for more on release to the customer.) Major activities for the IV&V phase include the following:

1. Integration, verification, and validation planning: It is important to have an overall strategy for IV&V. While plans and procedures are made for specific parts of IV&V, the effort should be planned as a whole, including the following:
   (a) Review system requirements and analyze test and evaluation aspects of requirements.
   (b) Define test requirements.
   (c) Update the TEMP. The TEMP serves as a top-level test management document.
   Note that while IV&V planning is listed as an activity for the IV&V phase, much of it is performed in earlier phases.

2. Integration and integration testing: Integration testing occurs after unit testing and before verification and validation. Code should be integrated incrementally in small pieces with each piece tested as it is integrated. Integration almost always reveals some unexpected incompatibilities. Regression testing may be required if some stage of the integration testing fails. Sub-activities include the following:

   (a) Plan the integration and its testing. Integration depends on the manner in which the system components and subsystems are developed.
   (b) Perform integration in the planned steps and test each step.
   (c) Document the integration test, test results, and all anomalies and associated conditions and actions.
   (d) Perform retesting and regression testing as necessary for the anomalies and apply the associated procedures for anomalies as detailed in the integration plan.

3. Software verification: Software verification is performed to determine if the software meets its requirements. If the software fails a particular requirement and the requirement is truly necessary, the software is changed and re-verified. Regression testing may be required due to verification failures. The following are potential steps in verification with an emphasis on verification by test:

   (a) Develop a verification plan and associated procedures. The plan and procedures include, among other things, specifying the required facilities and personnel, acceptance criteria, and how to handle test failures. Software modules are modeled to determine how they are expected to perform and this expected behavior is compared with test results. Perform risk analysis to ensure that the most critical requirements are verified with a high level of assurance. Non-test methods, such as formal methods, can also be used to supplement the testing of critical software, but testing should be considered the primary verification approach. Although functional requirements tend to receive the most attention, be sure to cover the interface and non-functional requirements also.

   Also develop test cases using the SRS, SDD, and software code. Develop individual test plans describing objectives, procedures, test data, test scenarios, expected results, and acceptance criteria. Make sure that the test scenarios cover the entire operating environment and all system operating modes.

   (b) Before verification testing (also known as system testing), ensure that the requirements have been validated and analyzed for traceability, completeness, consistency, correctness, and testability. Make sure that the design has been verified for compliance to the specifications and the code has been checked for coding standards compliance and best practices, that code metrics have been analyzed, and that the code has successfully passed unit tests, integration and integration testing.

   (c) Perform the tests, documenting the test set-up and test results in one or more test reports.

   (d) Report test failures and apply the associated procedures for test failures as detailed in the verification plan.

4. System validation: Validation is performed after software verification and ties back to the operational requirements. For validation, the software has been integrated into the target system and is operated in fully realistic and representative operational environments.

Again, regression testing may be required because of the validation test failures. Typical validation steps include the following:

(a) Develop a validation plan and associated procedures. The plan and procedures include, among other things, required facilities and personnel, acceptance criteria, and how to handle test failures. Where developing the plan and procedures, review previous verification and validation results. Also perform risk analysis to prioritize validation test activities.

(b) Develop test plans describing objectives, procedures, test data, test scenarios, expected results, and acceptance criteria.

(c) Inspect the installation documentation that documents how the software is installed.

(d) Perform the tests, documenting the test setup, and test the results in one or more test reports. Include all identifying information such as test and software configurations and test personnel.

(e) Report test failures and apply the associated procedures for test failures as detailed in the validation plan.

(f) Prepare the final test results documentation. The final documentation should include a description of the validated configuration.

Products for the phase include the following:

1. Integration, verification, and validation planning: The main product of this planning is a detailed TEMP.

2. Integration and integration testing: Products for this activity include integration test plans and procedures, integrated and tested software, and test reports documenting the integration testing. The software product may also be modified because of defects found during integration.

3. Software verification: As with integration testing, products associated with this activity include verification plans and procedures, tested (and corrected) software, and test reports documenting the testing.

4. System validation: Again, products associated with this activity include validation plans and procedures, tested (and corrected) software, and test reports documenting the testing.

5. Update products from previous phases: As the project matures, new information allows us to produce better versions of products created in previous project phases. Products from previous phases, such as the BOM, project schedules, and the risk assessment of the project should be reviewed and updated as needed.

Other products more directly related to software reliability that may be developed or updated in this phase include the following:

1. Perform reliability growth testing, reliability demonstration testing, and other software tests.

2. Update the FDSC as needed.

3. Update the reliability program plan as needed.

4. Update the OP as needed.

5. Continue assessing reliability metrics, adding new metrics or removing ineffective metrics as appropriate.

6. Update the reliability models and predictions as needed.
7. Continue and expand the DRACAS and DRB activities as necessary.
8. Perform reliability trade studies as needed.

Test personnel should work with requirements personnel to make sure that the requirements are understood and testable, with designers to ensure that the software is testable, and with the maintainers to make sure that the software is maintainable. There may also be software prototyping to reduce risk by constructing and testing immature and higher-risk subsystems or components.

4.5.2 Defects Typical for the IV&V Phase

Integration, verification, and validation each have the possibility of errors and the introduction of defects. Integration errors can result in a faulty product, such as the integration of the wrong software into the system or attempting to integrate the right pieces but doing it incorrectly and thereby adding defects. Verification and validation errors can result in software that does not satisfy the requirements or does not meet the needs of the customer. Test errors can prevent defects from being found and therefore allowing them to be in released software. If a defect is found, correcting the defect may inadvertently introduce new defects. Examples of defects for this phase include the following:

1. Integration, verification, and validation planning: If the planning for the integration, verification, or validation is defective or incomplete, the resulting activities may introduce defects into the product or leave defects in the product that otherwise would have been found and removed. Poor planning may also cause schedule, resource, and budget impacts affecting product release and profitability. The aspects of inadequate planning include the following:
   (a) Not understanding the system to be tested and the project objectives: Without this understanding of the requirements, customer needs, and design, the resulting plan is likely to be insufficient or poorly designed.
   (b) Not planning for the right tasks: This part may seem basic but doing the right tasks is critical to successfully running integration, verification, or validation activities. For example, there are a variety of different types of tests that can be performed: integration tests, system or verification tests, validation tests, reliability demonstration tests, robustness or stress testing, regression testing, and others. The project must choose the right types of tests to achieve the right results in a cost- and schedule-efficient manner. As another example, releasing a patch without adequate testing, including regression testing, may ease a schedule problem but result in the release of a defective product. The right combination of tests should be chosen to achieve the objectives. The "right tasks" also include activities such as performing the right types of analysis, documenting and archiving the results, and follow-through for identified anomalies and problems.
   (c) Not specifying adequate schedules or resources: Planning should account for task dependencies as some tasks must be started or completed before others. Scheduling should also accommodate adequate time to perform tasks. A good plan describes

what constitutes adequate staffing, resources, schedules, and budgets. Not specifying and planning for the right skill levels and skill mix can result in, as a minimum, schedule delays. It also increases the likelihood that too few testers or poorly skilled testers are assigned, resulting in tasks that are improperly performed. Inadequate resources, schedules, or budgets can all result in incorrectly performed tasks and possibly more defects in the software.

(d) Not being flexible enough: Complex sequences of tasks often do not go as planned. A good plan allows for alterations and re-planning. Try to avoid "single points of failure" with critical tasks and instead put "alternate routes" into the plan.

2. Integration and integration testing: Defects for integration and integration testing may be due to poor planning, but other defect sources exist as well. Specific types include the following:

(a) Test setup: Not having the right test environment for all requirements, operational environments, and operational modes can cause test setup failures. Another problem is not understanding the hardware, software, test plan, or other aspects of the test and test set-up. Also, not automating repetitive test cases can increase the likelihood of human error.

(b) Equipment failures: Test equipment can fail resulting in a failed test. The test facility should be tested at least to the level of the system under test.

(c) Poorly designed test procedures: Poor test procedures increase the likelihood of software defects by adding confusion and increasing the probability that needed tests will not occur or will be performed incorrectly. For example, following a poorly designed test procedure may result in a "successful" test when in reality, the software should not have passed the subject test objectives.

(d) Human error: Human errors include not following the plan or following it poorly. Human errors have the potential to be a significant source of defects, particularly if humans are to perform routine tasks many times or to perform complex tasks that can be confusing.

(e) Poor test analysis and follow-through: Any test should come with a collection of expected results and test analysis is required to determine if these expected results have occurred. If the results are not as expected, follow-through is required. Not recognizing or inadequately addressing anomalies can negate the purpose of the entire test. If the software fails part of the integration test, do a root cause analysis and relate the root cause to unit testing or some other technique performed earlier to see if the root cause could have been found earlier. If not, consider adding one or more techniques that will catch such defects in the future.

(f) Inadequate documentation and logging: The loss of information due to inadequate documentation and logging may mean the loss of information crucial to determining the root cause of an anomaly.

3. Software verification: Defects in software system/verification testing are often similar to those mentioned above for integration testing.

4. System validation: As with software verification, software validation testing defects are often similar to those mentioned above for integration testing.

4.5.3 Techniques and Processes for the IV&V Phase

Long recognized as a major project activity, there are many techniques that can be used to prevent or detect and remove defects in the IV&V phase. We start with general techniques and then cover anticipatory and feedback techniques. This is a sampling of techniques and not a complete or mandatory list.

1. General techniques for the IV&V phase:
   (a) Requirements reviews: It is critical that the people developing the tests understand the requirements. These test developers should be a part of requirements development from early in the project, not only to understand the rationale behind the requirements but also to ensure that the requirements that are to be verified by test are testable and that the test activities can be performed precisely and efficiently. In the IV&V phase, each member of a test team should be briefed on the meaning of the requirements that the team will test.
   (b) Software testing: Software testing is the main way that much of the software is verified and validated. There are various types of software tests, each with its own purpose. These include the following:
       i. Integration testing: Integration tests are used to see if the individual units of code work correctly together.
       ii. Interface testing: The purpose of interface testing is to ensure that the software-to-software and software-to-hardware interfaces are designed and coded correctly. Often, interfaces for a software module are initially tested using a simulated interface and then re-tested with the real interface after the interfacing software modules have been developed or obtained.
       iii. Regression testing: A regression test is the re-running of a previously performed test to determine if, after changes are made to the code, the previously tested behavior is still performed correctly.
       iv. Reliability demonstration testing: The purpose of reliability demonstration testing is to determine, to within a prescribed statistical confidence, if the software meets its reliability goals or requirements. Reliability demonstration testing is different from robustness testing in that reliability demonstration testing uses the system in as close to the expected real-world situations for the system as possible in order to obtain data that closely represents real applications. Robustness testing uses a high percentage of stressing and off-nominal situations to try to find and test "corner cases" and other places that defects may "hide." Note that when performing multiple runs of the software for reliability demonstration testing, we often consider each run as statistically independent of all other runs. In reality, this is not truly the case. Runs are rarely truly independent, but dependencies between runs should be minimized.
       v. Requirements testing: Requirements testing (also known as system testing or verification testing) is used to verify requirements.
       vi. Robustness testing: Robustness testing of software is analogous to accelerated life testing of hardware. With robustness testing of software, we test the software against a wide variety of test situations in an effort to find software defects. For this type of testing, it is important to have a method for determining how long

to perform the testing, such as the use of a capture–recapture technique to esti-mate the remaining defects. We then can test and fix until the estimated total number of defects is below some number. Code coverage metrics can also be used to ensure that a sufficient percentage of the code has been tested. Finally, design of experiment (DOE) techniques can be used to make the testing more cost-efficient.

Related to robustness testing, a project can use bug bounties. With bug bounties, the project sends a version of the software to "ethical hackers" outside the com-pany. The ethical hackers are paid for each valid bug or defect that they are the first to find and report.

Another similar concept is exploratory testing where testers use heuristics and intuition to try to find hard-to-find defects.

vii. Unit testing: Unit testing is a type of software test but is typically performed by the code developers during code development and as a result, it is covered in the DC phase.

These and other types of software tests are considered in Section 6.3.

(c) Software test considerations: Several considerations for successful software testing are as follows:

  i. Pay attention to code coverage in testing.

  ii. Use incremental testing to find incorrect assumptions and defects early.

  iii. Use good configuration control. It is critical in test activities.

  iv. Perform tests as early as feasible. The sooner that a defect is detected and cor-rected, the less impact it will have and the better for the project.

  v. Use test automation to reduce the workload and reduce errors.

  vi. Good logging and detailed documentation of test results, along with all of the test conditions and any anomalies, are critical. Success in troubleshooting a test or in finding the root cause of a defect may depend on a small detail.

(d) Consider the use of formal methods: Although software tests are the main way that software verification and validation are performed, formal methods are becoming more available and can add significant benefits to the process, especially for critical code. Formal techniques are discussed in more detail in Section 6.4.

(e) Use a software reliability advocate: A software reliability advocate continuously assesses the project for potential software reliability issues and advises project personnel on the impacts and mitigations of these issues.

2. Anticipation techniques for the IV&V phase:

(a) Critical path analysis: Test activities can result in complex schedules and they do not always go as originally planned. As a result, knowing the critical path can help manage resources and increase the likelihood of successfully completing on time.

(b) Checklists: Human error can be a significant source of test defects and the use of checklists is a proven technique for reducing such errors.

(c) Test automation: Another way to reduce human error is to automate as much of the testing as practical.

(d) Software FMECA: A software FMECA can help plan and prioritize tests. Testing can be planned to ensure software functions or situations that may result in a critical failure are adequately covered.

(e) DOEs: DOE techniques can be used to design effective but also efficient testing.

(f) Use reliability estimations: The IV&V phase is typically the first project phase that has sufficient test data to make useful reliability estimations from test data rather than predictions from software characteristics. If the data are from tests that are using a production-representative system in ways that represent the expected real applications of the system and its environment, the data can be used with any of several reliability estimation models to estimate the software reliability. These estimates are statistical, so confidence bounds must be included. Compare these results with the reliability predictions from the earlier phases and account for any inconsistencies. A number of useful techniques are found in [1, 6], and [15].

(g) Use reliability growth models based on the test data: Again if the data are from realistic tests, reliability growth models can be used to determine if the software is likely to meet its reliability goal by a certain time in the future. A number of useful techniques are found in [1, 6], and [15]. The use of reliability growth models for the project software should be documented in a software reliability growth plan.

(h) Always be aware that fixing a defect may introduce new defects. Consider more testing for software modules that have undergone more changes, whether it be code changes, design changes, or requirements changes.

3. Feedback techniques for the IV&V phase:

(a) DRACAS and DRB: Use a DRACAS and DRB when defects are found. This process should ensure that the defects undergo the proper recording, analysis, corrective action, and results verification loop.

(b) Peer reviews and inspections: Peer reviews or inspections can be used to detect defects in test cases, test plans, test procedures, and test reports, as well as defects and errors in their application.

(c) Statistical process control and trend tests: Consider the use of statistical process control and trend tests to assess peer review, inspections, and test results. These approaches may enable a more precise assessment of the implications of these feedback results.

(d) Daily stand-up meetings: Consider the use of a daily stand-up meeting to coordinate different test teams or to improve communication between the test team and other teams.

4.5.4 Metrics for the IV&V Phase

As with other phases, there are many metrics to choose from in the IV&V phase of a project. We consider process metrics and product metrics. Project metrics are covered at the start of Section 4.1. Several examples of process and product metrics applicable to this phase are listed as follows.

Process Metrics A number of applicable process metrics are listed at the beginning of chapter. Examples of activities that these may be applied to in the IV&V phase include the following:

1. Planning
2. Performing analysis for tests

3. Documenting analysis, the TEMP, test plans and procedures, and test results
4. Performing tests
5. Conducting peer reviews and inspections of products
6. Performing configuration control
7. Using DRACAS and DRB
8. Using metrics

Note that these metrics should indicate how well the process is working. They are not specifically designed to assess the quality of the product itself.

Product Metrics Examples of product metrics for the IV&V phase include the following:

1. IV&V planning: Key to this activity is the TEMP. Potential metrics useful for producing this plan and for assessing it include the following:
    (a) Size of the plan in terms of number of tests, number of requirements to test, or other metrics for the quantity of test work to be performed
    (b) Test time metric: A commonly used test time metric is the Halstead testing time metric. The Halstead test time metric, or test effort metric, is calculated as the Halstead effort metric divided by 18. This result is an estimate of testing time. The Halstead effort metric is the Halstead volume V times the Halstead difficulty D. The value 18 should be adjusted to better represent the values obtained by the organization using the metric. See [12] for more on these metrics. This metric can be compared with the test time estimated from the test schedules in the TEMP.
    (c) Test metrics based on requirements: Requirements issues can impact the ease and quality of software test planning. Several potential requirements metrics that can provide information on software test planning are as follows:
        i. Requirements clarity: Unclear requirements can negatively impact software test planning. Requirements clarity can be measured by counting the number of weak phrases, the number of options, or the specification reading ease as measured by the Flesch reading ease test score or other approaches.
        ii. Requirements defect estimate: Defective requirements can also negatively impact software test planning. One way to estimate the number of defects remaining in a specification is to use a capture–recapture technique with reviews of the specification.
        iii. Requirements volatility: Several examples of metrics for requirements volatility are covered in the section of metrics for the Requirements and Interfaces phase.
    (d) Test planning metrics based on software: The test effort can be expected to be dependent on the size and complexity of the software. Several software size and complexity metrics are covered in the metrics section for the DC phase.
    (e) Number and severity of test plan defects found with reviews and inspections
    (f) Defects in the software that, per root cause analysis, are traced back at least partially to defects in the TEMP
2. Integration and Integration Testing: Along with cost and schedule metrics, potential integration and integration test metrics are as follows:
    (a) Metrics for integration size and complexity: As with IV&V planning, metrics for requirements and for the software may be appropriate to provide indications of the

effort involved and the potential for issues. For requirements, potentially useful metrics include requirements clarity, volatility, and estimated number of defects. For the software, these types of metrics include the number of lines of code, number of modules to integrate, number of interfaces to integrate, and any of several complexity metrics.

(b) Number of test plans and procedures

(c) Number of test objectives per test plan and per test procedure

(d) Defects found in test plans and procedures with reviews and inspections

(e) Halstead test time metric: See [12] for more on this metric.

(f) Integration testing status: Test status includes such metrics as follows:

    i. Number of test cases completed

    ii. Number completed on schedule

    iii. Number completed successfully

    iv. Number failed

    v. Number of defects corrected

    vi. Defect repair cycle time

(g) Defects in the software that, per root cause analysis, are traced back at least partially to defects in the performance of the integration or integration testing

(h) Code coverage: Metrics for code coverage measure the degree that the software test exercises the code. Automated tools are usually used to support this type of metric. There are several different types of coverage, including the following:

    i. Function coverage: Function coverage measures the fraction of functions or subroutines that are called.

    ii. Statement coverage: Statement coverage measures the fraction of executable statements that have been invoked.

    iii. Branch coverage: With statement coverage, an "if A then B" statement may be exercised by letting condition A be true and never considering what happens if it is false. Branch coverage, also called decision coverage, considers both the true and the false conditions of the statement. Branch coverage measures the fraction of logic branches that are tested.

    iv. Condition coverage: Condition coverage, or branch condition coverage, is more detailed than branch coverage. With branch coverage, with an "If A then B" statement, we consider the case where statement A is true and the case where statement A is false. Sometimes, "statement A" is a complex Boolean statement with Boolean subexpressions. For example, the statement "$((x < 3$ AND $y <= 18)$ OR $z > 0)$" has three subexpressions, "$x < 3$," "$y <= 18$," and "$z > 0$," separated by logic operations. For condition coverage, each of these subexpressions must take a value of true and of false for complete coverage.

    v. Parameter value coverage: For software that takes parameters, all "common" values of each parameter are tested.

(i) Number of retests and effort per retest

3. Software verification: Metrics for software verification may be similar to those for integration testing although with an emphasis on requirements and system/requirements testing. For example, a metric for requirements traceability during testing may be used.

This metric measures the percent of the requirements that are tested and is given by

$$RT = \frac{R_1}{R_2} \times 100\%$$

where R_1 is the number of requirements that the testing covers and R_2 is the total number of requirements that the software under test is designed to cover. A value close to 100% is desired.

4. System validation: As with software verification, metrics for software validation may be similar to those for integration testing but with an emphasis on validation and validation testing.

5. Software defects: When defects are found during testing, use a DRACAS and DRB to find the root causes and corrective actions and as a minimum report:

   (a) The number of defects of each type. Keep a time record of these defects to determine if the situation is improving.

   (b) The number of repeated defects.

   (c) The number of backlogged defects.

   (d) The time required to determine the root cause of the defect.

   (e) The time required to successfully correct the defect.

   (f) The corrective action effectiveness, which is the fraction of corrective actions that correct the defect or reduce its likelihood or severity to an acceptable level divided by the total number of defects addressed.

   (g) The regeneration rate, which is the rate at which new defects are introduced into the software because of corrective actions. Note that the regeneration rate does not include defects because of the addition of new features. See [6] for more details.

   Faults found in testing are often used to estimate the remaining defects or faults in the software and the time required to detect those remaining. These results can then be used to estimate the resources required to reduce the number of defects to a specified level.

6. Additional defect metrics: A few useful defect metrics that are sometimes not considered in the DRACAS include the following:

   (a) Defect density: The total number of defects found divided by the total number of lines of code covered in the testing. Defect density is typically in terms of defects per thousand lines of code.

   (b) Requirements defect density: Requirements defect density is a metric for the impact of requirements and standards compliance on the software. It is defined as the number of defects because of specifications and standards divided by the SLOC.

   (c) Correctness: Correctness is a metric of how well the software conforms to specifications and standard. It is defined as the total number of requirements met divided by the total number of requirements.

   (d) Defect potential: The sum of all requirements, design, coding, document, bad fix, test plan, and test case defects

   (e) Defect discovery efficiency (DDE): The ratio formed by dividing the total number of defects found before product release by the total number of defects found, including defects found after product release

   (f) DRE: The ratio formed by dividing the total number of defects removed before release by the sum of the total number of pre-release defects and post-release defects removed

Examples We end this section by continuing the pair of examples introduced in the CDP section and continued through the Requirements and Interfaces and DC sections. Example 4.5.1 below continues Example 4.4.1 to the IV&V phase. This example is of a small company that produces a commercial software product for hospitals. Example 4.5.2 continues Example 4.4.2, which is of a large company that works on large government contracts.

Example 4.5.1   As stated above, this example continues Example 4.4.1 from the DC phase through the IV&V phase. Recall that Company A is a small software company that produces software for hospitals. It follows an agile philosophy and uses agile processes such as BDD to produce its products.

The company elicits a new list of desired behaviors for its software before the start of a sequence of build sprints. It then decides when to start a sequence of sprints to produce a new version of their product, with the first sprint being a planning sprint and the last being a transitional sprint. Recall that the main tasks for the IV&V phase include the following:

1. IV&V planning
2. Integration and integration testing
3. Software verification
4. System validation
5. Update products from previous phases

IV&V for each software version follows the same basic plan. Any additional planning individualized for a specific version is performed in the planning sprint.

Company A uses BDD, so integration and integration testing is performed incrementally in each of the build sprints and documented incrementally.

Initial software verification is performed incrementally in the build sprints and final verification is performed in the transitional sprint. Continuous verification is performed through the BDD process. Final verification starts with a review of the requirements traceability and allocation paths. Verification testing is largely automated and is incrementally documented at the end of each build sprint and the transitional sprint. Automated regression testing is performed for behaviors from previous versions. Verification also includes code reviews and reviews of the path from user stories through meeting minutes with diagrams and sketches through to end documentation and from each behavioral description in the user documentation being traceable through the testing and design implementation to a user story and customer need.

System validation is performed at the end of each build sprint and in the transitional sprint. This validation includes a customer assessment of the software at the end of each sprint to make sure that the product provided is the product desired. If the customer cannot support each sprint, then it is hoped that they at least support validation of the sequence of sprints. If customer support for validation is not feasible, then Company A assigns someone to act in the customer's interest for validation. The team also performs a peer review of the processes that have been followed during the BDD sprints to see that there is evidence that they were followed.

Other testing includes the following:

1. Reliability testing: Reliability testing using the OP is performed incrementally during each build sprint, but it is performed with the most emphasis during the transitional sprint. Reliability testing is mostly automated and is used to assess the software MTBF and whether the software is ready for release. If the testing indicates that the software has too many defects, release is delayed until this problem has been corrected. The release decision considers a comparison of the predicted and the estimated MTBF values and if there is a significant difference, why. Code coverage is measured incrementally, sprint by sprint.

2. Stress testing: Company A has a standard test suite that it uses for stress testing, although it may be modified during the planning sprint. The purpose of the stress testing is to test loading, resource use, and items of concern from previous software versions or identified in this version.

As noted in Example 4.4.1, products such as characterizing the software, OPs, and MTBF predictions are assessed throughout the sequence of sprints and the resulting products are updated as needed.

As with the previous phases, Company A also uses metrics to assess progress and alert for potential problems in the IV&V phase. Metrics useful for this phase sometimes overlap those of the CDP, Requirements and Interfaces, and DC phases. Examples of product metrics more specific to the IV&V phase are as follows:

1. Estimated test time
2. Number of behaviors to test
3. Predicted MTBF
4. Estimated MTBF
5. Software size
6. Number of defects found during testing
7. Code coverage of the testing
8. Number of defects found without root causes
9. Percent of defects found without root causes
10. Time required to determine a root cause
11. Time required to successfully correct a defect
12. Defect density
13. DRE

Example 4.5.2 This example continues Example 4.4.2 from the DC phase through the IV&V phase. Company B is a large company working on a large government contact and must deliver a hardware and software system at the end of the multi-year contract. They are also required to develop and deliver a collection of documents and meet a series of milestones, such as a PDR and a CDR. Company B uses an incremental develop approach where tasks from the CDP and Requirements and Interfaces phases are performed early, followed by several DC and IV&V phases. Each of these iterations produces a version of the final product with more features and capabilities than the previous version, or build. At the end of the contract, the product should meet the full set of requirements and be ready for limited production. Recall that the main tasks for the IV&V phase include the following:

1. IV&V planning
2. Integration and integration testing
3. Software verification
4. System Validation
5. Update products from previous phases

The initial IV&V planning for the project is performed as a part of the proposal. Since contract award, test planning has been updated based on the results of each project phase. For example, test personnel are an integral part of requirements development to make sure that the requirements to be verified by test are written in such a way that they are verifiable and that the test personnel understand the requirements and what it will take to verify them.

Project tests are guided by the TEMP. This plan is updated as the project has progressed. Since testing is time-consuming and expensive, it needs to be carefully planned to get the required information efficiently. DOE techniques are frequently used. Unexpected events often occur during testing, and test schedules are frequently compressed because of delays and impacts from other phases, so critical path analysis is performed and carefully monitored. Daily stand-up meetings are held to keep everyone informed of progress, required retesting, and potential schedule conflicts and changes.

There are a variety of tests that have to be planned and coordinated. Integration testing, verification or requirements testing, and validation testing are three notable types, but there is also reliability demonstration testing and reliability growth testing as the builds progress, soak testing performed on each build, robustness or stress testing that is used for high-risk critical LRUs, and regression testing is performed as a minimum with each new build to test that previously-passed requirements are still acceptable. Also, formal methods are used for safety critical pieces of software.

Integration testing for software takes software that has passed unit testing and has been through peer reviews and integrates the pieces into a functioning whole. It must be successful before verification testing on the system can be performed. It is performed incrementally to find issues early and to relieve the schedule. Regression testing is performed if issues are found.

Software verification is performed after the requirements have been validated and analyzed for traceability, completeness, consistency, correctness, and testability. Verification testing is a part of the overall verification process, where verification testing is just one of the verification methods. Verification tests are carefully planned to get the required information with minimum cost and schedule impacts. There are also situations when tests must be run to provide data for verification by analysis. Finally, some verification is performed through formal methods.

System validation testing is performed by the customer with contractor support. This testing is very visible, so it is preceded by reviews, simulations, and pre-validation tests to reduce risks.

The software FMEA may be updated because of defects found during testing. Also, reliability estimations are produced based on these defects and are compared with the reliability predictions and if there are significant differences, they are analyzed, corrections are made as needed, and results are used for lessons learned. All defects go through the DRACAS and DRB in an effort to remove the root cause of the defect. Software fault trees are often used for finding defect root causes, and defect detection rate monitoring is used to provide an early indication that defects are not being found at an expected rate.

Early in the CDP phase, the test organization performs a premortem to anticipate issues and problems and uses the results to plan testing activities. Lessons learned and process improvement techniques are used between builds to improve testing. Software reliability engineering (SRE) reviews test plans and procedures, particularly with respect to reliability demonstration testing and reliability growth testing. SRE and the testing organization work together to plan these tests. Additionally, the software FMEA may be used to prioritize tests.

Test plans, test cases, test procedures, and test reports all undergo peer reviews or inspections. Checklists are standard for test performance and automation is used where possible. Standard guidance in developing tests includes ensuring adequate code coverage, performing the tests early and incrementally when possible, and creating strategies for how to perform good logging of test results. There is also guidance for what constitutes adequate documentation to make sure that tests are as repeatable as possible and to help with root cause analysis. Automation is emphasized and there are strict configuration control processes and requirements.

Metrics for tasks that are of particular use for activities in the IV&V phase include the following:

1. Plan size, such as number of tests and number of requirements to test
2. Requirements metrics, such as requirements clarity, requirements volatility, and requirements defect estimates
3. Number and severity of test document (such as plans and procedures) defects
4. Defects found in the software that can be traced at least partially to a defect in test documents
5. Test status metrics, such as number of tests completed, number of tests completed on schedule, number of successful tests, and number of failed tests
6. Code coverage of the tests
7. Number of defects and categories for these defects, such as those traceable to a requirements defect, or a test defect
8. Defect regeneration rate (the rate at which new defects are introduced because of corrective actions)
9. Defect density
10. DDE
11. DRE
12. Defect containment

4.6 Product Production and Release

In the Product Production and Release phase of the project, we determine when a version of the software design and its documentation are ready for release to the customer and then produce that version of the product and transfer it to the customer.

4.6.1 Description of the Product Production and Release Phase

Software release occurs after the project has verified and validated the software and its documentation and has determined that it is suitable for release to the customer. This release process consists of a collection of predetermined tests, analyses, inspections, and decisions required by the project and sometimes by the customer before the decision to release the new or updated product. The release process also serves as an important gate, ideally preventing a substandard product from being released and therefore serving as an important software reliability function. Software release is often tied to key schedule goals and as such, software release is often the main project schedule driver.

In addition to being ready to release the product, it is important to accurately reproduce the software product onto quality media or to properly integrate it into the target hardware and ensure that the correct product reaches the customer. Activities in support of the Production and Release phase include the following:

1. Develop a system: It is important to develop a release, production, and delivery system for the product. Tasks for this activity include the following:
   (a) Create a product release process with the associated sequence of criteria that the product (including documentation and installation software if applicable) must meet before release approval. This set of release criteria must be constructed carefully. Just as the true meaning of a requirement is determined by the pass/fail criterion used to verify it, the release criteria of the product go a long way toward determining how good the product will be. Release criteria should be unambiguous and repeatable. The release process should be produced and documented well before time to use the process and can therefore serve as an early warning for possible project trouble. Also, pressure to release the product often increases as the scheduled release date approaches. Defining the process early reduces the risk that the process will be defined in such a way as to meet the schedule.
   (b) Develop or gain access to a product production process and facility to reproduce the product to the required quality standards and in the required quantities.
   (c) Develop acceptance criteria for the produced product. The product may be designed to be highly reliable but if the production of the product is defective, what the customer receives may not be highly reliable. For example, configuration control issues going from the design to produced copies of the design can result in a defective product.
   (d) Determine or develop the product packaging and delivery system to be used to deliver the product and to monitor the delivery to ensure that it is successful.
2. Release, produce, and deliver the product: Representative tasks for the release, production, and delivery of the product are as follows:
   (a) Implement the release process for a developed and tested product.
   (b) Once there is a releasable version of the product, determine when to release it. This decision is often a business, not a technical decision.
   (c) Implement the production processes in the production facility.
   (d) Apply the acceptance criteria to the produced product. After passing the release process criteria, the design of the product is considered to be acceptable for release.

Acceptance criteria are criteria applicable to each instance of the design to make sure that each instance that has been produced is this approved design.

(e) Use the approved packaging and delivery system to transfer the product to the customer.

(f) Monitor these processes and seek customer feedback to ensure that the system is working as desired and, if not, use the monitoring for feedback to improve it.

3. Update products from previous phases: As the project matures, new information allows us to produce better versions of products. Products from previous phases, such as the BOM, reliability program plan, FDSC, reliability models and predictions, the risk assessment of the project, and OP, should be reviewed and updated as needed in support of ongoing and future product lines.

Products for this phase include the following:

1. Develop a system: The product for this activity is a release, production, and delivery system for the product, along with the following:
   (a) Documentation of the system and processes used to release the product
   (b) Documentation of the system to produce the product
   (c) Documentation of the system to deliver the product
   (d) Documentation of the acceptance criteria for the product
   (e) Training material as needed to successfully use the system
2. Release, produce, and deliver the product:
   (a) Develop or otherwise obtain one or more production facilities to produce the product
   (b) Develop or otherwise obtain packaging for the product
   (c) Develop or otherwise obtain delivery capabilities and facilities
   (d) Collect metrics to assess how well the system is performing
3. Update products from previous phases: Examples of updated products from other phases include the BOM, reliability program plan, FDSC, reliability models and predictions, the risk assessment of the project, and OPs.

4.6.2 Defects Typical for the Product Production and Release Phase

Errors in the release phase can result in the release of software before it is sufficiently mature. These errors can also introduce defects, such as an error in configuration control resulting in the wrong version of a software module being used for a release. Software production and shipping defects can also introduce defects, even to the point of shipping the wrong product. Examples of defects in this phase are as follows:

1. Product release: Releasing a product too soon may mean releasing an unreliable product and releasing it too late may result in lost business opportunities. Potential types of errors and defects for product release include the following:
   (a) Incorrect release decision process: The release process and subsequent criteria should be based on what is needed to ensure that the product will satisfy the customer while preventing unnecessary release delays. If the process is incorrect, it is more likely that the product will not satisfy the customer.

(b) Not following the release decision process correctly: Pressure to meet a schedule may encourage shortcuts. Also misunderstandings or simple human errors may result in the process not being followed. Any of these cases increases the risk of defects.

2. Product production: Defects can enter the product because of product production. Plan the production process early and do so with one of the goals being to not introduce defects into the product. Examples of defects when performing product production include the following:

   (a) Inadequate production processes or facilities: Plans and procedures should be made for producing the product and sufficient facilities made, leased, or otherwise obtained.

   (b) Poor version control: Version control throughout the phases of the project is critical to successful production. Configuration control is an important part of product production and should be planned and implemented early in the life of the project.

   (c) Inadequate quality control: Production monitoring and quality control are important to the quality of the product. Produced versions of the product should be sampled and assessed. Statistics of these assessments provide feedback as a part of the determination of whether the processes are adequate or not.

3. Product delivery: Product packaging and shipping should be well planned. Configuration management and good knowledge of the customer can be critical to successful product delivery. For example, there may be different versions of a product for different operating systems or operating environments. The delivered product may interact with other third-party software, and this third-party software may be updated at different times by different customers. If the delivered product is not "approved" for the customer's operating environment, defects are much more likely. Knowing which version of the product applies to each customer and delivering that version sounds basic but can easily be performed incorrectly. Delivering the wrong product, delivering a damaged product, or delivering the product late or to the wrong location, all reduce customer satisfaction.

4.6.3 Techniques and Processes for the Product Production and Release Phase

As with the previous project phases, this section starts with general techniques and then covers techniques that anticipate defects and aim to prevent them, and techniques that are used to find the existing defects and remove them. This is a sampling of techniques and advice and is not a complete or mandatory list.

1. General techniques for the Product Production and Release phase:

   (a) Start development of the production and release system and infrastructure early in the project. This system should be based on customer needs while preventing unnecessary release delays. Determine the release criteria well ahead of the release date so that the release is not rushed and a substandard product is released to meet a schedule.

   (b) Check the quality of the product: Aspects of this step include, but are not limited to, the following:

      i. Check that contributing products meet their quality requirements. These contributing products include everything from the statement of needs and the CONOPs through the specifications and interface documents, software architecture design, detailed design, and on to the coding, integration and testing, and configuration management.

ii. Check the critical items list (CIL) from the SFMEA to ensure that there are no unresolved critical items.

iii. If third party products contribute to the final product, review the quality of these products. From the time of the decision to use a particular third party product to the time of release of the project's product, more information about the quality of the third party product may be available. If this information indicates that the third party product is risky, incorporate this information into the release decision.

iv. Review other tests, such as robustness or stress testing and timing and performance testing to ensure that they meet the minimum acceptable standards for product release.

v. Check the stability of the product. As software testing progresses, defects are found and corrected. If the rate of defect discovery is decreasing and appears to be in the tail of the Rayleigh curve (or is projected to be by the end of testing) and the estimated failure rate is acceptable, then the product is likely to be stable. See [6] for more details.

vi. Check the results of the reliability demonstration testing. A reliability demonstration is normally performed after the product is estimated to be stable and reliable. The results of the demonstration test, along with the associated confidence bounds, should be acceptable and the test coverage and requirements coverage should each be close to one. Also check that the reliability demonstration testing is performed per the OP.

vii. Use pre-release testing to "simulate" a release and see if the release is likely to have issues. The simulated "release" can be in-house or to trusted beta testers.

viii. Check that the remaining defects are not going to be an issue for future releases and that the next release will not be faced with correcting an excessive number of defects.

ix. Check to see if it is cost-effective to release the product. Release of a product is a business decision and cost plays a significant role. If the software has too many defects at release, it will be expensive to maintain. As an example of this consideration, assume that the software undergoes software reliability growth testing supported by a nonhomogeneous Poisson process software reliability growth model. Let c_1 be the cost of removing a defect during testing, c_2 be the cost of removing a defect after software release, and c_3 be the cost per unit of testing. One of the outputs of the software reliability growth model is the failure intensity function $\lambda(t)$. With this information, if

$$(c_2 - c_1)\, \lambda(t) > c_3,$$

it is more cost-effective to continue the reliability growth testing (and defect removal) and not release the software yet. If instead

$$(c_2 - c_1)\, \lambda(t) \le c_3,$$

it is more cost-effective to release the software. This example gives some of the flavor of tying the software release decision to reliability testing and cost. See [16] for more on this approach. There are many other approaches for estimating the cost-effectiveness of software release, and each project should tailor its approach to its needs.

(c) Consider using a continuous delivery approach. For many industries, the release of a product needs to be carefully timed. Continuous delivery is an agile concept for releasing a product in response to market demands.

(d) Use a software reliability advocate: A software reliability advocate continuously assesses the project for potential software reliability issues and advises project personnel on the impacts and mitigations of these issues.

(e) Other considerations include the following:

i. Make small incremental changes to the software and use regular release cycles. Keep the code in a releasable state as much as practical.

ii. Keep the design and code modular.

iii. Use a collaborative work environment to keep the various different groups involved in the production and release of the product informed.

2. Anticipation techniques for the Product Production and Release phase:

(a) Analyze the production and release system for bottlenecks. Look for where critical tasks are performed in series and determine what to do if problems occur with these tasks. Consider performing a process FMEA/FMECA on the production and release system.

(b) Automate and standardize as much of the process as practical, thereby reducing human error and adding consistency.

(c) Use production readiness reviews.

(d) Use checklists: Checklists are a simple and inexpensive technique for reducing defects in rote activities.

3. Feedback techniques for the Product Production and Release phase

(a) Use QA checks on the product. For example, inspect the packaged products for defects or sample the software for the correct versions.

(b) Peer reviews: Perform peer reviews on the release, production, and delivery processes and procedures.

(c) Inspections: Perform formal inspections on the release, production, and delivery processes and procedures.

(d) Beta test the product, including the documentation. Get independent people to use the product and the documentation. Beta testing can find defects in the product and determine how clear, complete, and accurate the documentation is. If defects are found, use the DRACAS and determine how the defects originated and how to prevent such defects in the future.

As a final note, Chapter 10 of [15] and Chapter 10 of [17] discuss techniques for optimizing the time of software release under various conditions.

4.6.4 Metrics for the Product Production and Release Phase

There are various metrics applicable to product production and release. Project metrics are covered at the start of Section 4.1. Examples of process and product metrics applicable to this phase are listed as follows.

Process Metrics A number of applicable process metrics are listed early in this chapter. Examples of activities that these may apply to in the Product Production and Release phase include the following:

1. Planning
2. Performing analysis for release, production, and delivery systems
3. Documenting analysis and resulting systems
4. Performing release, production, and delivery of products
5. Performing peer reviews and inspections
6. Performing configuration control
7. Using DRACAS and DRB
8. Using metrics

Typical metrics for the process of releasing the product include the following:

1. Backlog (the number of changes waiting for a future system release)
2. Number of failures due to a release
3. Number of successful changes in a given release
4. Number of failed changes in a given release
5. Percentage of releases that are delivered to production on time
6. Percentage of releases that are released on time
7. Percentage of releases by type (such as priority)

Note that these process metrics should indicate how well the process is working. They are not designed to directly assess the quality of the product itself.

Product Metrics Product metrics typically look at the quality of the product being produced. Examples of product metrics for the Product Production and Release phase include the following:

1. Documentation: Examples of documents for this phase are release procedures, production procedures, acceptance criteria, and delivery procedures. Potential metrics include the following:
   (a) Configuration control metrics
   (b) Estimated remaining defects based on techniques such as Capture–recapture from reviews and inspections
   (c) Number of defects found after document approval
   (d) Document reading ease: Metrics such as the Flesch reading ease test score, the Flesch–Kincaid grade level, or the Fog index can be used. Note that a higher score

(meaning that the document is on a lower reading level) is not necessarily better. The idea is to be on the reading level of the target audience.

2. Production facility: Potential metrics include the following:
   (a) Set-up cost
   (b) Set-up schedule
   (c) Operating cost per produced item
   (d) Percent on-time delivery
   (e) Time to make changeovers
   (f) Customer returns due to production errors and resulting product defects
   (g) Production cycle time
   (h) Production capacity
   (i) Downtime due to scheduled maintenance
   (j) Downtime due to unplanned maintenance
   (k) Delivery or quality risk due to production facility adequacy, capability, or schedule
3. Produced products: Some potential metrics applicable to the results of production and delivery include the following:
   (a) Number of defects in released products at least partially due to production, release, acceptance, packaging, or delivery
   (b) Downtime due to these errors or defects
   (c) Type and priority of releases
   (d) Number of on-time releases
   (e) Number of delayed releases
   (f) Number of delayed deliveries
   (g) Number of lost deliveries

Examples We end this section by continuing the pair of examples introduced in the CDP section and continued through the IV&V section. Example 4.6.1 continues Example 4.5.1 to the Production and Release phase and is of a small company that produces a commercial software product for hospitals. Example 4.6.2 continues Example 4.5.2, which is of a large company that works on large government contracts.

Example 4.6.1 This example continues Example 4.5.1 from the IV&V phase through the Production and Release phase. Recall that Company A is a small software company that produces software for hospitals. It follows an agile philosophy and uses agile processes such as BDD to produce its products.

The company elicits a new list of desired behaviors for its software before the start of a sequence of build sprints. It then decides when to start a sequence of sprints to produce a new version of their product, with the first sprint being a planning sprint and the last being a transitional sprint. Recall that the main tasks for the Production and Release phase include the following:

1. Develop a system
2. Release, produce, and deliver the product
3. Update products from previous phases

Company A typically uses the same release, produce, and delivery system for each software version, but during the planning sprint checks if any modifications are needed. As they apply continuous integration and delivery techniques, they often have small, incremental releases. Company A has its own software production capability and per their business model, mass production is not needed. The production and release processes are documented and make extensive use of checklists. Automation is used whenever practical.

Because tasks associated with each phase of the project affect the quality of the end product, release criteria come from all phases of the project. The company uses a checklist for release that includes the following criteria:

1. Process checklists indicate that all processes have been followed and, if not, the deviations have been justified.
2. All new behaviors have been verified and validated.
3. All user story behaviors are allocated to the design and to a test.
4. All design updates and tests are traceable to a user story or to a documented and needed change.
5. The software passes the reliability demonstration test.
6. The reliability testing indicates that the rate of defect discovery is not increasing.
7. The estimated remaining defects are manageable from a maintenance perspective.
8. The software passes the stress testing.
9. The reliability risk assessment indicates that the reliability risks are low.
10. Requirements, design, test, and configuration control representatives sign off on the product version.
11. The product version lead signs off on the product.

Production is limited because each version of the software is tailored to a specific customer. However, a hospital chain may have several hospital locations, each requiring the software and each potentially needing the software to run on a different system. Configuration management is very important. The company has a standard set of acceptance criteria for each product instance which includes a check of the versions for the various software components of the release and some sample software runs.

For initial delivery to a new customer, Company A sends an installation and training team to install the software and to train hospital personnel to use and maintain the software. For updates to the existing customers, the new version is usually downloaded and self-installs. Any such download is pre-coordinated with the customer.

As noted in Example 4.5.1, products such as characterizing the software, OPs, and MTBF predictions are assessed throughout the sequence of sprints and the resulting products are updated as needed.

Metrics mentioned for previous phases are often important in the Product Production and Release phase. Some of the key metrics for this phase are as follows:

1. Code test coverage
2. Requirements traceability
3. Requirements allocation
4. Estimated MTBF
5. Estimated number of delivered defects

6. Corrective action effectiveness
7. Product version defect backlog
8. Estimated defect density
9. DRE

Example 4.6.2 Our next example continues Example 4.5.2 from the IV&V phase through the Production and Release phase. Company B is a large company working on a large government contact and must deliver a hardware and software system at the end of the multi-year contract. They are also required to develop and deliver a collection of documents and meet a series of milestones, such as a PDR and a CDR. Company B uses an incremental develop approach where tasks from the CDP and Requirements and Interfaces phases are performed early, followed by several DC and IV&V phases. Each of these iterations produces a version of the final product with more features and capabilities than the previous version, or build. At the end of the contract, the product should meet the full set of requirements and be ready for limited production. Recall that the main tasks for the Production and Release phase include the following:

1. Develop a production and release system
2. Release, produce, and deliver the product
3. Update products from previous phases

Company B is on a development contract and will produce production-like prototypes in preparation for limited production, followed by full-scale production. The contract requires that Company B plan for production and release to ensure that the customer will receive a design that is producible on an acceptable schedule and at acceptable costs. At the end of Company B's current contract, there will be an FCA and a production readiness review as a part of the process of determining if the product is ready for limited production. If limited production is considered successful, full-scale production is expected. Ideally, the system for limited production is to be the same system as for full-scale production. Production processes, schedules, and facilities are planned. Acceptance criteria for each instance of the product are developed. Packaging and delivery of the product are detailed. Training material for how to use the system is developed. Documentation for all of these tasks must be developed and approved by the customer before approval to enter limited production. The overall plan for production includes a list of critical milestones, success criteria, and metrics.

The current contract only has scheduled releases of prototypes of the product. However, these prototypes are to be as close to production units as feasible. For follow-on contracts, delivery of produced units is on a schedule that is set by the customer. All produced units will be per the approved plans and processes. Automation is used whenever practical. Production is monitored and metrics are recorded for both production and delivery.

There are also updates to previous products. In particular, there are estimates of the product's reliability based on customer experiences with the prototypes, and these estimates are compared to previous predictions and estimates. Significant differences are analyzed to see if the tests or predictions were not performed correctly or if the OP is not accurate or if other factors are involved.

The release process and production system and its infrastructure are designed early in the project so that it can impact the system design and, potential issues can be addressed early. There are various checks of the quality of the product before release:

1. Check that all contributing items, such as specifications, drawings, and software, meet their quality requirements. For example, there are checks that process checklists indicate that all processes have been followed and, if not, the deviations have been justified.
2. Check that there are no unresolved critical items from the FMECA.
3. Check third-party items for quality.
4. Review that IV&V, the reliability demonstration tests, and other tests meet the minimum acceptable standards for product release.
5. Check the stability of the product and make sure that the rate of defect discovery is not increasing.
6. The reliability risk assessment indicates that the reliability risks are low.
7. Make sure that the remaining defects will not be an issue.

The contract requires the contractor to prepare plans for Production Readiness Reviews and other important production and release milestones. A trial run of the release process is used to "simulate" a release. A process FMEA is performed on the production and release systems, and they undergo peer reviews and inspections.

Metrics carry a particular urgency in this phase because of the key decisions that are to be made. Some of the key metrics for this phase are as follows:

1. Configuration control metrics
2. Estimated remaining defects based on techniques such as Capture–recapture from reviews and inspections
3. Facility cost and schedule metrics
4. Production cycle time
5. Production capacity
6. Operating cost per produced item
7. Time to make changeovers
8. Downtime due to scheduled maintenance
9. Downtime due to expected unplanned maintenance
10. Delivery or quality risk due to issues with production facility adequacy, capability, or schedule
11. Corrective action effectiveness
12. Product defect backlog
13. Estimated product defect density

4.7 Operation and Maintenance

There is more to software than design, production, and release. Released software will require updates due to defects, product improvements, or changes in the software environment (such as an operating system update). Even with good documentation and installation software and procedures, customers expect high-quality product support.

This post-release interaction with the customer should be considered to be an opportunity to learn more about the current product and about the wants and needs of the customer. It is also an opportunity to build customer good will.

4.7.1 Description of the Operation and Maintenance Phase

The Operation and Maintenance (OM) phase involves supporting the customer use of the software through general help with software installation and usage and with software updates. Usage help involves both documentation on how to install and use the software as well as technical support addressing issues that the user may have. Good customer support requires planning, training, and proper staffing. Decisions must be made as to the level of support, such as "24/7" support or "normal working hours" support, with a support system developed accordingly. Customer support should also consider both day-to-day support and support for exceptional situations.

Customer support also implies software maintenance. The project should account for system administration activities, data logging and analysis of logged data, routine maintenance requirements, system backups, and other activities. Plans and procedures must be made for software maintenance. A maintenance system needs to be established that allows for customer feedback. Staff and resources to identify, verify, correct, and verify corrections of defects must be planned. The report of a defect requires DRACAS and DRB activities and may involve project development personnel. Establishing this "reach-back" mechanism may be critical to a successful maintenance system. The proposed change should be thoroughly tested at the developer's site prior to release and installation. Defect correction typically requires regression testing, so plans, facilities, and staffing for this activity are required. Also any software change must go through rigorous configuration control. Finally, a determination of whether operation and maintenance documentation must be updated needs to be made and acted on accordingly.

Per [18], software updates generally fall into one or more of four categories:

1. Adaptive updates: Adaptive updates are made to software to accommodate changes in the software environment, such as an updated operating system.
2. Perfective updates: A perfective update is made to software to implement new or changed user requirements or to remove features that are not needed.
3. Corrective updates: Corrective updates are made to software to fix software defects.
4. Preventive updates: A preventive update is performed to prevent future problems. For example, after several updates, the code may become complex and difficult to maintain, so code refactoring may be needed.

Adaptive and perfective updates are for enhancements and corrective and preventive updates are for corrections. Both perfective and preventive updates are proactive updates. Adaptive and corrective updates are classified as reactive updates.

OM activities begin early in the project. A few examples of these early activities are as follows:

1. CDP phase:
   (a) The statement of needs and operational requirements require the project to have an understanding of how the customer wants to operate the product and wants it to be maintained.

(b) A determination of what service agreements to provide and how they are to be structured and priced are also considered in this phase. These agreements can have significant cost impacts and so the product should be developed with these agreements in mind.

(c) Maintenance plans are initially developed in this phase and updated in later phases.

2. Requirements and Interfaces phase:

(a) Requirements for system administration, backups, data logging, and other "routine" operations are needed.

(b) Maintainers are involved in the development of requirements to ensure that requirements reflect maintenance needs.

(c) Interfaces to backup systems, for data logging, and for other "routine" operations are required.

3. DC phase:

(a) Ease of installation and of system administration should be designed into the product.

(b) Maintainers should work with designers to ensure that the developed software will be maintainable and easy to debug.

Some basic tasks for the OM phase include the following:

1. Operations:

(a) Design for operations: Before producing any software, the product and its software should be designed to support operations. This design process starts with understanding what the customer wants and needs, including administrative requirements.

(b) Develop installation guides and installation software, operating procedures, and user documentation: The software installation and software documentation are often the first interactions that a customer has with a software product and as such can strongly influence customer satisfaction. Good documentation, whether on-line or a "hard copy," improves the customer's product experience and potentially reduces the amount of customer support required.

(c) Provide customer support to help with installation and usage of the software: Even with user-friendly software and high-quality documentation, customers will still have questions concerning the software product. Good customer support is a requirement for building and maintaining a loyal customer base.

(d) Develop and implement service agreements: Customers often expect service agreements to ensure some minimum level of support. These agreements are an important part of customer satisfaction.

2. Maintenance:

(a) Develop a software maintenance plan: A good software maintenance plan is critical for successful software operations and maintenance. Reference [18] provides details on such plans. A maintenance plan should be developed alongside the development plan and the configuration control plan and should cover topics including (but not limited to) the following:

i. How are software updates handled?

ii. How does the customer submit defect reports, desired changes to requirements, suggestions for new features, and other means of potentially triggering a software update?

iii. What resources are required to support customer software maintenance needs, such as staffing, budget, facilities, equipment, tools, and training?

iv. How to determine when to retire the software, i.e., when and how to announce that the software will no longer be supported. Note that sometimes old software exhibits increasing numbers of defects, much like the right-hand side of a reliability bathtub curve. One cause of this is that the architecture of the software may no longer be able to adequately support the maintenance changes made to the software. It may also be that maintenance updates have been hurried and are inconsistent. Finally, customer needs may have evolved over the life of the software, resulting in the customer using a different OP than the one that the software was originally designed for.

(b) Maintain the software: Perform software maintenance per the software maintenance plan, including implementing software updates per the approved processes and procedures, monitoring software maintenance performance, dealing with security and vulnerability issues, and providing suggested improvements to the processes.

When maintaining the software, update the software reliability estimates based on defects found after delivery of the software to the customer. Compare the postdelivery estimates with the corresponding predelivery estimates and predictions and account for differences. Table 45 of [6] lists several reasons that the predicted reliability and the postdelivery reliability estimate may differ. Be sure to retain all relevant data for use in future projects.

(c) Release updates and deliver the software to the customer: Release an update per the approved update procedures and monitor the release and delivery to ensure that the appropriate processes are being followed and improved when needed.

Products important for this phase include the following:

1. Operations: Products developed for operations include the following:
   (a) Customer support services
   (b) Installation software and guides
   (c) Operating procedures
   (d) User documentation
   (e) Service agreements
   (f) User support plans and procedures
2. Maintenance: Products developed for maintenance include the following:
   (a) Maintenance plans
   (b) Maintenance procedures and processes
   (c) Software maintenance as required
   (d) Software updates as required

Other products more directly related to software reliability that should be developed or continued in this phase include the following:

1. Update previous reliability products as needed.
2. Track customer satisfaction with the performance and reliability of the product.

3. Continue assessing reliability metrics, adding new metrics or removing the old ones as appropriate.
4. Continue the DRACAS and DRB activities as necessary.
5. Perform reliability trade studies as needed in support of product maintenance, product improvements, and new products.

4.7.2 Defects Typical for the OM Phase

Defects in the OM phase can result in incorrect software or system operation, poor customer support and ensuing ill-will, and badly maintained and therefore defect-prone software. These defects can also mean missed opportunities to collect both software defect information and customer feedback, both extremely useful for product improvement and customer satisfaction. Some typical defects for the OM phase include the following:

1. Operations: The operations aspect of a software product is critical for customer satisfaction. Defects in operations not only risk losing customer base, but also mean potentially losing information that can be used to improve the current product and possibly develop new products. Potential operations defects include the following:
   (a) Poorly planned software operations, such as not understanding the customer needs, not designing the software with operations in mind, or software operations not consistent with hardware operations
   (b) Defective installation procedures or installation software
   (c) Failure to properly verify and validate the operations aspects of the product
   (d) Poor customer support
   (e) Poor user documentation
   (f) Inadequate consideration to life-cycle costs
   (g) Inadequate plans and procedures for retiring software
   (h) Poorly designed or implemented service agreements
2. Maintenance: Software maintenance is an important aspect of a software product and has a strong impact on customer satisfaction. Examples of potential maintenance defects are as follows:
   (a) An insufficient maintenance concept or maintenance plan
   (b) Failure to design for maintenance
   (c) Inadequate customer support, such as poor plans and procedures, inadequate staffing, facilities, and resources, or other failures to support the customer
   (d) Poor debugging procedures
   (e) Inadequate configuration control of changes
   (f) Inadequate software maintenance records
   (g) Poor understanding of maintenance and change impacts, including lack of regression testing when needed

4.7.3 Techniques and Processes for the OM Phase

As with other project phases, it is important to make a judicious choice of approaches and techniques to apply in the OM phase. This section starts with general techniques and then covers some techniques that anticipate OM phase defects to prevent them, and techniques that are used to find the existing defects and remove them. This is a sampling of techniques and advice and not a complete or mandatory list.

1. General techniques for the OM phase:
   (a) Start OM planning at the start of the project. Make sure that OM personnel are embedded in the project and are an integral part of project activities.
   (b) Make a test area for testing software patches and upgrades. By having a separate test area for maintenance, software changes can be thoroughly tested and maintenance turn-around times can be better controlled.
   (c) Life-cycle cost analysis: Perform a life-cycle cost analysis to determine what support will be cost-effective and how to best structure and schedule software maintenance.
   (d) Configuration control: Configuration control is critical for various reasons. For example, third-party software may be updated at virtually any time, and different customers may differ as to when they update their versions of it. Also, different customers may prefer different versions of a software product, again making good configuration control important.
   (e) Maintenance records: In addition to good configuration control, keep good maintenance records. Use a maintenance database and automation where practical.
   (f) Make sure that the maintenance team understands the software and has access to designers when needed.
   (g) Use a software reliability advocate: A software reliability advocate continuously assesses the project for potential software reliability issues and advises project personnel on the impacts and mitigations of these issues.
2. Anticipation techniques for the OM phase:
   (a) Checklists: Use checklists based on best practices and previous experience
   (b) Process FMECA: Use a process FMECA to analyze the operations and maintenance plans and procedures.
   (c) Throughout the software development of the software, ensure that software maintenance is given sufficient attention. For example, make the software easy to de-bug.
   (d) Plan operational techniques for software aging. Software aging is the accumulation of error conditions in the software over time. For example, Chapter 14 of [1] discusses software rejuvenation, a technique of occasionally stopping the software, cleaning its internal states, and then restarting the software.
3. Feedback techniques for the OM phase:
   (a) Regression testing: Thorough testing of any software patch or upgrade is critical if the software reliability is to be maintained or improved.
   (b) Software testing: Installation software should be well tested.
   (c) Peer reviews and inspections: Operations and maintenance plans and procedures and any user or maintenance documentation should undergo reviews and inspections to reduce defects. Use peer reviews and inspections on the OM products, plans, processes, and other documents.
   (d) Institute customer satisfaction assessments: Use surveys, interviews, web searches, or other means to determine how the customer rates the product. Any such feedback should be assessed for use in improving the product and product support. Also use

survey, interviews, questionnaires, or other techniques to elicit what the customer wants and expects in customer support.
(e) Quality circles: Consider the use of quality circles to allow OM personnel to identify, analyze, and correct project problems.
(f) DRACAS and DRB: Apply DRACAS to defect reports from the customer and for internally found defects.
(g) Statistical process control and trend tests: Consider the use of statistical process control and trend tests to assess test results and customer feedback results. These approaches may enable a more precise assessment of the implications of these feedback results.
(h) Apply Pareto analysis on defect data and on customer feedback to ensure that high-priority issues are given adequate resources.

4.7.4 Metrics for the OM Phase

As with the other phases, metrics for the OM phase can provide early warnings of potential trouble and also inform the project of issues. Project metrics are covered at the start of Section 4.1. Examples of process and product metrics applicable to this phase are listed as follows.

Process Metrics A number of applicable process metrics are listed at the start of this chapter. Examples of OM phase activities that these may apply to include the following:

1. Planning
2. Performing analysis for operations and maintenance
3. Developing documentation: examples of documents in this phase include documentation of analysis, installation guides, operating and maintenance procedures, user guides, service agreements, and other artifacts.
4. Engaging in customer support and maintenance
5. Performing peer reviews and inspections
6. Performing configuration control activities
7. Performing DRACAS and DRB activities
8. Using metrics
9. Using feedback from quality circle participants
10. Analyzing of trends in Pareto analysis of defects

Product Metrics Here, the product metrics that we are interested in typically look at the quality of the product being produced. They vary depending on the product being produced, but examples include the following:

1. Operations: Although operations is mostly about performing activities and hence are in the purview of process metrics, there are several documents that are produced. These include the following:
   (a) Installation software and guides
   (b) Operating procedures

    (c) User documentation

    (d) Service agreements

    (e) User support plans and procedures

    Typical document metrics can be applied to these documents.

2. Maintenance: As with operations, maintenance involves performing activities, although a number of documents are produced, such as maintenance plans, procedures, and processes. These documents may benefit from the typical set of document metrics. Other products in this phase are as follows:

    (a) Software updates as required: Metrics for software have already been covered. Both software quality metrics and cost and schedule metrics may be appropriate.

    (b) Halstead maintenance effort metric: More on this metric can be found in [12].

    (c) Customer satisfaction: There are various ways to assess customer satisfaction. One approach is to encourage and assess customer comments. Another is to actively survey customers with surveys or interviews concerning their experience with the project's maintenance.

    (d) Fielded defect density: The fielded defect density is calculated as the number of fielded defects divided by the actual size of the code in thousand lines of the software code.

    (e) Number of backlogged customer defects

    (f) Incoming defect rate: This metric is the number of customer-reported defects per time interval.

    (g) Defect fix rate: This metric is the number of defects that are fixed per time interval.

    (h) Software reliability estimate based on postdelivery defects

Examples We end this section by continuing the same pair of examples that we have been considering throughout the previous phases. Example 4.7.1 below continues Example 4.6.1 to the Operation and Maintenance (OM) phase and is of a small company that produces a commercial software product for hospitals. Example 4.7.2 continues Example 4.6.2, which is of a large company that works on large government contracts.

Example 4.7.1 This example continues Example 4.6.1 from the Production and Release phase to the OM phase. Recall that Company A is a small software company that produces software for hospitals. It follows an agile philosophy and uses agile processes such as BDD to produce its products.

    The company elicits a new list of desired behaviors for its software before the start of a sequence of build sprints. Depending on the nature of the software maintenance, maintenance may be a part of new sequence of sprints that is adding new features, or it may be an expedited effort to fix a critical problem. Recall that the main tasks for the OM phase include the following:

1. Operations:

    (a) Design for operations

    (b) Develop installation guides and installation software, operating procedures, and user documentation

    (c) Provide customer support to help with installation and usage of the software

    (d) Develop and implement service agreements

2. Maintenance:

    (a) Develop a software maintenance plan

    (b) Maintain the software

    (c) Release updates and deliver the software to the customer

As with the release, produce, and delivery system, Company A typically uses the same system to support operations for each software version, but in the planning sprint, they check to see if any modifications are needed. The software allows for both regular users and administrative users. The administrative users have more privileges and can change certain settings, such as what results are logged and the frequency of data logging and system backups.

Company A's product is designed to self-install and tailoring each version of its product to a specific customer makes this aspect of the design easier. The product has its own database of product instructions and user help along with paper copy summaries. The company also provides customer support to help with installation and usage of the software.

The company has several different levels of service, each priced accordingly. However, service agreements are often tailored to the customer. The company has a standard maintenance plan for each level of service. If a standard service agreement is tailored to the customer, then the corresponding maintenance plan is also tailored.

Company A maintains the software, releasing updates and delivering the software to the customer through its continuous delivery process. The company coordinates with the customer to prioritize and schedule maintenance actions. Emergency fixes are handled through an expedited process. The customer pays for adaptive, corrective, and preventative updates through the maintenance agreement. The frequency and size of perfective updates depend on the service agreement.

The company uses automation where possible. Regression testing is automated and uses the BDD tests. Maintenance records are stored on a database that is automatically updated when various maintenance steps are taken. This database is used for configuration control and to update the traceability and allocation of the software behaviors. Software maintenance is aided by keeping the design modular and by frequent refactoring. They monitor software maintenance performance and make an effort to stay on top of potential security and vulnerability issues. While maintaining the software, several reliability-related tasks are performed:

1. Update the software size, defect density estimate, and other reliability prediction and estimation parameters

2. Monitor the MTBF of the fielded software

3. Compare the fielded MTBF with pre-delivery predictions and estimates and if the differences are significant, determine why and adjust the prediction and estimation data and processes accordingly

4. Archive data for future use

5. Update the OP

6. Update tests as necessary to keep the test coverage high

Metrics mentioned for previous phases are often important in the OM phase. Some of the key metrics for this phase are as follows:

1. Indications of accuracy of software size, defect density, and other assumptions
2. Ability to support the fielded system
3. Defects found by the customer in the first six months of operation after delivery
4. Number of backlogged defects
5. Time to release an update
6. Maintenance response time
7. DRE

Example 4.7.2 This example continues Example 4.6.2 from the Production and Release phase to the OM phase. Company B is a large company working on a large government contact and must deliver a hardware and software system at the end of the multi-year contract. They are also required to develop and deliver a collection of documents and meet a series of milestones, such as a PDR and a CDR. Company B uses an incremental develop approach where tasks from the CDP and Requirements and Interfaces phases are performed early, followed by several DC and IV&V phases. Each of these iterations produces a version of the final product with more features and capabilities than the previous version, or build. At the end of the contract, the product should meet the full set of requirements and be ready for limited production. Recall that the main tasks for the Production and OM phase include the following:

1. Operations:
   (a) Design for operations
   (b) Develop installation guides and installation software, operating procedures, and user documentation
   (c) Provide customer support to help with installation and usage of the software
   (d) Develop and implement service agreements
2. Maintenance:
   (a) Develop a software maintenance plan
   (b) Maintain the software
   (c) Release updates and deliver the software to the customer

Company B's current contract only includes limited operations and maintenance of the product. The contract requires delivery of several production-like prototypes to be used by the customer to determine if limited production, and possibly full scale production, will be contracted. Company B must provide electronic and physical operation and maintenance manuals for these prototypes and provide training for operators and maintainers. They also initially maintain the product and then support the customer maintainers as they become trained.

Company B realizes that operations and maintenance are cheaper and more trouble-free if they are designed into the system from the start of the project, so experts in these topics are a part of the conceptual design, requirements, and design efforts. Features such as fault tolerance, fault detection, and fault prediction are carefully explored and designed into the product. Software debugging is considered when designing and coding the software. The contractor also ensures that the verification of operation- and maintenance-related requirements is performed correctly. Periodic software upgrades are expected throughout the life of the system, so the software is designed to be modular, and installation guides and installation software are peer reviewed and tested for errors and ease of use. Process FMEAs are performed on operations and maintenance processes and contractor- and customer-found defects go through the DRACAS and DRB.

The company must also provide evidence that the final product will be easy and cost effective to operate and maintain. In the current contract, the contractor provides operations and maintenance plans describing how to meet the customer's operations and maintenance requirements. Cost and schedule information, such as a life-cycle cost analysis, is provided along with proposed facility locations and training plans. A suggested service agreement is provided based on customer information and feedback. A support configuration control plan is also provided to show how different versions of the product will be controlled and maintained. The maintenance plan also describes testing of any software changes, the use of peer reviews, and inspections of product changes, and how DRACAS and DRB are to be applied throughout the support contract.

Finally, the contractor continues to track defects and update reliability estimates, comparing predictions and earlier estimates with the current estimates based on fielded units. Significant differences are investigated and process improvements are made when warranted. They also monitor how well their operations and maintenance processes are performing and recommend changes when appropriate.

For the current contract, metrics for operations and maintenance are mostly metrics on processes, such as those for the performance of these tasks. Additionally, there are cost and schedule metrics for these items. There are also defect metrics for the operations and maintenance plans and procedures and for the support that the contractor provides the customer for operations and maintenance of the prototypes. Finally, metrics on any defects found with the products are collected and analyzed through the DRACAS and DRB and are used to update reliability estimates, such as product reliability and fielded defect density, and to assess the project's defect processes with metrics such as the number of backlogged customer defects, the incoming defect rate, and the defect fix rate.

4.8 Management

Project management plays a critical role throughout all of the project phases. In most organizations, management makes or approves the bulk of the "critical" decisions. Management also controls much of the work environment, having a significant impact on whether it is productive or not.

4.8.1 Description of Management

Project management works with and through other people to set and accomplish certain objectives. Typical management tasks include the following:

1. Setting objectives
2. Creating high-level plans for the organization

3. Staffing the project
4. Assigning responsibilities (objectives and tasks) and authority to project personnel
5. Setting up project controls
6. Monitoring project progress
7. Motivating personnel
8. Providing resources sufficient for the assigned tasks
9. Developing personnel
10. Coordinating with outside organizations

Management plays a major role in the reliability of the project's software. Management has the opportunity to establish the "vision" for the project. This vision includes the core values for the project, such as "customer focused," "quality first," "process focused," or "safety first." In Chapter 31 of [1], total dependability management is discussed. In this approach, these visions influence the set of methodologies used by the project. These methodologies are overall ways to work within the project to achieve certain goals. One example of such a methodology is risk analysis. Another is design for reliability. To be successful with a methodology, the project must choose a set of "tools" or techniques to use. For the example of risk analysis, brainstorming, affinity diagrams, project premortems, and reliability risk assessment are all potential tools or techniques to be used to achieve the methodology goals. By coordinating the project from vision to individual techniques, the project can focus on constructing its systems based on its core vision.

Software reliability is often considered to be costly and time-consuming. While it is true that there are costs and scheduled activities associated with it, these cost and schedule impacts need to be compared with the cost and schedule impact of not considering software reliability properly. Chapter 32 of [1] discusses this topic in the context of total quality for software engineering management.

4.8.2 Defects Typical for Management

Management errors can result in a poor work environment through under staffing, inadequate resources, poor direction, "office politics," or employees feeling underappreciated. Bad management decisions can adversely affect project objectives and the direction of the project. A significant number of activities affect software reliability, either directly or as secondary factors, so managing a project well increases the likelihood of reliable software. A few types of management defects and their results include the following:

1. Poor work environment: A poor work environment makes human errors much more likely. Several associated defects of this type are as follows:
    (a) Underappreciated employees: Keeping morale high is a major function of management for any project. Software is designed and written by humans and humans perform better with good morale.
    (b) Inadequate resources for the assigned tasks: Employees need the right tools for the assigned tasks. For example, the inability or unwillingness to get the right software tools or to fix the printer in a timely manner adds to the perception of a poor work environment.

(c) "Office politics": Few things are more demotivating than to be blamed for something that you did not do, for someone else to receive credit for your work, or to be more qualified for a promotion than the person who gets it. Employees need to feel that "management has their backs," not that management has favorites or is apathetic to them. Office politics destroys *esprit de corps* and the possibility of a great team.

(d) Project chaos: A well-run project has a few key objectives and tries to keep focus on them. If a project has constantly changing priorities and goes from one crisis to another, employees will have trouble doing quality work and morale will suffer. Change is inevitable, but change needs to be properly managed.

2. Staffing issues: Management is responsible for hiring, developing, and maintaining the personal who performs the project tasks. Management should
   (a) ensure adequate staffing.
   (b) ensure that all members of the project have at least a basic competency for their assigned tasks.
   (c) ensure that employees have the authority required to do their jobs.
   (d) invest in training, but make sure that the training is relevant and effective. Training can be disruptive and if it is performed largely to satisfy some administrative desire, it can have a negative effect on morale.
   (e) avoid frequent employee turn over.

3. Schedule issues: Management also controls the schedules for the project tasks. Potential related guidance includes the following:
   (a) Make sure that the schedules are adequate for the tasks based on the resources provided and personnel assigned.
   (b) Ensure schedule consistency. Inconsistency causes confusion and a feeling of chaos. Schedule consistency also helps to prevent imbalanced priorities such as schedule-driven priorities at the expense of product quality.
   (c) As with schedule consistency, task and task priority consistency are important.
   (d) Avoid being excessively and unrealistically success oriented. A frequent project problem worthy of special mention is reducing the schedule and budget for later phases of the project to make up for earlier delays and budget shortfalls.

4. Poor project controls: Poor project controls can result in poor communication, poorly defined roles and responsibilities, constantly changing priorities, and a sense of chaos. Project control issues include the following:
   (a) Loss of focus: Always know what the main objectives are and channel actions toward these. It is too easy to get side-lined with low-impact activities. For example, we may "optimize" a sub-process at the expense of the overall process. Keep the basics in sharp focus at all times. Focus on what affects the product. When adding or changing activities, ask "Is this impacting the product, getting a high-quality product produced on schedule and in budget?" If the answer is "No," then reconsider the need for the activity. If you are concentrating on too many things, you are not concentrating.

   Another sign of loss of focus is excessive documentation. Make sure that each document produced has a purpose that over-rides the impacts of producing the document. Some documentation is needed, but a project can be heavily burdened by too much.

(b) Poorly defined roles and responsibilities: One of the major tasks of management is to assign roles and responsibilities. If these are not clearly defined for the employees, confusion, excess work, missed tasks, and a general feeling of a bad work environment are likely.

(c) Not understanding or ignoring the limits of the processes and methods: It is important to understand how to perform the processes and methods, but it is also important to understand why they are in place and what their limitations are. Without this understanding, there is the risk of excessive rigidity. Have processes but do not let processes become an excuse for not doing smart things. Highly process-oriented organizations often become highly bureaucratic organizations.

(d) Relying on the wrong numbers, measures, or feedback: Management relies on feedback of various kinds to keep the project on track. If the feedback is incorrect or not providing the information that is needed to understand the true situation, discovering and correcting this early may be paramount. Always be checking feedback for relevance. If a problem occurs and the current feedback did not provide a warning, decide if it should have and, if so, make corrections. If the current feedback should not have been expected to provide a warning, decide if additional feedback should be put in place and if so, what type.

(e) Poor communication: Poor communications includes either management communication to other project personnel, between project personnel, or between the project and the customer. Make sure that each employee
  i. knows who they report to
  ii. knows what they are doing
  iii. knows why they are doing it
  iv. knows how to do it
  v. knows when it needs to be done
  vi. knows what success looks like
  vii. knows who needs it

4.8.3 Techniques and Processes for Management

There are many techniques and processes for the management of a project, with new ones created almost daily. These techniques and processes tend to become "trendy" and fall out of favor as other techniques grow in popularity. Several standard techniques and processes are listed below; however, it is important to determine which techniques are the best fit for the project and its personnel.

1. General techniques for management: Some general management techniques are as follows:

(a) Setting the tone: Project management has the responsibility to set the tone for the project and how it emphasizes the reliability of the software. If early in the project, management lets it be known by words and actions that the reliability of the software is important, then reliable software is much more likely to be produced. Make sure that project personnel feel responsible for quality work. This tone setting is a balancing act in that ultra-reliable software is expensive and if the customer wants reliable software but not very expensive software, too much emphasis may be contrary to the customer's wishes. However, it should also be noted that properly managed, software reliability activities can reduce the project cost and ease the development schedule. (It is cheaper and quicker to do it right the first time.) If that trend does not happen, investigate to see why.

(b) Staffing: Having the right people on a project is perhaps the biggest factor toward project success. Important considerations in choosing project personnel are as follows:

    i. Make sure that the person's skills and experience match the needs of the position.

    ii. Look for good judgment and intelligence. Both are important and neither are easy to train people for.

    iii. Make sure that the person has integrity and will work for and value the success of the project. People who play "office politics" to gain advantages can cause severe harm to a project.

    iv. Attitude is important. Choose people with a "can do" attitude who have a drive to get things done.

(c) Planning: Schedules and budgets should be based on the size of the project and the complexity and reliability required of the products. Ideally, historical information is used to make estimates and plans. Aspects of planning and scheduling may include the following:

    i. Establish a project baseline: The baseline consists of the tasks, resources, and schedule available for the project. A WBS may be useful in this task.

    ii. Set up the processes: There may be existing processes that, after analysis, are determined to be applicable to the project and adequate for its needs. If not, use techniques such as process-flow diagrams to understand and analyze the required processes.

    iii. Schedule the tasks: Techniques such as program evaluation review technique (PERT), critical path method, and Gantt charts are often applicable. Knowing the critical path can be very important and it should be monitored carefully.

    iv. Set up metrics: It is sometimes said that we should "measure everything." Although "everything" is an exaggeration, early and frequent feedback is critical to a well-managed project. However, some metrics are more important than others. For example, the metrics should include key performance indicators (KPIs). KPIs are metrics with a particularly high priority. They can indicate that a very important aspect of the project may be in jeopardy.

(d) Setting the level of authority: Let project personnel have as much authority as they can safely control. Micromanaging not only hurts morale, it also produces inferior products.

(e) Day-to-day project operations: A few examples of day-to-day management techniques and processes include the following:

    i. Project notebook: Keep a project notebook, which is a complete record of all project events.

    ii. Monitor the schedule regularly and update it when necessary. Pay special attention to when the critical path needs updating. Note that schedule updates may be necessary, but try to avoid doing it too often as a "constantly changing" schedule can cause project confusion.

iii. Project communications: Make sure that the right people have the right information in a timely manner. There are numerous ways to disseminate information, such as daily stand-up meetings or e-mail notices. Examples of important project information that can be distributed or available online are a project glossary and a project acronym list. Note however that too much information, especially information that is not relevant to a given employee, can be counterproductive.

iv. Reward quality work: Behavior follows rewards and punishments. "You get more of what you reward and less of what you punish."

v. Periodically review the tasks and processes of the project and ask "Is this impacting the product, getting a high-quality product produced on schedule and in budget?" If the answer is "No," then reconsider the need for the activity. A related question from Peter Drucker is "If we were not already doing this activity, would we start it now?" If the answer is not an emphatic "Yes," it probably should be eliminated.

2. Anticipation techniques for management: Management techniques and approaches for anticipating and avoiding potential issues include the following:

(a) Management commitment: Creating reliable software within project constraints requires management commitment and fortitude. Resist the pressure to push reliability-related issues further down the schedule. Management needs to fully understand that spending upfront costs on reliability is likely to prevent much larger downstream costs.

(b) Defined expectations: Management should ensure that all project personnel know what is expected. This setting of expectations includes setting expectations for the reliability of the software, including any reliability goals. In doing so, management needs to keep expectations in line with staffing, schedule, and resources.

(c) Clearly defined quality: Quality can mean different things to different people. Ensure that everyone is working to the same goals.

(d) Software reliability practices: Early in the life of the project, establish a set of software reliability practices and techniques that cover all major sources of software defects. Evaluate the effectiveness of these practices and techniques and update them accordingly.

(e) Project premortem: The project premortem technique is described in the CDP phase and also in Section 6.5. For this technique, project stakeholders meet and pretend that the project has failed and then list reasons why. The list of reasons is prioritized and solutions are developed. Actions are then assigned based on chosen solutions.

(f) Checklists: Checklists are a proven method of preventing or reducing many types of human errors.

(g) Risk management: Always be looking for potential risks so that the project can address them proactively. It is usually advantageous to have a risk management plan and set up risk management processes.

(h) Process FMECAs: A process FMECA anticipates the mistakes that may be made with a process, the effects that these mistakes could have on the project as a whole and on the software specifically, how to detect them, and what to do to prevent them.

(i) Defect education: Introduce task performers to the types of defects likely to occur in their tasks. Do this early in the project to train the employees and make them sensitive to these errors and defects and to teach them how to recognize them and prevent them.

(j) Formal entry and exit criteria: A useful technique to help manage a process is to develop and use formal entrance and exit criteria for the process. After the entrance criteria for a particular process have been met, the process is ready to commence. The process finishes after the exit criteria are met.

3. Feedback techniques for management: A few techniques for obtaining actionable feedback are as follows:

(a) Employee feedback: Set up a system that allows employees to provide feedback to management without fear of retribution. Employee feedback may be the first indication of a project problem.

(b) Management feedback: Just as management needs feedback from project personnel, project personnel need feedback from management. This feedback includes general information on project status and events, on changes to the project that have a special impact to a given employee or group of employees, and on an employee's performance.

(c) Customer feedback: Customer feedback can provide valuable information to project management. For example, customer dissatisfaction with the interaction with an employee or group may indicate a morale problem, poor project communication, or other management-influenced issues.

(d) Daily stand-up meetings: These short meetings allow feedback from project personnel. They also give management an opportunity to provide project personnel with a quick project status. Note that meetings are often viewed negatively, so keep them short and focused. If employees do not see them as useful, the meeting will be considered to be a burden.

(e) Quality circles: Quality circles are a way for project personnel to identify, analyze, and correct project problems.

(f) Metrics: After collecting a meaningful set of project, process, and product metrics, it is important to analyze them to determine if they indicate that problems exist or that the project is trending toward a problem.

4.8.4 Metrics for Management

Project management is typically the aspect of a project most associated with metrics. The project metrics listed at the start of this chapter indicate how well the project is running. Management reviews them and possibly changes project direction based on them. Other potential metrics for management are listed below as follows:

1. Work environment: A good work environment can be defined in various ways. Here we focus on employee motivation, adequate tools and resources, project information, and feedback.

(a) Motivation can be measured, to some degree, through surveys. Employee surveys can ask about the employee's motivation and let the employee rate the group's motivation. They can also ask about an employee's commitment to project or company goals as poor commitment to these goals may indicate employee dissatisfaction.

These survey results can be compared with surveys of customer satisfaction. Other indications of satisfaction level are percent of employee targets met (employees who meet their stated goals tend to have higher satisfaction), training alignment (when employees request training that is line with their responsibilities, it could indicate job satisfaction), and personal goal alignment with project goals. Employee discipline issues and frequent employee turnover may indicate a lack of motivation, as can frequent employee complaints.

(b) Adequacy of tools and resources can be assessed by employee surveys. Another way to assess tools and resources is to determine if what has been provided is similar to what has been successfully used for the same or similar tasks on previous projects. A third indicator is the number of defect root causes that can be traced back to, or negatively influenced by, poor tools or resource issues.

(c) It is important for each project member to have basic project information, such as definitions of project terms, a good acronym list, a statement of project goals and objectives, a definition of what quality means on the project, a schedule of significant project milestones, and a summary overview of the project. Determine how accessible and up-to-date this information is and if the project personnel are using and understanding it.

(d) A good work environment allows for employee feedback in a comfortable, low-risk manner. Metrics of the confidentiality of the feedback, ease of access, and responsiveness of management to the feedback can be important in assessing the overall work environment.

2. Staffing: Adequate staffing involves, among other things, good workload assessments, low turnover, and good training.

(a) Workload can be measured in several ways. One way is to use historical information, although this approach may not account for skill levels and project changes, such as new tools and added automation. (Note that adding automation may increase the workload before it reduces it.) Other approaches are to use employee feedback, such as employee surveys, and surveys of internal customers of the product. These tend to be subjective but can still provide useful information. Finally, consider employee morale, schedule slippage, and hours worked as workload indicators.

(b) Retention is an important staffing metric. Frequent employee turnover or project members requesting to be moved to another project can indicate project difficulties.

(c) Employee feedback can be used as a metric for training. For example, a survey could include items such as relevance to their job, clarity and completeness of the training, and overall assessment. Another way to assess if training is effective is to determine what aspects of a group's work should be improved by the training and measure it before and after the training. Finally, determine how many defects are attributed to, or negatively influenced by, poor or inadequate training and if the trend is not improving, there may be a training issue.

3. Scheduling: Project scheduling is an important management function and can be measured with the quality of task estimating and the schedule change rate.

(a) The quality of a schedule depends on the likelihood of it being successful. If a schedule is made based on solid historical data of similar tasks performed under similar conditions by similarly qualified personnel, there is a higher likelihood of schedule success than if the schedule is based on rough estimates or even worse, "wishful thinking." A collection of scheduling metrics can be produced that ranks time allotted for a given task by the quality of the estimate source and considers the fraction of tasks with each ranking. Various metrics applicable to project scheduling are listed early in Section 4.1.

(b) Schedule change rate can be measured as the number of schedule changes per time period and can indicate potential project problems. Not all schedule changes are of equal importance, so several such metrics may be in order.

4. Project controls: Several project control metrics for management span of control, roles and responsibilities, and changes in priority are listed as follows:

(a) Management span of control is the number of employees directly reporting to a given manager. The more employees reporting, the heavier the workload for the manager. A manager should have neither too many nor too few reporting employees and the ideal number depends on the manager, employees, and types of tasks performed.

(b) Employees should have clearly defined roles and responsibilities. Ideally these are written and readily available for each employee. Employees should be encouraged to ask questions to ensure clarity, particularly as aspects of the project change and an employee's responsibilities may seem to have changed. Surveys can be used to determine if employees know their roles and responsibilities, know how to perform them, and are given the resources and authority to do them. It can also ask if management assigns them tasks outside their assigned roles and responsibilities.

(c) Priority changes may also be indicative of project issues. As project management issues new "top priority" statements to personnel, a metric of the number of changed priorities per time period can be used to indicate possible project trouble. Note that it is likely a positive indicator if the priority change is due to the associated task having been completed.

Examples We end this section by covering the last of the pair of examples based on the examples introduced in the CDP section and continued in each of the sections describing a project phase. Example 4.8.1 continues Example 4.7.1 to management activities and is of a small company that produces a commercial software product for hospitals. Example 4.8.2 continues Example 4.7.2, which is of a large company that works on large government contracts.

Example 4.8.1 This example continues Example 4.7.1 from the OM phase to management activities. Recall that Company A is a small software company that produces software for hospitals. It follows an agile philosophy and uses agile processes such as BDD to produce its products.

The company elicits a new list of desired behaviors for its software before the start of a sequence of build sprints. It then decides when to start a sequence of sprints to produce a new version of their product, with the first sprint being a planning sprint and the last being a transitional sprint. Recall that some of the main tasks of management are as follows:

1. Setting objectives
2. Staffing the project

3. Setting up project controls and monitoring progress
4. Motivating personnel
5. Providing resources sufficient for the assigned tasks
6. Developing personnel
7. Coordinating with outside organizations

For Company A, setting objectives starts with developing a clear vision statement for the project. This statement includes what the project should accomplish, why the project is valuable, and what the success criteria for the project are. Per the agile manifesto, their highest priority is to satisfy the customer through early and continuous delivery of valuable software. The emphasis is on value to the customer, and each vision statement relates to this emphasis.

In the planning sprint, customer-desired behaviors are placed in the backlog for the sequence of sprints. The backlog is the list of all of the user stories and related tasks that will be included in a given sequence of sprints. Each sprint has a specific goal and the user stories chosen for that sprint reflect that goal. Large tasks are broken into smaller, more manageable tasks. Prioritization is done as a team.

During a sprint, the team performs these assigned and agreed-to tasks and nothing more. Each sprint has one or more slack tasks, such as reducing technical debt (the "less-than-perfect" aspects of the design and code), or employee education. Slack tasks are useful but if finishing the behaviors of a sprint takes longer than planned, these slack tasks can be deferred without impacting the main goal of the sprint. Remember that finishing a behavior for a sprint means, among other things, building tests, coding the behavior, testing the behavior, refactoring, integrating the new code into the rest of the code, updating build scripts, database scripts, and install scripts as necessary, reviews, fixing defects, and validation. Checklists are used to ensure that everything is covered before a behavior is considered to be "done."

As Company A tries to follow an agile philosophy and is therefore heavily relying on agile techniques, it values collaboration and empowered teams. Teams are given objectives and time boxes and are allowed considerable leeway on how to achieve those objectives. Teams are allowed to be self-organizing, meaning that the team members decide who will be doing what tasks. Teams are collocated so that communication is face-to-face, and pair programming is used when needed. Also, assignments to a project are full-time assignments.

Company A uses the same project controls and monitoring for each customer and updated version of software, although they are reviewed at the end of each sequence of sprints and possibly at the end of an individual sprint. The main "positions" for the sequence of sprints are a project manager, a product manager, and the build team. The project manager (sometimes called a Scrum Master) makes sure that things run smoothly, supports the team and keeps it going and supplied, and runs interference for the team as needed. The product manager (or product owner) supports the customer and makes sure that the resulting product is what the customer wants. The product manager works with the customer and maintains and promotes the product vision. The build team has all of the expertise required to create the product. The team consists of five to nine people who are responsible for coding the software, testing it, debugging it, integrating it, demonstrating it, and all other aspects of making the product. One person is designated the "programmer coach." This person performs the same function as other team members but also provides expertise on the processes, standards, and techniques being used.

As the team members are collocated, communication is as simple as an in-person conversation or a face-to-face meeting. Short daily stand-up meetings are used to keep everyone on track and informed. These meetings last five to ten minutes. Progress is tracked with a task board, which is a highly visible set of tasks and their states (to do, in process, done). Also burndown charts are used to show tasks remaining as a function of days into the sprint. At the end of a sprint, the team has a sprint review to discuss what was done. They also have a retrospective meeting to discuss what went well and what did not, and they make suggestions for improvements. The workspace has visuals that show everyone the vision statement, the status of the project, and process improvement charts.

Company A puts a premium on keeping employees motivated. Part of the motivation comes from people working together in empowered teams with each member treated as an important member. The project vision statement is posted so that it is visible to all team members. There is collective ownership of the resulting product and a team spirit is intentionally encouraged. The company tries to set up a challenging but manageable and rewarding environment. Workloads are managed to avoid overtime and everyone is encouraged to go home on time.

The first line of defense against insufficient resources is to correctly assess what resources are needed and getting a commitment for these resources before committing to a given sequence of sprints. As the company regularly updates its products by adding, modifying, or removing behaviors, it has historical data on what resources will be needed and uses these for resource planning. The project manager is the second line of defense for making sure that the project team has needed resources. The project manager has direct access to Company A's upper management and can appeal to them as a last resort. Finally, upper management makes a major effort to avoid negatively impacting an ongoing sequence of sprints. The company encourages keeping a sustainable development pace and adding slack tasks to each sprint. The team is empowered to determine how much work to do for each sprint and unless there is obvious abuse of this empowerment, is trusted to make good determinations.

Developing personnel is a part of keeping employees motivated. It helps employees feel fulfillment in their jobs and gives them a sense that the company values them. Company A develops personnel on the job through their team approach to developing their software. Here, less experienced personnel can learn from the more experienced ones. Pair programming also helps train people. In addition to learning-while-doing employee development, slack tasks are often used for employee education.

As stated earlier, Company A's highest priority is to satisfy the customer through early and continuous delivery of valuable software. To do this, they must know what the customer needs and wants. Upper management maintains good communications with customers and potential customers, and each sequence of sprints has a product manager whose purpose is to make sure that the customer's needs and concerns are foremost in the product development. In addition, customer tests are encouraged. These are customer-generated ideas for product tests which typically are tests of behaviors, or sequences of behaviors. In addition, some customer suggestions result in ideas for exploratory testing. Customer feedback is encouraged throughout the process and the earlier that a customer concern is raised, the sooner it can be addressed, meaning that it has less impact downstream.

Company A also has a long-term effort to look for new markets and product lines and to consider changes to the core structure of its existing product lines.

Other reliability-related aspects of Company A's project management include the following:

1. Setting the tone: Company A puts plenty of emphasis on the quality of their products and that everyone is responsible for that quality. For example, when a defect is found, the whole team is involved in finding the root cause and preventing that type of defect in the future.
2. Setting the level of authority: With the team approach that the company uses, programmers, testers, and other "technical" people have considerable authority over many aspects of their work.
3. Clearly define quality: Part of the vision for the project is clearly defining what quality means for the product.
4. Software reliability practices: Company A has put in place a set of practices and techniques that addresses the main sources of reliability defects in their software.
5. Checklists: Checklists are used throughout the project to reduce human errors.
6. Defect education: Each type of job has its own set of "typical" errors and associated defects. Part of on-the-job training is to familiarize employees with the defects that they may encounter or produce.
7. Formal entry and exit criteria: Although there tends to be a smooth flow from task to task in a sprint, there are very definite entrance and exit criteria for sprints and tasks performed in sprints. Satisfying these criteria make it more likely that the project is ready to perform a task and that the task is performed completely and satisfactorily.
8. Employee feedback: Employees know what is going on "in the trenches" and non-retaliatory employee feedback is one way for the company to discover this information. It also is a part of letting employees know that they are valued.
9. Customer feedback: Company A always seeks customer feedback and uses it to develop products that the customer wants.
10. Daily stand-up meetings: Daily stand-up meetings enable quick communication of status and new information of interest to the team.

Metrics mentioned for previous phases are often important to management. One obvious example is the number of defects found after the code is considered to be "done." Some of Company A's metrics that are more specific to management are as follows:

1. Employee mix
2. Employee turnover rate
3. Sprint burndown report
4. Sprint planning based of successful sprint completions
5. EV, AC, and PV
6. Project churn
7. Customer sentiment
8. Number of employee safety incidents

Example 4.8.2 This example continues Example 4.7.2 from the OM phase to management activities. Recall that Company B is a large company working on a large government contact and must deliver a hardware and software system at the end of the multi-year contract. They are also required to develop and deliver a collection of documents and meet a series of milestones, such as a PDR and a CDR. Company B uses an incremental develop approach where tasks from the CDP and Requirements and Interfaces phases are performed early, followed by several DC and IV&V phases. Each of these iterations produces a version of the final product with more features and capabilities than the previous version, or build. At the end of the contract, the product should meet the full set of requirements and be ready for limited production. Recall that some of the main tasks of management are as follows:

1. Setting objectives
2. Staffing the project
3. Setting up project controls and monitoring progress
4. Motivating personnel
5. Providing resources sufficient for the assigned tasks
6. Developing personnel
7. Coordinating with outside organizations

Company B uses a project-wide statement of objectives for the project so that priorities are clearly stated and available to everyone. The list of key objectives is kept small so that the project can concentrate on what is most important and these objectives are made simple and to the point. Each functional organization then expands on the statement of objectives to state how the organization fits into these overall objectives.

Management realizes that one of its jobs is to set the tone for the whole project. The project has, and clearly displays, a vision statement that helps to set this tone by clearly stating, in simple language, what the project is about and what is important for it. It also clearly defines what quality means for the project. In a sense, this is where software reliability begins. Management lets the project know that high-quality, reliable software is one of the key considerations for the project.

As cost is one of the main factors in the contract award decision, competing contractors try to keep costs down, and this effort means that staffing levels and skill levels are carefully estimated. Throughout the performance of the contract, Company B carefully monitors the required staffing levels and skill mixes, adjusting as needed. Many of these staffing and employee mix changes are expected and employees are told of them upfront. Company B is a large company with numerous projects, so when a reduction in staffing is needed, every effort is made to find a suitable replacement position for each affected employee. Employee morale is extremely important to a well-run project, so these matters are handled carefully.

Controlling the project requires a solid understanding of what the project is and needs to do. One tool to this end is a project plan detailing the what, when, why, how, and other critical aspects of the project. Various organizations in the project are used to handle important but specialized management tasks. There is an organization that tracks the schedule and budget. There is also an organization that manages project changes, particularly those that are out of scope and require cost and schedule estimates. Risk analysis, performed by the risk assessment group, is used to anticipate and plan for potential issues that might derail or otherwise negatively impact the project.

Controlling the project also requires good communications and setting up lines of communication from management to employees and employees to management as well as between management personnel and between employees. Management encourages employees to fit the type of communication to the need, whether it be a short "instant message," a concise e-mail, a phone call, or perhaps a short, focused face-to-face meeting. Long or monopolized meetings are discouraged.

Another aspect of controlling a project is setting up processes, procedures, and standards. Most of the ones used by Company B on this project are some of the company's "standard" processes and procedures, or are adapted from these. The company also realizes that while processes, procedures, and standards are important, they are not perfect and so the project has methods to bypass them in a controlled manner. Processes are frequently under review for improvement through process FMEAs and other techniques, and process improvement metrics are prominently displayed.

Each member of the project has one or more specific roles and is informed of these in writing. This informal document explains the following:

1. Who the employee reports to
2. What the employee is tasked to do
3. Why this tasking is important
4. What if any constraints there are on how to do the tasks (while avoiding micro-managing)
5. When it needs to be done
6. Who needs it and
7. What success with their tasks looks like

Sometimes the employees write part of this document, but the manager or supervisor signs the final document in agreement. The employee's supervisor or manager also discusses expectations for the employee and what authority the employee has in performing the assigned roles and tasks. Employees are also educated on the types of defects that are likely to occur with their job. The objective is for each employee to feel empowered and to know how he or she provides value and ultimately supports the customer. The discussion also covers training needs and career direction. Reducing job confusion and providing a career path are two of the ways that Company B attempts to keep employees motivated. Management also continually tries to establish and maintain a challenging yet rewarding work environment. Team spirit is encouraged and employee feedback is sought and valued.

Resource management is initially based on the resource estimates contained in the proposal. If possible, these estimates are based on historical information from similar projects and tasks under similar conditions. As the understanding of the project matures, these estimates are updated and if a shortfall is expected, they are coordinated with the customer. Again, risk analysis comes into play with resource estimation.

Another aspect of providing employees with needed resources is information. For example, the project produces and makes readily available lists such as definitions of project terms, a good acronym list, the project vision statement, a statement of project goals and objectives, a definition of what quality means on the project, a schedule of significant project milestones, and a summary overview of the project.

Coordinating with outside organizations mostly means coordinating with the customer and subcontractors. As the customer is embedded with the contractor, contractor coordination is usually easy. For example, the reliability engineering lead has frequent contact with the customer's reliability engineering lead. There are also more formal activities for this coordination. For example, there are regular status reports that management sends to the customer.

Errors in almost any of these tasks can cause significant project disruption and increase the likelihood of software defects. For example, schedules should provide adequate time to perform the tasks correctly and not be overly optimistic. They should emphasize the key tasks and keep the critical path in mind. Also schedule consistency is important. Otherwise, time will be used inefficiently and project morale will suffer, increasing the risk of human errors and risking the loss of key project personnel.

Management is also constantly on the lookout for "office politics," knowing that if such tactics are successful, they will be repeated and soon, the work environment will become toxic.

Many of the metrics used by management are listed in Example 4.2.2; however, some of the metrics more specifically tailored to how well management is working are as follows:

1. Confidential employee surveys
2. Customer feedback
3. Employee turnover rates
4. Number of project plan changes per month
5. Reasons for the changes and their percentages
6. Average advanced notice of a project plan change
7. Number of times that the same aspect of the plan is changed

References

1 Pham H, editor. *Handbook of reliability engineering*. Springer-Verlag, London, 2003.
2 McConnell S. *Code complete: A practical handbook of code construction*. Microsoft Press, Redmond, Washington, 2004.
3 Tomar A, Thakare V. The CMM level for reducing defects and increasing quality and productivity. *Int. J. Comput. Sci. Mob. Comput.*, 5(2):132–138, February 2016. Available via https://ijcsmc.com/docs/papers/February2016/V5I2201629.pdf. Accessed 22 Aug 2020.
4 Briand L, Emam K, Freimut B, Laitenberger O. A comprehensive evaluation of Capture–recapture models for evaluating software defect content. *IEEE Trans. Softw. Eng.*, 26(6):518–540, June 2000.
5 IEEE Standard 12207. *System and software engineering - software lifecycle processes*, 2017. Software and Systems Engineering Standards Committee of the IEEE Computer Society.
6 IEEE Standard 1633. *Recommended practice on software reliability*, 2017. Software Engineering Technical Committee of the IEEE Computer Society.
7 Hooks I. Writing good requirements. *Proceedings of the 3rd international symposium of the NCOSE*, 2, 1993. Available via https://reqexperts.com/wp-content/uploads/2015/07/writing_good_requirements.htm. Accessed 22 Aug 2020.

8 Nuseibeh B, Easterbrook S. Requirements engineering: A roadmap. *ICSE proceedings of the conference on the future of software engineering*, pages 35–46, 2000. Available via https://www.cs.toronto.edu/ sme/papers/2000/ICSE2000.pdf. Accessed 22 Aug 2020.

9 Fagan M. Design and code inspection to reduce errors in program development. *IBM J.*, 15(3), 1976. Available via https://link.springer.com/chapter/10.1007/978-3-642-59412-0_35. Accessed 22 Aug 2020.

10 IEEE Standard 1008. *IEEE standard for software unit testing: An American national standard*, 2003. Software Engineering Technical Committee of the IEEE Computer Society.

11 Kazman R, Klein M, Clements P. *ATAM^{SM}: Method for architecture evaluation*. Technical report CMU/SEI-2000-TR-004, ESC-TR-2000-004, Carnegie Mellon Software Engineering Institute, Pittsburgh, PA, 2000. Available via https://resources.sei.cmu.edu/asset_files/TechnicalReport/2000_005_001_13706.pdf. Accessed 22 Aug 2020.

12 Halstead M. *Elements of software science*. Elsevier, New York, 1977.

13 McCabe T. A complexity measure. *IEEE Trans. Softw. Eng.*, SE-2(4):308–320, 1976. Available via http://literateprogramming.com/mccabe.pdf. Accessed 22 Aug 2020.

14 Etzhorn L, Delugach H. Towards a semantic metric suite for object-oriented design. *Proceedings on the 34th international conference on technology of object-oriented languages and systems*, pages 71–80, 2000. Available via https://ieeexplore.ieee.org/document/868960. Accessed 22 Aug 2020.

15 Pham H, editor. *System software reliability*. Springer-Verlag, London, 2006.

16 Defense Technical Information Center. *Handbook of software reliability and security testing*, 2011. CSIAC report number 519193.

17 Kapur P, Pham H, Gupta A, Jha P. *Software reliability assessment with OP applications*. Springer-Verlag, London, 2011.

18 IEEE Standard 14764. *Software engineering – software life cycle processes – maintenance: An American national standard*, 2006. Software Engineering Technical Committee of the IEEE.

5

Roadmap and Practical Guidelines

We have covered a lot of competing suggestions and it is easy to get lost in the details. This chapter provides a summary and roadmap for using this book as well as giving some overall guidance for keeping the big picture in mind.

5.1 Summary and Roadmap

The following is a short summary of the material covered earlier in this book and a roadmap for how to use it in situations that frequently occur in projects. We start with Chapter 2 and understanding software defects. To efficiently and effectively handle defects, we need to understand them. This understanding includes knowing where they enter the system, their effects on the system, how to detect them, and what causes them. As a brief summary:

1. Where they enter the system: Defects affecting software can enter the system anywhere that there are activities and products that affect the software. For example, fixing a defect may introduce one or more additional defects.
2. Effects of defects: Some software defects are very subtle and hard to find, even with a rigorous test program. For example, the behaviors of some defects are changed by our efforts to find them. As a result, it is almost never cost-effective to rely solely on software testing to produce quality software. This consideration is also one of the reasons that it is best to prevent defects from occurring and if that is not possible, to find and correct them as early in the project as possible.
3. Detection of defects: As stated in Section 2.3, detection of defects should
   (a) find errors and defects early
   (b) be complete
   (c) not miss very many errors or defects
   (d) be cost and schedule efficient
   Section 2.3 also outlines a number of techniques for detecting defects in processes and in products. More on this topic is covered in Chapter 4, the chapter on phases of a project.
4. Causes of defects: To prevent or eliminate a defect, it is important to know the cause or causes of the defect. This knowledge helps us design processes to reduce the number of defects in a product. It also helps us to efficiently remove or reduce the likelihood of a given type of defect from occurring again and not just remove a single instance of the defect. Other considerations include the following:

Software Reliability Techniques for Real-World Applications, First Edition. Roger K. Youree.
© 2023 John Wiley & Sons Ltd. Published 2023 by John Wiley & Sons Ltd.

(a) Defects typically involve a chain of events, initiated by one or more root causes. Anticipating the chain of events that results in a defect gives us the earliest and easiest opportunities to prevent the defect.

(b) Section 2.4 lists and describes the following high-level issues that can result in defects:

    i. Not producing or monitoring the right things

    ii. Poor processes for producing or monitoring a product or process

    iii. Not following the processes

    iv. Following the process poorly or monitoring poorly

    v. Non-human errors

The section then lists a number of causes of human errors.

(c) Section 6.5 has techniques for finding root causes.

The first step for using this book is to understand defects and how they can affect software. This understanding forms the basis for nearly all of the decisions that will be made later for handling defects. Of course, simply understanding defects is not enough. We must know how to handle them. To this purpose, we address three situations:

1. At the start of a project when we want to plan for cost- and schedule-efficient software reliability.
2. As a member of an organization on a project and we want to ensure that we and our organization make positive contributions to software reliability.
3. As a member of a troubled project with software reliability issues and the objective is to turn the project around.

Each of these situations is common enough and important enough to merit its own piece of the roadmap.

5.1.1 Start of a Project

Suppose that a project is being planned and software reliability is important to the success of the project. How should the project plan and implement the software reliability effort to achieve the desired goals? To some degree, the answer depends on the starting point. If the project is similar to other projects that the company has had and these projects successfully achieved software reliability goals that are similar to those of this current project, it may be best to see what they did and adapt their approaches and techniques to this current project. If this is not the case, we have to develop a new software reliability plan for the project and decide how to implement it.

At a high level, we handle errors and defects by preventing them, by detecting and removing them, by designing the system to tolerate them, and by forecasting them to better manage for them and develop confidence in the system. Handling defects requires good planning and execution, and these call for well-defined objectives, careful planning, and well-thought-out implementation, monitoring, and feedback to account for the dynamics of the situation. In Chapter 3, we presented a simple four-step approach for developing and executing a software reliability plan. We start by determining our objectives. We then go through a list of steps to develop a plan. Next, we determine how to implement the plan and how to monitor progress and supply feedback when needed. We summarize the chapter as follows:

1. Objectives: At a high level, most software projects have the following two reliability objectives:
   (a) Create a highly reliability software product on schedule and within budget constraints.
   (b) Know with a high level of assurance that the software is sufficiently reliable.
2. Plan: To achieve these objectives effectively and efficiently, we need a plan suitable to the nature of the project and its personnel. Typical steps include the following:
   (a) List steps that the project will perform to produce the software product.
   (b) List what can go wrong in each of these steps.
   (c) List how we can prevent these defects and errors or at least significantly reduce their likelihood and impact.
   (d) List ways that we can quickly know if something goes wrong, i.e. list what monitoring is needed.
   (e) List when the information from the monitoring indicates that we should do something different and what it should be.
   (f) List how we will know if our processes and corrective actions are effective, and if they are not, list what we should do. We need to know how confident we can justifiably be in our product.
3. Implementation: A good plan must be implementable by the people assigned to the project. Establish good processes to help project personnel implement the plan. Section 3.4 lists properties that a good process should have. It also notes that the project should control the processes and not let processes control the project. The goal is not the processes but rather the products. A few additional implementation guidelines and considerations include the following:
   (a) Task people who are qualified for the job.
   (b) Determine the critical milestones and their schedules and build around these.
   (c) Determine what resources are needed for tasks and when they will be needed, then ensure that they are available at the right time.
   (d) Set up clear lines of communication for the project.
   (e) Make sure that project personnel have a clear and current understanding of all matters pertinent to their work.
   (f) Make sure that people feel ownership of their tasks.
   (g) Start software reliability processes early.
   (h) Instill a "quality mindset" in project personnel.
   (i) As much as possible, be prepared for the unexpected.
4. Monitor and Feedback: Implementation needs to be monitored and changes made as needed. Changes may be needed if the current approach is not working well, if there is evidence that a significant improvement can be made, or if there is a change in the project. Monitoring and feedback is how systems adapt and improve. Other advice on monitoring and feedback from Chapter 3 includes the following:
   (a) All other things being the same, the monitoring that finds a defect earliest is the most effective monitoring.

(b) The right monitoring is based on making correct estimates of what can go wrong.

(c) Monitoring has "noise" in it, so allowances must be made for these potential errors and defects by creating a robust monitoring and feedback system.

(d) Feedback is the main source of reliability growth.

(e) Feedback must be appropriate and timely to be useful.

(f) The right feedback requires the right monitoring.

(g) The right feedback, consistent with other project constraints, is based on the root cause (and its precursors) of the monitored problem.

(h) From a cost and schedule perspective, the closer the feedback is to the creation of the defect or error, the more effective the feedback is.

The above summary is a good start to a roadmap. First, we need to have a good understanding of what our objectives are. Even if software reliability is not a priority, defect prevention and control should always be an important project consideration as defects negatively affect project cost and schedule. At the start of a project, make a list of objectives for software reliability, and for defects, make these objectives measurable and make sure that the project personnel know them and know why they are important.

Next, we make a plan for how to achieve these objectives. The six typical planning steps listed above are a good start. To make our plan, we need to understand our product and the chain of internal products used to develop and produce it. Remember that part of the overall product is installation, maintenance, and other support. Also recall that the final software product is typically the result of several intermediate internal products. Defects in these internal products can result in defects in the software, so the plan should account for preventing and controlling defects in them as well. Know the strengths and weaknesses of the processes the project uses and know how we can monitor both these processes and the products they produce for indications that something may be wrong. Note that we are not just looking for process failures. No process is perfect, so even sound, well-executed processes can result in defects in products. As a result, always have techniques for detecting defects in products. In short, choose techniques to prevent defects or control defects in the final product, the intermediate products, and in the processes used to produce them. As the project probably uses products and processes implemented by several different organizations, these different organizations should be a part of the development of the plan. Chapter 4 contains material related to these intermediate products and ways to control defects in them.

It is important to have a plan, but it is the results of the plan that matter. If the plan cannot be implemented or is poorly implemented, expect poor results. The next step is to implement the plan. Planning for implementation begins when we are setting objectives: make sure that the objectives are achievable. During the second step, making a plan, continually consider how the various parts of the plan will be implemented. Just because Company A does a task in a particular way does not mean that all companies should do it that way. Consider the available resources, skill sets, and schedules. Also consider how things are currently done. Too much change in a short amount of time may be disruptive. The change may be necessary, but consider the impacts of the change. Make sure that project personnel understand the need for the changes and try to get them to accept the changes.

Finally, we need to monitor the key aspects of the project and make changes as needed. The sooner we know of a problem, the easier it is to correct and the less damage that it causes. Determine what should be happening and what to do if it is not. Make sure that the things that are monitored are important and provide useful information. If we have developed a good plan, we have determined what to monitor, but project environments are dynamic, so if the current plan is not working well or if significant improvements can be made without causing a major project disruption, we may need to change the plan.

5.1.2   As a Member of an Organization

Another situation is when an organization that is supporting or is a part of a project wants to make sure that it creates and propagates as few defects affecting the software as possible. This situation is common at the start of a project.

We noted in Section 5.1.1 that organizations that produce products or processes supporting the final software product should be involved in the development of the overall project plan. If so, this situation can be considered to be a part of the one covered in Section 5.1.1. However, it may be that the project has not created such a plan, but an organization wants a plan for their aspect of the project. We can view this situation as a miniature version of the previous one. In both cases, we want to choose processes and techniques to produce products supporting the overall project objectives. For this situation, a single organization considers its products and processes within the context of the overall project objectives. If the project has not developed these objectives, the organization should encourage the project to do so, and if necessary, create its own objectives.

The organization can use the same steps outlined in Chapter 3 and Section 5.1.1 to make and implement a plan. Also, Chapter 4 may be a particularly helpful resource. Chapter 4 breaks a project down into phases that are typical to many projects. Even if a project does not explicitly employ these phases, many of the activities considered in the phases are likely to be performed. For each phase, we consider typical activities for the phase, defects that are common for them, techniques often used for these activities, and applicable metrics. Under techniques, we include general techniques, anticipation techniques, and feedback techniques. The phases considered are as follows:

1. Concept Development and Planning
2. Requirements and Interfaces
3. Design and Coding
4. Integration, Verification, and Validation
5. Product Production and Release
6. Operation and Maintenance

There is also a section on project management. When an organization is creating its objectives and plan for supporting software reliability, it can make a list of its activities and find material in Chapter 4 on techniques and monitoring of the associated processes and products.

5.1.3   Troubled Projects

Our last situation is when we are on an existing project and realize that it is likely that, unless we make significant changes, we will be producing and delivering unreliable software. How do we rectify the situation? There is no single answer, but one approach is to use the following five steps:

1. Recognize that there is a problem
2. Determine why there is a problem
3. Decide what to do about the problem
4. Implement the solution
5. Monitor the situation and apply feedback as needed

We elaborate on these steps below.

Recognize that There Is a Problem

The first signs of a problem may occur almost anywhere and may be obvious or subtle. Some examples include the following:

1. Aspects of the project are over budget
2. Aspects are behind schedule
3. Values of metrics are outside their expected ranges (Recall from Section 3.3 that the monitoring for software reliability should include "triggers" that indicate that some form of action may be required and should state what that action should be.)
4. Products (including internal products) have too many defects
5. The project is receiving poor-quality external products
6. There is excessive project churn
7. Project personnel are unhappy
8. The customer is unhappy

Sometimes, there is a potential sign of a problem that, upon further investigation, does not indicate a problem, but the default position should be that there is a problem that needs correcting. The sooner that it is recognized that the project has a problem, the better the prospects for a solution.

It often seems that when a project becomes troubled, it occurs at the worst possible time. A potential risk reduction approach is to identify critical periods and events for the project and to perform project reviews timed to identify potential issues before each critical time period and in time to mitigate the issues before these periods and events.

Determine Why There Is a Problem

Recognizing that a problem exists is the first step. The next step is determining why there is a problem. Start with the first indication of a problem and do a root cause analysis using Section 2.4 and the root cause analysis topic in Section 6.5 as guides. These sections list and describe a number of techniques that can be used for this analysis. Note that there may be more than one root cause.

In addition to investigating the indication of a problem, review the entire project. A problem in one area may mean that there are problems in other areas that have not risen to the level of being noticed. Go back to the software reliability plan from Chapter 3 and review the objectives, plan, implementation, and feedback. Are the objectives reasonable and tied to customer needs and wants? Is the overall plan complete and is it being followed? Is the plan being implemented effectively? Is the feedback covering key potential problem areas and aspects of the project that provide insights into the state of the project? What has been learned since the start of the project that may affect the plan and its implementation? A few ideas for performing this review are as follows:

1. Perform one or more brainstorming sessions. Given the current state of the project, what "hidden" issues may exist?
2. Do a premortem to create a list of potential issues.
3. Consider and analyze the metrics. Look for values that are out of place. Also, look at trends in metrics.
4. Do interviews with project personnel. What issues have they seen or what concerns do they have?
5. Look for products and processes that may not be adequately monitored and check them in more detail. For example, do a peer review on an important document or a process failure modes effects and analysis (FMEA) on a critical process.

Use Chapter 4 to check each phase of the project. Is each product that affects the software adequately checked? Are the metrics sufficient? Are the processes satisfactory and are they being followed correctly?

There are many possibilities for what may be the problem or problems. It may be that the project did not adequately plan for software reliability. It may have chosen to use the same processes and techniques that have been used in the past and hoped for better results. The root cause of the problem may not be directly related to the software. For example, the schedule may be unrealistic, resulting in quick but poor-quality work. The project may be using the wrong people or inadequately trained people. Project communication may be poor. Morale may be bad. Goals may be inadequate, poorly communication, or not accepted. Processes may not be appropriate or may be poorly implemented. It is also important to realize that the problem may not be with the reliability of the software but rather some other aspect of the project. In such cases, software reliability needs protecting, but the software reliability approach may not need to be changed.

Finally, in determining why there is a problem, avoid making the problem worse. Do not focus on blame but rather stick to the more impersonal process and product evidence.

Decide What to Do About the Problem

Deciding what to do about the problem is a matter of making an implementable plan and fitting this new plan into the existing project plan. More explicitly:

1. Decide what stop-gap measures need to be taken now to meet near-term commitments.
2. Determine what to do about the problem, preferably at the root cause level.
3. Determine how to interleaf this new plan into the existing plan as smoothly as possible.

Just because a problem has occurred in a project does not mean that everything on the project can stop so that the problem can be addressed. The project has commitments. Because of this, it is often the case that stop-gap measures are needed. The project may need to fix the immediate issue to meet a commitment and then finish addressing the root causes of the problem. Stop-gap measures are obviously not ideal and should only be used as a last resort and then only with a plan on how to "fill in" any issues that the stop-gap measures have not addressed.

Once you have determined the root cause or causes of the problem, they need to be corrected. We also need to see if these issues have propagated to other, as-yet unseen places in the project and address these as well. At this point, we make a plan for addressing the problem based on the root cause or causes. This plan is, in a sense, an "addendum" to our original plan. Start with objectives. The objective or objectives should be consistent with the objectives of the overall software reliability plan. Material in Section 3.2 may serve as a starting point. Then, material in Section 3.3 may help with the actual plan. Sometimes, it is obvious how to proceed and sometimes it is not. Often, the more obvious approach puts additional pressure on the schedule or budget. If the defects are in a specific product area, ensure that adequate resources are there to help and consider defect-reduction and defect-detection techniques for those products. Consider bringing in additional help, either additional in-house personnel or outside expertise. Correcting the problem may mean additional budget, extending the schedule, or both. Changing the course of an ongoing project is difficult and a very big change may cause more harm than good. Look for the smallest, most targeted changes that will correct the problem. Finally, remember to include metrics to help monitor progress and use expected results and trends to determine when feedback is needed.

At this point, the project has its original plan and this newly created "addendum." These two plans need to be combined into a single implementable plan. Combining plans can be challenging. There are still budget and schedule constraints, and the morale of the team needs to be protected or possibly improved. Some parts of the original plan can possibly be replaced by a part of this issue-correcting plan, but often, new tasks are required. As stated above, look for the smallest, most targeted changes that will do the job.

Implement the Solution

Project personnel will be needed to implement the change or changes. They will need to be aware of what the changes are, why they are needed, and how the changes are to be applied. Try to keep this education objective and avoid blaming people or departments. Be sure to set clear goals and milestones. The same implementation guidelines stated in Section 3.4 apply here, particularly the list of key implementation considerations near the end of that chapter.

Monitor and Apply Feedback

As stated in Chapter 3, the final part of the process is monitoring progress and applying feedback as needed. If metrics deviate sufficiently from desired values or trends are unacceptable, interventions may be required. As much as possible, anticipate in the plan what should initiate action and what that action or actions should be. Unfortunately, unexpected situations often occur, and project personnel have to determine what to do "in real time." In such situations, keep the overall objectives in mind, and as stated earlier, try to make the smallest, most targeted changes that will do the job.

5.2   Guidelines

While any set of guidelines will leave out important information, the following set covers some of the key points that people interested in producing reliable software should always keep in mind.

1. Key general principles inherent in the processes that we have covered are as follows:
    (a) Objectives: Know your few key objectives and stick with them. Always be looking for value added. Eliminate valueless goals, processes, and procedures and change the ineffective ones. However, always understand the goal, process, or procedure before changing it.
    (b) Feed-forward error and defect control: Errors and defects that do not occur are the best kind. Try to prevent defects and errors by anticipating where and why they can enter the system. Anticipate what can go wrong and plan for these events. This antic-ipation requires a mental model of the potentially defective process, and this model is also subject to errors. Therefore, we should continually monitor for other sources of defects and errors.
    (c) Feedback error and defect control: Realize that errors and defects will occur. Plan how to monitor for such events and how to correct them as early as practical to min-imize their impacts. Find the defect as soon as practical after it is introduced into the system. The longer that it is in the system, the more damage it has an opportunity to cause and often times, the more embedded it becomes in the system and the harder it is to remove. Errors that are quickly mitigated and do not result in defects or result in defects that have little or no impact are the second best kind to have.
    (d) Implementation: Continually improve these defect control processes. Even a well-designed set of defect control processes is subject to current conditions, and conditions change continuously. Always be looking for improvements.
    (e) Responsibility: Software reliability is everyone's responsibility. Projects that produce reliable software tend to be well-run projects staffed by motivated and talented per-sonnel that take software reliability and quality in general seriously.
2. Maxims: Sometimes, advice can be summarized with an insightful maxim. A few such maxims that apply to software reliability are as follows:
    (a) Make it easy for the human to do the right thing.
    (b) People first. The most important factor in the project's success is the team of people on the project. Processes, organizations, and techniques can all help, but it is the people involved that make it work even when things do not go as planned.
    (c) Avoid "paralysis by analysis." Doing a few things well is better than just planning for a lot of things. There are many options and techniques available, but it is okay to start with a few key approaches and techniques and add others as needed.
    (d) Understand the types of errors and defects that are likely to occur.

(e) Monitoring and feedback are critical for reliability growth.
(f) Nearly all activities affect software reliability. Some have a primary impact and others a secondary.
(g) Keep it simple. When things start getting complicated, drop back and reassess the situation and see if there is a more simple solution.
(h) Try to measure everything that is important. "That which is measured is what is improved."
(i) A requirement is not truly defined until a "pass/fail" verification for the requirement is defined.

More on the subjects covered in this chapter may be found in the referenced chapters and their references. In addition, see [1–5]

References

1 Lyu M. *Handbook of software reliability engineering*, Computer Society Press and McGraw-Hill Book Company, New York, 1996.

2 Lyu M. Software reliability engineering: A roadmap. *Future of software engineering*, pp. 153–170, IEEE Computer Society, 2007. https://www.researchgate.net/publication/4250863_Software_Reliability_Engineering _A_Roadmap. 22 Aug 2020.

3 McConnell S. *Code complete: A practical handbook of code construction*. Microsoft Press, Redmond, Washington, 2004.

4 Musa J. *Software reliability engineering: More reliability software faster and cheaper*. AutherHouse, 2004.

5 Neufelder A. *Ensuring software reliability*. Marcel Dekker, Inc., New York, 1993.

6

Techniques

Chapter 4 considers techniques by project phase, where each technique directly or indirectly improves software reliability. In this chapter, we consider 67 techniques. They are introduced as a whole in Section 6.1; then, we consider these techniques by organization. The techniques are divided into techniques typically used by systems engineering (SE), software, reliability engineering, or any project organization. More specifically, Section 6.2 describes techniques typically associated with SE activities, Section 6.3 describes techniques usually used by software personnel, and Section 6.4 describes techniques for reliability engineering organizations. We also consider techniques in Section 6.5 that can be used by any project organization, although many of these techniques are most often associated with quality assurance (QA). Each technique has a short description and often references for further information.

6.1 Introduction to the Techniques

Throughout this book, we have considered the nature of defects and many techniques to prevent, tolerate, or find and remove them. We have also discussed techniques to determine how well we have performed these tasks. In this final chapter, these and other such techniques are described in more detail. The description typically provides enough information that the reader can apply the technique, although references are often included if more information is needed. There are also numerous examples to demonstrate the potential uses of the techniques. The techniques described in this chapter, along with the associated project organization, are provided in Table 6.1.

When interested in what techniques are covered, or if information about a specific technique is required, Table 6.1 lists the techniques and where to find a detailed explanation of it. To understand what techniques are associated with a given organization, go to the section for that organization.

6.2 Techniques for Systems Engineering

There are many techniques that are not traditionally considered to be a part of reliability engineering, but that can significantly improve the reliability of software. For example,

Software Reliability Techniques for Real-World Applications, First Edition. Roger K. Youree.
© 2023 John Wiley & Sons Ltd. Published 2023 by John Wiley & Sons Ltd.

Table 6.1 Techniques for Software Reliability.

Technique	Main Organization
Affinity diagrams	6.5 Program-Wide
Algorithm description document	6.3 Software
Architecture decision document	6.3 Software
Architecture tradeoff analysis method	6.3 Software
Bayesian techniques	6.5 Program-Wide
Behavior-driven development	6.3 Software
Benchmarking	6.5 Program-Wide
Brainstorming	6.5 Program-Wide
Checklists	6.5 Program-Wide
Cleanroom methodology	6.3 Software
Continuous delivery, Continuous delivery Pipeline	6.3 Software
Data collection for metrics	6.5 Program-Wide
Decision-making techniques	6.5 Program-Wide
Defect detection rate monitoring	6.4 Reliability Engineering
Defect education	6.5 Program-Wide
Design of experiments (DOE)	6.5 Program-Wide
DRACAS, DRB	6.4 Reliability Engineering
Dynamic analysis of code	6.3 Software
Failure definition and scoring criteria	6.4 Reliability Engineering
Fault tolerance	6.3 Software
Formal methods	6.4 Reliability Engineering
Formal specification languages, Restricted languages	6.2 Systems Engineering
FRACAS, FRB	6.4 Reliability Engineering
Inspections and reviews	6.5 Program-Wide
Lessons learned	6.5 Program-Wide
Models and simulations	6.5 Program-Wide
Operational profiles	6.4 Reliability Engineering
Orthogonal defect classification	6.4 Reliability Engineering
Pair programming	6.3 Software
Pareto analysis	6.5 Program-Wide
Process FMEA/FMECA	6.5 Program-Wide
Process improvement techniques	6.5 Program-Wide
Project best practices	6.5 Program-Wide
Project premortem	6.5 Program-Wide

Table 6.1 (Continued)

Technique	Main Organization
Prototypes	6.5 Program-Wide
Quality circles	6.5 Program-Wide
Quality function deployment	6.2 Systems Engineering
Rationale documentation	6.5 Program-Wide
Reliability risk assessment	6.4 Reliability Engineering
Reliability trade studies	6.4 Reliability Engineering
Requirements analysis techniques	6.2 Systems Engineering
Requirements elicitation techniques	6.2 Systems Engineering
Requirements validation techniques	6.2 Systems Engineering
Root cause analysis	6.5 Program-Wide
Semantic analysis of code	6.3 Software
Software defect root cause analysis	6.4 Reliability Engineering
Software fault tree analysis	6.4 Reliability Engineering
Software FMEA/FMECA	6.4 Reliability Engineering
Software rejuvenation	6.3 Software
Software reliability advocate	6.4 Reliability Engineering
Software reliability and availability allocations	6.4 Reliability Engineering
Software reliability and availability objectives	6.4 Reliability Engineering
Software reliability and availability predictions	6.4 Reliability Engineering
Software reliability casebook	6.4 Reliability Engineering
Software reliability estimations	6.4 Reliability Engineering
Software reliability growth plan	6.4 Reliability Engineering
Software reliability program plan	6.4 Reliability Engineering
Software reuse	6.3 Software
Software testing	6.3 Software
Static analysis of code	6.3 Software
Statistical process control	6.5 Program-Wide
Style guides	6.5 Program-Wide
Test-driven development	6.3 Software
Trend tests	6.5 Program-Wide
Use cases	6.2 Systems Engineering
User scenarios	6.2 Systems Engineering
User stories	6.2 Systems Engineering

the seeds for a software defect may be planted in a conceptual design or a requirement specification. The SE organization typically has the prime responsibility for the early conceptual design and the requirements and interface documentation, and as such, techniques that they use to prevent or remove defects from these products can improve software reliability. In Table 6.2, we list techniques that SE can use to reduce defects in their products and do so in cost- and schedule-efficient ways.

The eight SE techniques described in this section are primarily associated with SE activities that are performed early in a project, although different projects may apply them at different times. Other organizations are typically involved with these techniques and may in some instances play the lead role with them. For example, user stories are often created by software but are included here under SE because they can serve as requirements for the software and requirements are traditionally an SE task. It is also a common situation for the bulk of the effort with these techniques to be applied early in a project, but updates and refinements occur in later phases or in different cycles or iterations of the project. For example, most of the effort in producing requirements is early in a project, but customer needs may change or become better understood after this time period, or as the software design, coding, and software tests progress, issues may mandate that changes be made to the requirements.

Table 6.2 lists eight SE techniques that can improve the reliability of a software product. Table 6.3 lists other techniques that are related to SE activities, the organization that the technique is described under, and how the technique is related to SE.

Next, we describe each technique listed in Table 6.2.

Formal Specification Languages, Restricted Languages

The use of formal specification languages reduces the ambiguity in the meaning of requirements when writing specifications. These languages are an important part of formal methods but can be used solely to clarify requirements. Although they reduce ambiguities, these languages have the disadvantage of reducing expressive freedom, and they require that all participants understand the new language. An example of such a restricted requirements language is the language Z. For more information, see the Formal Methods topic in Section 6.4.

Quality Function Deployment

Quality function deployment (QFD) is a customer-driven process for determining customer "must haves," "wants," and "wows" and prioritizing these so that they can be converted into engineering requirements and assigned to the appropriate organizations. QFD seeks to add "positive quality" by adding value to the customer throughout the marketing, development, and support processes and not just reduce the number of defects as traditional quality programs tend to do. It starts with identifying the qualities that the customer desires and then finds the functions required to provide these qualities. Finally, it identifies the means to deploy the available resources to provide the resulting product or service. While not normally associated with software reliability, QFD can help a project reduce churn, and project churn can be a major source of errors and defects.

There are different versions of QFD. Traditionally, a well-known part of QFD is a technique called house of quality (HOQ), which is used to identify customer desires and

Table 6.2 Techniques for Systems Engineering.

Technique	When Used	Notes
Formal specification languages, restricted languages	Early	This technique can reduce requirements ambiguity
Quality function deployment	Early	Quality function deployment is a technique that helps to determine what requirements a customer needs
Requirements analysis techniques	Early	Requirements analysis techniques are used to take user needs and decompose them into clear, doable requirements
Requirements elicitation techniques	Early	Requirements elicitation is the process of discovering requirements for a system
Requirements validation techniques	Early	These techniques are used to determine if the requirements specify the right product
Use cases	Early	Use cases analyze functional requirements from a user's perspective
User scenarios	Early	This technique documents the process the user may take to use the product and helps determine what requirements a customer needs
User stories	Early	A user story is a high-level description of a software feature or behavior and is an agile approach for requirements writing

translate them into engineering targets to be met by the product. Other versions of QFD have been adapted to more modern lean development processes and may not include HOQ. These versions use tools such as customer segments tables to identify the chain of customers, customer process maps to understand what the customer likes or dislikes, and customer voice tables to better understand the "whys" of customer wants and therefore focus on the true needs of the customer. Another QFD tool is the analytic hierarchy process (AHP) for prioritization and decision-making. AHPs use pairwise comparisons to reduce complex decision-making to a clear rational result. See Section 6.5 for more details on AHP. Affinity diagrams (see Section 6.5), hierarchy diagrams, and maximum value tables are other tools that may be used. See [1] and [2] for more details.

Requirements Analysis Techniques

Although requirements engineering techniques may not immediately come to mind when discussing techniques to improve software reliability, requirements defects are a major source of software defects, and techniques to reduce these defects deserve careful consideration. Requirements analysis techniques are used to take user needs and decompose them into clear, doable requirements. Part of this process is to determine if the resulting requirements are necessary, consistent, complete, feasible, unambiguous, affordable, and verifiable. Requirements analysis is a very important part of preventing or reducing the

Table 6.3 Techniques Related to Systems Engineering.

Technique	Main Organization	Notes
Algorithm description document	6.3 Software	This technique can support requirements development
Behavior-driven development	6.3 Software	Behavior-driven development is a software development process and as a part of the process develops requirements for the software
Continuous delivery, Continuous delivery pipeline	6.3 Software	Continuous delivery, continuous delivery pipeline is an agile software development approach emphasizing release on demand and as a part of this process requirements are developed
Formal methods	6.4 Reliability Engineering	A part of formal methods is developing precise requirements for the software
Style guides	6.5 Program-Wide	Style guides are an important tool for writing good requirements
Test-driven development	6.3 Software	Like behavior-driven development, this technique is a software development process and as a part of the process develops requirements for the software

number of requirements defects and therefore preventing software defects. There are many techniques that can be used. Some of the more useful are as follows:

1. Requirements traceability and allocation analysis: The goal of requirements traceability and allocation analysis is to ensure that every requirement is traceable to a source and that every requirement is in a path that allocates to design elements. Also, each requirement should also be allocated to verification and validation elements. There are tools and databases to help automate and enforce this process.
2. Use cases: Use cases are a modeling technique used to identify and analyze functional requirements from a user's point of view and to document and communicate these clearly. They are used to define the interactions between one or more actors and the system so as to accomplish certain goals. Use cases are explained more fully in their own topic later in this section.
3. Models: A use case is an example of a type of model. Its value in requirements analysis is such that we have given it its own item number. Other types of models are also important. These include flowcharts, Gantt charts, storyboards, Unified Modeling Language (UML), Systems Modeling Language (SysML), user interface models, and data flow diagrams, to name a few. More on models can be found in Section 6.5.
4. Prototypes: Prototypes can provide information of the feasibility and desirability of select subsets of potential requirements. More on prototypes can be found in Section 6.5.

5. Restricted languages: As mentioned in the item on formal specification languages and restricted languages, the use of a formal specification language can reduce specification ambiguity. Restricted languages are covered in more detail above and in Section 6.4.

6. Peer reviews and inspections: Peer reviews and inspections are proven ways to reduce specification errors. A useful way to apply these techniques is to have several reviewers, each with a different viewpoint. For example, one reviewer can be reviewing the specification to determine if each requirement is necessary, another reviewer considers requirement consistency, and so forth. Peer reviews and inspections are covered in more detail in Section 6.5.

More details may be found in [2–4] and [5].

Requirements Elicitation Techniques

Requirements elicitation is the process of discovering requirements for a system. In requirements elicitation, we determine the performance, features, and behaviors required to meet the functional, non-functional, and interface requirements. The process involves activities such as identifying the relevant stakeholders and setting boundaries for the system under development to scope the problem. As with requirements analysis, requirements elicitation is not usually considered as a software reliability technique, but it is an important part of producing good requirements and therefore good software. Techniques often used with requirements elicitation include the following:

1. Analysis of existing systems: Analysis of existing systems can be an important requirements elicitation technique when building an improved version of an existing system or producing a product to compete with a competitor's product. Knowing the strengths and weaknesses of these systems may provide clues to behaviors and features that separate the proposed new product from its competitors.

2. Traditional data gathering techniques: Traditional data gathering techniques include surveys, interviews, focus groups, and reviews of existing documentation and systems.

3. Customer meetings: A customer meeting for requirements elicitation typically consists of a relatively informal discussion of user needs and wants.

4. Product tracking: Product tracking is the process of keeping track of what is being said about a given product, as well as for competing and related products. Product tracking can be used to gain insights into the relative strengths and weakness of products from the viewpoints of product users.

5. Use cases and user scenarios: Use cases and user scenarios are used to identify functional requirements from a user's point of view. Each is covered in more detail in its own topic later in this section.

6. User stories: A user story is written by a stakeholder and consists of a high-level description of a feature or behavior. User stories are covered in more detail later in this section.

7. Quality function deployment (QFD): QFD is a customer-driven process for determining customer "must haves," "wants," and "wows" and for prioritizing these. The results are used to develop requirements. As with use cases, user scenarios, and user stories, QFD is covered in more detail in its own topic in this section.

8. Prototyping: A prototype is a mock-up or partial implementation of a system. Prototyping for requirements elicitation includes techniques such as storyboarding, screen mock-ups, and pilot systems. Prototyping is covered in more detail in Section 6.5.
9. Brainstorming: Brainstorming is a well-known group technique for generating ideas. Brainstorming is covered in more detail in Section 6.5.
10. Stakeholders lists: A stakeholders list identifies all of the requirements stakeholders. The list is then used to support other elicitation techniques, such as interviews or brainstorming.

More details may be found in [2–4] and [5].

Requirements Validation Techniques

For requirements validation, we ask if the requirement specifies the right product. Each requirement should ultimately address one or more customer needs or wants. (Note that some requirements address "business non-value-added activities," such as regulatory requirements, that do not add value from the standpoint of the customer but are still needed.) Requirements validation also considers whether the requirements are consistent, do not conflict, and are verifiable, complete, unambiguous, and accurate.

When validating requirements, be sure to examine the requirements as a whole to ensure that they meet the operational requirements and therefore the customer needs. Stakeholder and customer involvement can be anywhere from very useful to critical in this process. Requirements validation may use any of a number of different techniques, including the following:

1. Reviews and inspections: Reviews and inspections are probably the most common requirements validation techniques. These typically compare the requirements against earlier artifacts such as needs statements, operational requirements, use cases, or user scenarios. Reviews and inspections are covered in more detail in Section 6.5.
2. Test cases: Designing tests for requirements is a good way to find requirements defects in the early phases of a project. A requirement should be clear and unambiguous, and a way to check for this is to build a test case for the requirement. Also, test cases for user requirements should provide an early understanding of what the acceptance testing will look like and if the requirements are addressing the user needs. Section 6.3 covers software testing in more detail.
3. Prototyping: Prototypes can serve as an intermediate between requirements and the finished product, making it easier to validate the related requirements. Section 6.5 has more on prototyping.
4. Modeling: As with prototypes, models increase the level of information about what product will result from the requirements. Models are addressed in more detail in Section 6.5.
5. Validation checklists: A useful technique for reducing the likelihood of requirements validation issues is to use validation checklists as the requirements are being developed. Section 6.5 further addresses checklists.
6. Requirements traceability and allocation checks: Part of making sure that the requirements meet the customer needs is making sure that every requirement is traceable to a source and that every identified customer need and requirement is in a path that allocates to design elements. That way, we can check that each customer need is

addressed in the requirements and in the design and that if there are any requirements not needed to address a customer need, they are needed for "business purposes" or removed.

More details may be found in [2] and [5].

Use Cases

Use cases and use case analysis are used to identify and analyze functional requirements from a user's point of view and to document and communicate these clearly. A use case is a modeling technique that is used to define the interactions between one or more actors and the system so as to accomplish certain goals. Informally, a use case is a "story" describing how actors interact with the system to achieve a goal. "Actors" in the use case may not be humans, but they can make decisions. A use case typically has the following:

1. A title
2. A short description
3. A list of actors, often denoted as a primary actor and other stakeholders
4. The initial conditions describing the environment before the events of the use case
5. The end conditions describing the environment after the events of the use case
6. Triggers that initiate the sequence of events
7. The use case sequence of events

A use case may also have other features. A use case describes a nominal or normal sequence of events and may also be expanded to describe off-nominal sequences. Interactions between use cases are shown in use case diagrams. These diagrams often use UML or other modeling languages, and they show the activities of the associated actors, with each diagram showing a separate sequence of activities and events. For more details, see [2].

User Scenarios

A user scenario is a narrative of how a potential user of the software may use the product to accomplish some goal. It documents the process the user may take to use the product. User scenarios do not generally cover all such scenarios but rather attempt to cover the most common ones.

A user scenario tries to understand the motivations of the users with respect to the product and is used early in the project to develop initial design ideas and improve the usability of the product. It answers questions such as

1. "Who is the user of this product?"
2. "What does the user want to accomplish with this product?"
3. "When might the user perform these tasks?"
4. "How is the user going to achieve this goal?"
5. "Why does the user choose this product over other products?"

Typically, user scenarios answer each of these or related questions in short, succinct paragraphs. User scenarios can also be used to develop test objectives.

Example 6.2.1 As an example of a user scenario, we consider an elderly woman interested in using a bank's on-line account information and management site:

Maggie is an 82-year-old retired teacher who lives alone. She has little computer experience but is bright and engaged with her community. She has a small, inexpensive computer with Internet connectivity at her home. She has a pension that automatically deposits a monthly amount into her checking account at the local bank and considers it to be important to carefully monitor her finances. She has two accounts with the bank. One is a checking account that bears no and the other is an interest bearing savings account.

The bank that Maggie uses has an on-line account information and management site, so she wants to use her home computer to connect to the site and monitor and manage her accounts. Her typical routine is to check how much money is in her checking account and, if it is below a certain amount, to transfer money from her savings account to her checking account. She also monitors her savings in the savings account to see that it is above a certain amount and, if it is not, replans her spending to reduce her expenditures and save more of her money. She performs these tasks weekly, usually on a Sunday afternoon, but may monitor her account at any time during the week if a bill has come in or some other financial event has occurred.

Maggie is intelligent but not computer savvy. First, she wants the system to be secure and does not want to worry about someone stealing money from her accounts. She also wants simple, easy-to-understand processes to find the information that she wants and to manipulate the money between the accounts. She wants to see how much she currently has in her account as soon as her checks clear, and she wants to be able to move any amount of money from one account to the other, with the transfer taking effect immediately.

User Stories

A user story is a high-level description of a software feature or behavior. The stakeholders, not the software developers, write the stories. These stories contain just enough information that software developers can produce a reasonable estimate of how much effort is required to produce the implementing code. They are informal and use natural language. They also tend to be shorter and more concise than a user scenario. A popular format for a user story is the "As a [description of the user] – I want [feature] – so that [benefit of the feature]" format. A user story also typically has an acceptance criterion. Acceptance criteria are often in a "Given [state at the beginning] – When [some action is taken] – Then [some observable outcome]" format.

User stories are often an important part of agile programming and extreme programming and are usually created with a minimum of tools. Stories may be hand written on index cards or other easily managed writing surfaces. There are typically multiple iterations of them. The next step after creating a user story is often creating a high-level sketch of how the story will be implemented. There are many variations and techniques to get the most from user stories. For more details, see [6].

Example 6.2.2   In this example, we are interested in software for a website for an investment company or firm. The website allows clients of the investment firm to log on and see the status of their accounts and to make changes to their accounts. Clients of the firm invest in order to make money, so website users often want to see the status of their accounts to see if their investments are growing. For this example of a user story, the stakeholder is a client of the investment firm who uses the firm's website. The particular thing that the client wants is to see her account status. The reason that she wants to see it is to determine if her account is making money for her. A possible user story for this situation in the "As a – I want – so that" format is

> **As a** [client of the investment firm],
> **I want** [to see the current status of my account]
> **so that** [I can see the annual percent growth of each of my assets.]

In addition to the basic user story, it is useful to have an acceptance criterion. The situation starts with a client of the investment firm being logged onto her account on the firm's website. As clients do not have to be computer savvy, the website should be simple and intuitive. An example of an acceptance criterion for this user story is:

> **Given** [that I am a client at the investment firm and that I am logged into my account on their website],
> **When** [I click the "View My Account" button on the screen of the main page]
> **Then** [the computer screen shows the current status of all of my assets at the firm and the percent change of each from the same quarter last year.]

6.3 Techniques for Software

The software design and coding organization, or "software" for short, is often referred to simply as "design." This organization takes the conceptual designs and the requirements and interface documentation and produces a software design consistent with these items. It then codes the software to produce the design. As with the techniques associated with SE, software techniques can be used to prevent or remove defects in the design or software well before they are found by tests. As a result, we can have more reliable software at a lower cost and in less time.

Below, we list a collection of 15 techniques that can be used to prevent or remove defects and that are generally used by the software organization. Five of these techniques are usually used early in a project, five are used later in a project, and five are typically used throughout the project time frame. As with the SE techniques, when a technique is most appropriate depends on the project. Also, a technique that is primarily associated with being used early in a project may also be of use throughout the project, such as when the bulk of the effort is applied early in a project, but then updates and refinements occur in later phases. Another important note is that techniques that are typically used later in a project, such as when analyzing or testing code, are typically planned for early in the project.

As just noted, the list contains five techniques for software that are often used throughout the project time frame. These five are techniques such as behavior-driven development (BDD) and test-driven development (TDD), which are techniques for overall software development. They begin early in the project with the effort to understand what the customer needs and continues through to deliverable software. Other techniques included in these five are for preventing or detecting defects and apply throughout the project time frame.

Table 6.4 lists 15 software techniques that can improve the reliability of a software product. Table 6.5 lists seven other techniques that are related to software activities, the organization that the technique is described under, and how the technique is related to software

Table 6.4 Techniques for Software.

Technique	When Used	Notes
Algorithm description document	Early	This document helps bridge the gap between requirements and design by documenting the algorithm agreement between requirements engineering and software
Architecture decision document	Early	This document helps communicate what the software architecture design is and why it is designed as it is
Architecture tradeoff analysis method	Early	This technique is an approach for analyzing the software architecture
Behavior-driven development	Throughout	This technique is an agile-inspired behavior-focused software development process that emphasizes interaction between stakeholders
Cleanroom methodology	Throughout	Cleanroom methodology is a highly structured software development process that focuses on the prevention of defects
Continuous delivery, continuous delivery pipeline	Throughout	This technique is an agile-inspired software project approach that emphasizes product release based on market opportunities
Dynamic analysis of code	Later	This technique involves running executable code to find software defects
Fault tolerance	Early	Fault tolerance techniques allow successful software performance in spite of software defects
Pair programming	Later	Pair programming uses two programmers working together on the same software to produce better code
Semantic analysis of code	Throughout	This technique analyzes software for runtime errors without having to run the code
Software rejuvenation	Later	The technique helps mitigate software aging, the process by which software errors accumulate with time or load
Software reuse	Later	Software reuse takes advantage of proven software
Software testing	Later	The technique is a mainstay for software projects. There are many types of testing with various purposes
Static analysis of code	Early	Static analysis tools find certain types of defects in code without executing the code
Test-driven development	Throughout	This technique is an agile-inspired feature-focused software development process that uses small incremental cycles of testing, coding, and refactoring

Table 6.5 Techniques Related to Software.

Technique	Main Organization	Notes
Formal methods	6.4 Reliability Engineering	Formal methods may be used in the design and development of the software
Orthogonal defect classification	6.4 Reliability Engineering	This technique can help software allocate resources
Software fault tree analysis	6.4 Reliability Engineering	This technique is often used by software to troubleshoot code
Style guides	6.5 Program-Wide	Style guides are an important tool for writing consistent and readable software
Use cases	6.2 Systems Engineering	Use cases are often a part of an overall agile software development process
User scenarios	6.2 Systems Engineering	Like use cases, this technique is often a part of an overall agile software development process
User stories	6.2 Systems Engineering	Like use cases and user scenarios, this technique is often a part of an overall agile software development process

activities. Some of this overlap between organizations is due to different organizations being assigned different tasks in one project than in another. Other overlaps are due to emphasis. For example, a style guide for SE may describe how a requirement specification or interface document should look, whereas for the software organization, a style guide may outline how to write readable software code.

Next, we describe each technique listed in Table 6.4.

Algorithm Description Document

An algorithm description document (ADD) is a technique used by requirements and software design personnel to better organize the process of going from requirements to software design and code. As a result, it can reduce the number of defects in the software. Rather than taking requirements and starting to design and write code, many problems can be prevented by taking the intermediate step of first focusing on designing and documenting the algorithms that will be coded. To meet the collection of requirements assigned to the software, certain functions must be performed by algorithms, and these functions must be performed in a certain order and meet certain requirements. A useful hierarchical way to design code is to lay out a high-level list of these functions and the algorithms used to perform them and use a flowchart or equivalent techniques to indicate which functions feed which other functions. This effort is often documented in an ADD. An ADD has some or all of the following information:

1. A narrative to describe the overall purpose of the software
2. A list of functions required of the software
3. A list of algorithms that are to perform the functions
4. An arrangement of the algorithms to indicate which algorithms feed which other algorithms
5. Requirements for each algorithm, such as timing, speed, accuracy, stability, and so forth
6. A description of each chosen algorithm, typically using mathematics or pseudo-code
7. A description of interfaces between the algorithms
8. Trade studies and decision processes supporting the chosen algorithms and ensuring algorithm accuracy, throughput, timing, memory requirements, and other considerations
9. The validation of the algorithms, using analysis, simulations, or other techniques

The ADD documents the algorithm agreement between requirements engineering and software and describes the life cycle of the algorithms. Requirements are potentially updated based on the results found producing the ADD. The ADD may be updated as the software matures to better document the algorithm and its place in the software. It is important to keep the requirements, ADD, and software synchronized. The resultant document should be peer reviewed or inspected and used to develop the code. More on ADDs may be found in [7].

Architecture Decision Document

Software architectures tend to be costly to change, and decisions made to design them should be carefully thought out, well understood, and documented. An architecture decision document describes the software architecture and how it satisfies the software and system requirements. It usually employees architecture diagrams. In addition to describing the architecture, it describes the major architecture decisions that were made, what alternatives were considered, and the rationale for the choices that were made. By documenting these choices, the project has a better understanding of why the architecture is designed as it is, and if at some time in the future there is a desire to change the architecture, critical decision information will still be available. More on documenting software architectures may be found in [8].

Architecture Tradeoff Analysis Method

Architecture tradeoff analysis method^{SM}(ATAM) is a technique for analyzing a software architecture. The method gathers stakeholders to analyze system goals and requirements, resulting in a collection of scenarios that are used to create software architecture tradeoffs. Risk analysis is performed on these tradeoffs resulting in architecture decisions. In more detail, the steps for this technique are as follows:

1. Present the method: In this step, the architecture team describes the ATAM process to the various stakeholders. These stakeholders include designers, customers, managers, maintainers, and other representatives affected by the software architecture.
2. Present the business drivers: This step is to ensure that everyone understands the system to be evaluated, which is presented from a business point of view. The most important functional requirements are presented, various constraints are outlined, and the overall goals are covered. Drivers for the software architecture are also presented.

3. Present the architecture: The architecture is presented by the lead architect at an appropriate level. The presentation includes the technical constraints such as operating system, hardware, and the other systems that the software architecture must interface with. Also, architecture approaches used to meet other requirements, such as quality requirements, are covered.

4. Identify architecture approaches: In this step, architectural approaches are identified but are not analyzed.

5. Create a quality attribute utility tree: Next, the team works with the stakeholders to identify, prioritize, and refine the most important quality attributes for the system. At the end of this step, the team has a prioritized collection of quality attribute requirements presented as scenarios where these scenarios state the quality attributes in terms of stimuli and responses. The intent is to focus attention on the aspects of the architecture that are most important to the success of the system.

6. Analyze the architecture approaches: In this step, the team analyzes the architecture approaches to obtain enough information about each architecture approach to perform an early analysis of its attributes to see if the approach holds promise for meeting the requirements. Outputs of this step include a list of approaches along with questions associated with these approaches with early answers to these questions. It also includes risks, sensitivity points, and tradeoffs associated with each approach. In addition, this step looks for and documents interactions and tradeoffs between the approaches.

7. Brainstorm and prioritize the architecture approaches: Here, stakeholders brainstorm use case scenarios and change scenarios. Use case scenarios represent the ways that the stakeholders expect the system to be used and change scenarios represent the ways that they expect the system to change in the future. After brainstorming, scenarios are prioritized, typically by a voting procedure. The technique of brainstorming is covered later in this chapter.

8. Analyze the architecture approaches: At this point, the highest ranking scenarios are mapped onto the architecture descriptions. Ideally, this will be mapping onto the previously discussed architecture approaches.

9. Present the results: Finally, the results are collected, summarized, and presented back to the stakeholders.

See [9] for more on this technique.

Behavior-Driven Development

BDD is a software development process that expands on TDD (TDD is explained below). Like TDD, BDD involves developing software in cycles, first developing a test, then creating just enough code to pass the test, then refactoring. However, TDD generally focuses on producing code one feature at a time, whereas BDD is behavior focused, developing code one behavior at a time. BDD emphasizes the interaction between stakeholders, both "business" and "technical." A major way that it does this is through the use of test cases written in a language that all of the stakeholders can understand.

BDD incorporates user stories (covered in Section 6.2) with short narratives to specify behavior. These stories give initial conditions, triggers, and the expected outcome. As a result, they also serve as acceptance criteria. With its emphasis on clear simple language, BDD is also an approach to improving communication between engineers, business people, and customers so that the end product meets the needs of the customer. In more detail, BDD uses the following:

1. Deliberate discovery: In deliberate discovery, the stakeholders create real-world examples of how the system should work. These are used to define the behavior of the system. Typically, BDD uses an issue tracker to write user stories in a common, or "ubiquitous," language, such as Gherkin or JBehave. These are structured languages that use natural language words. The main point of their use is to remove as many requirement ambiguities and misunderstandings as possible. It is also a step in the development of test cases through its extensive use of scenarios. These scenarios can then be automated.
2. Apply test-driven development: A scenario has a sequence of steps, typically one per line. At this point, we use TDD and create a test for the first step, observe that the current code fails the new test, create new code that should pass the test, repeat the test, and so on.
3. Pass the scenario: Continue these small incremental TDD steps until each step in the complete sequence of steps for the scenario has passed. Next, test the scenario.
4. Refactor: After passing the scenario, refactor. Refactoring code is the process of making the code clearer, and more clean, efficient, and maintainable without changing the external behavior or functionality of the code.
5. Repeat: Having successfully added the scenario and refactored, repeat with a new scenario until all scenarios are incorporated.

BDD has the disadvantage of requiring a certain amount of training and professional maturity. However, it can improve communication, increasing the likelihood that the product developed is the product that the customer wants. Proponents also claim (with some justification) that it produces very reliable software, partly because of the following considerations:

1. It creates fewer defects by using a coherent, systematic design approach.
2. The team finds and fixes defects quickly through frequent testing. Frequent use of the collection of tests also prevents the defects from reoccurring.
3. The team also refactors often to reduce the likelihood of poorly designed code.

More details may be found in [10].

Cleanroom Methodology

Cleanroom methodology refers to a highly structured software development process that focuses on prevention of software defects by ensuring clear requirements, producing precise functional and usage specifications and applying correctness verification before producing code. Pseudo-code is developed and goes through correctness verification. After this verification, the pseudo-code is converted to actual software code. The pseudo-code is so precise that translating it into the chosen coding language is relatively easy. Statistical testing using the operational profile (OP) is usually performed to determine the reliability of the resulting software.

Cleanroom methodology is very much focused on "doing it right the first time" and has some impressive results. The cleanroom approach is reported to have produced as low as 3 defects per 1000 lines of code in-house and 0.1 defects per 1000 lines of code in a released product (see [11]). However, it requires more technical expertise than the personnel of some projects have. See [12] for more details.

Continuous Delivery, Continuous Delivery Pipeline

A continuous delivery pipeline is an agile concept to allow for the release of a product in response to market opportunities, i.e. to release on demand. To do this, small changes to the product are continuously being thought of, integrated into a version of the product, deployed for production, and readied for when management believes is the right time for release. The idea is to make every change releasable so that releases can be made quickly. In more detail, continuous delivery typically uses at least these four processes:

1. Continuous planning: Continuous planning takes stakeholders from both business and software functions to produce plans that are open-ended and that evolve in response to the business environment. Ideas that are expected to create value for a customer are analyzed and a concept for a minimal viable product (MVP) is developed. The result is usually a change to an existing product to be implemented in a future program increment.
2. Continuous integration: With continuous integration, a product idea from continuous planning is designed, coded, and tested by an agile team and then validated. With continuous integration, code is integrated throughout development and is frequently committed to a version control repository, typically at least daily. This process reduces integration problems by forcing associated efforts to be more coordinated.
3. Continuous deployment: Next, continuous deployment takes changes from the staging environment and makes them available for production. In the continuous delivery pipeline process, deployment is performed continuously, even if some features are not ready or management is not ready to release the product. Version control is critical and the process is automated as much as possible.
4. Release on demand: Finally, management can decide to release the updated product immediately or wait for a more opportune time.

There are many versions of continuous delivery. They all focus on adding value to the customer, the use of lean and agile processes to accelerate the flow, and the use of automation. The approach can lead to more streamlined processes, faster product turnaround times, increased product predictability, improved productivity, improved product reliability and quality, and more stakeholder satisfaction. However, it can be challenging to implement. See [13, 14], and [15] for more details.

Dynamic Analysis of Code

Dynamic analysis of code involves running executable code to find defects in the code. Software testing, covered in a separate topic below in this chapter, provides examples of dynamic testing. Both dynamic analysis and static analysis find defects, but some types of defects are more likely to be found by one method than the other. Dynamic analysis finds symptoms of the defect but may require extensive debugging to find the actual defect. Dynamic analysis can rarely be exhaustive and so it may miss defects. As a result, dynamic analysis is usually supported by other techniques such as code reviews and inspections. Inspections can find defects before software testing so that test costs and schedules may be reduced. It also tends to find a different subset of defects than tests, so the released products may be more reliable by using multiple techniques.

Fault Tolerance

Throughout this book, we have focused mostly on preventing defects and for those not prevented, finding and removing them early. However, no matter how well we perform these tasks, we have to assume that there will be some defects in the software. It is therefore important to consider fault tolerance techniques. Fault tolerance is the ability to perform adequately in spite of the occurrence of faults. The two most common approaches for software fault tolerance are fault recovery and fault compensation. Fault recovery is where a fault-free state is substituted for an erroneous state. For example, defensive programming may be used to check input and output conditions and look for illegal operations and follow preplanned steps if the input is expected to trigger a fault condition. Defensive programming can prevent faults from manifesting themselves. Exception handling can be used to contain faults that do manifest themselves and prevent them from propagating further. Checkpointing and rollback mechanisms can be used to recover from faulty conditions. Fault compensation uses enough redundancy that the software continues to perform adequately in spite of the fault, such as the use of redundant software modules, each module being different in some way but each performing the same or similar tasks. Fault tolerance may allow fault-free performance in spite of a fault, or it may just allow for graceful performance degradation without a system crash or some other more serious response to the fault. See Chapter 12 of [16, 17], Chapter 14 of [18], Chapters 33 and 34 of [19], Chapter 11 of [20, 21], and [22] for more on fault tolerance.

Pair Programming

Pair programming is a software coding technique that uses two programmers working together at the same workstation. One programmer codes and the other observes, thinks, and plans. The person coding focuses on the current coding activity. The person observing suggests improvements, considers how to test the code, evaluates the design, and considers the direction of the coding activity and potential future issues. To make pair programming effective:

1. Use a well-understood set of coding standards and style guides so that the programmers can concentrate of coding.
2. Rotate the roles of the pair.
3. Make sure that the person who is not programming is still actively participating.
4. Choose pairs carefully. If both programmers in the pair are new to pair programming, there may be a steeper-than-needed learning curve. Also, avoid pairing people who do not work well together.
5. Make sure that both programmers can see the monitor.

Pair programming can result in higher quality code in less time. It can also help train a less-experienced coder when paired with a more experienced one. Finally, pair programming ensures that at least two people understand the associated code. See [11] for more on pair programming.

Semantic Analysis of Code

Semantics analysis is a mathematical approach to analyzing software code for runtime errors while not actually running the code. Semantic analysis looks for the meaning of the code, not the syntax or structure of the code. Metrics for code syntax are referred to as syntactic metrics and include metrics such as lines of code and cyclometric complexity. One approach to semantic analysis is to use semantic metrics, such as in [23] and [24]. Semantic metrics provide several advantages over syntactic metrics:

1. Unlike syntactic metrics, semantic metrics are usually independent of the programming language used.
2. Semantic metrics provide information on the functioning of the code directly, whereas syntactic metrics tend to provide this information only indirectly and their connection to software quality may in some cases be questionable.
3. Semantic metrics can often provide useful information before the code has been written. Syntactic metrics require the code.

However, there are disadvantages to semantics metrics:

1. Semantic metrics are often difficult to collect. Good tools are usually required.
2. The use of semantic metrics is much newer than that of syntactic metrics and so there is less of an experience base to draw upon. Syntactic metrics have a long history and have been extensively studied.

In practice, both types of metrics can be useful and can be applied complementarily.

Semantic analysis is more than semantic metrics. For example, semantic analysis can aid in locating software faults and is used in formal methods to prove that certain defects are not possible with the code. See the Formal Methods topic in Section 6.4 for more details.

Software Rejuvenation

Software reliability engineers typically think of how to prevent defects in delivered software and how to tolerate the ones that are present. However, software aging can occur in delivered software, degrading its performance and making it less reliable. Causes of this aging include data corruption, the accumulation of numerical errors, and resource depletion, such as memory leaks. Defects can also enter a software system postdelivery through maintenance errors or changes in the software operating environment, but these conditions are not considered to be software aging.

Software rejuvenation is a technique to mitigate software aging. It is a proactive and preventive maintenance technique. With it, the software is occasionally stopped and its internal state is cleaned. The software is then restarted. Examples of this process are garbage collection, flushing operating system kernel tables, and reinitializing internal data structures. A more extreme example is a hardware reboot. Software rejuvenation is in many ways analogous to hardware preventive maintenance. Chapter 14 of [19] provides a more in-depth discussion on software rejuvenation.

Software Reuse

Software reuse is a frequently used approach for preventing software faults. The approach takes advantage of proven software rather than creating new software that essentially does the same thing. It is important however to ensure that the reused software has the proper interfaces and operating characteristics for its new intended use. If not, the project must create adequate interface software to incorporate the reused software. It is also important to use properly certified software. See [11] for more on software reuse.

Software Testing

Software testing is a critical part of producing and releasing good software. There is currently no substitute for software testing, but it should not be the only reliability technique used. Testing is rarely exhaustive, and so defects are likely to remain even after a well-designed and well-implemented test program. Also, software testing can show the presence of a software defect, but unless the testing is known to be exhaustive, it does not show the absence of defects. In addition, different defect detection techniques tend to find different types of defects, so while testing is critically important, other techniques should be used to both improve defect detection and to reduce cost and schedule impacts. Finally, software testing requires executable code, so testing only finds defects later in the life of the project. Other techniques can find defects earlier, before they have become as embedded in various aspects of the project.

Software testing is sometimes categorized as either white box testing or black box testing. With white box testing, also called clear box or structural testing, the internal workings of the code are tested. Knowledge of these inner workings can be used to improve the coverage of the testing. White box testing requires that the testers have good knowledge of the code and use this knowledge in designing the tests. Black box testing considers the software functionality from a user point of view and does not require knowledge of its internal workings but rather requires knowledge of the software requirements or user needs and how to use the software. Note however that in practice these two categories may be mixed.

There are a variety of different types of software tests. A few include the following:

1. Acceptance testing: Acceptance testing is a set of tests used to determine if the system meets acceptance criteria. It is usually performed by an organization that is independent of the system developers. Acceptance testing is performed after integration and system testing.
2. Beta testing: Beta testing is the process of sending an advanced copy of the software to some trusted users and letting them run the software. They then report defects that they find and provide general user experiences.
3. Exploratory testing: Exploratory testing is a manual testing process (although it can be assisted with automation), typically associated with agile processes, for finding hard-to-find software defects that automated testing is likely to miss. With exploratory testing, the tester or test team takes some code and, using heuristics and intuition, creates a test for it and observes the behavior of the code. Usually, each test leads the test team to the next test. Often, the test team consists of a developer and a tester. Exploratory testing can be performed anytime that there is working code and is typically used at least once in each build sprint and ideally several times a week. Exploratory testing usually uses the following:

(a) Charters: A charter outlines the areas and functions that are of most importance for the testing and gives the test team an idea of what to explore and look for. The charter serves as a guide rather than a strict set of boundaries.

(b) Observations: Exploratory testers should always be looking for things that are not normal or not expected. These observations lead to other observations or to tests, resulting in more observations and potentially to tests that find defects.

(c) Notetaking: It is easy to get lost while doing exploratory testing. One observation leads to another under a different set of conditions, and soon, a test team may not remember how they got to where they are. That is why notetaking is important. The test team should take notes of the actions that they take and possibly take screenshots so that they can always repeat their actions.

(d) Heuristics: Over time, we all learn that certain aspects of code are potential trouble areas. A few example questions are: if we are supposed to input an integer, what happens if someone inputs a floating-point number? If inputs are supposed to be between 0 and 100, what if a negative number is entered? How does the code handle incorrect inputs? What if a given input is missing? If the input should be in the closed interval from 0 to 1 and the output should approach a value of 10 as the input approaches 1, how does the sequence of outputs look as we input values closer and closer to 1? Does it approach 10 smoothly or erratically?

(e) Sessions: Sometimes, exploratory testing is performed in time-bound sessions, usually no less than 30 minutes and no more than 2 hours. The idea is that if exploratory testing goes for too long, productivity diminishes.

When exploratory testing finds a defect, the test team learns about the software and also about the effectiveness of the team's processes. Both should be considered for improvement.

It is important to note that exploratory testing is not intended to be a form of regression testing. Exploratory testing focuses on insights and creativity. It is also mostly or completely manual, whereas regression testing should be as automated as possible. For more on exploratory testing, see [10].

4. Fault insertion: Fault insertion is the process of perturbing the execution of the software and observing the results. These perturbations represent randomly introduced off-nominal conditions. The purpose of fault insertion is to test the robustness of the software to such conditions. Fault insertion requires suitable tools. Note that fault insertion is different from fault seeding. With fault seeding, known faults are deliberately introduced into the code, and the success of the test team in finding these faults is used to estimate the number of faults remaining in the code after completion of testing. The accuracy of this estimate is dependent on how similar the seeded faults are to the typical (non-seeded) faults. Faults that are seeded are often easier to find than many of the non-seeded faults, so the validity of the results of fault seeding is often questionable.

5. Integration testing: Integration testing is performed on partially assembled code as the software units are gradually integrated together. This testing is used to see if the individual units of code work together correctly in progressively higher levels of integration.

6. Interface testing: The purpose of interface testing is to ensure that the interfaces are designed and written correctly. Often interfaces for a software module are initially

tested using a simulated interface and then retested with the real interface after the interfacing software module or modules have been developed.

7. Regression testing: A regression test is the rerunning of a previously performed test to determine if after changes are made to the code, the previously tested behavior is still performed correctly. Except for small, simple software products, automating regression testing is useful for both saving time and for reducing human errors. As even automated regression testing can be time-consuming, it is also advisable to plan the testing carefully and avoid unnecessary tests.

8. Reliability demonstration testing: The purpose of reliability demonstration testing, also called reliability qualification testing, is to determine to within a prescribed statistical confidence if the software meets its reliability goals or requirements. Reliability demonstration testing is different from robustness testing in that reliability demonstration testing uses the system in as close to the expected real-world situations for the system as possible in order to obtain data that closely represents real applications. Robustness testing uses a high percentage of stressing and off-nominal situations to try to find and test "corner cases" and other places that defects may "hide."

Reliability demonstration testing starts with the established reliability goals or requirements and the developed OP. Testing is performed in accordance with these profiles, and the software should be run on the target hardware with the same operating system, memory, drivers, load from other users, and other background processes that are expected in real-world operation. Testing is performed on a sufficient number of times to obtain statistically significant results. These results are used to drive decisions, such as whether the software is adequately reliable.

More specifically, suppose that we are interested in estimating the failure rate λ of a software line-replaceable unit (LRU) and have performed reliability testing under appropriate conditions. We assume that the time between failures is exponentially distributed. Per [12], we can estimate the failure rate as

$$\bar{\lambda} = \frac{n}{T}$$

where we have n failures in execution time T. Along with this estimate, it is also important to consider how confident we can be in this estimate. This calculation first requires considering whether the test is time terminated or failure terminated. The test is time terminated if it is run until a specific ending time. If, however, the test ends when a certain number of failures have occurred, then it is said to be failure terminated. Another consideration is whether we want one-sided confidence bounds or two-sided bounds. If we want two-sided bounds, we have an expression of the form

$$P(L < \lambda < U) = 1 - \alpha$$

where L and U are the lower and upper bounds, respectively, and α is used to express the level of confidence. For example, if we want to have 80% confidence in our result, then $\alpha = 0.2$ so that $(1 - \alpha) \times 100\% = 80\%$. For one-sided confidence bounds, we can have an upper bound, expressed as

$$P(0 < \lambda < U) = 1 - \alpha$$

or a lower bound

$$P(L < \lambda) = 1 - \alpha$$

The expressions for these probabilities, found in Tables 6.6 and 6.7, use the χ^2 (chi-squared) distribution. In the tables, $\chi^2_\alpha(2n)$ is the value at which the χ^2 distribution with $2n$ degrees of freedom gives a probability of α. More explicitly, for a random variable X with a χ^2 distribution with v degrees of freedom,

$$P(X > \chi^2_\alpha(v)) = \alpha$$

See [12, 25], and [26] for more details.

Example 6.3.1 There are many considerations that go into designing a reliability demonstration. One such consideration is that we need to know how long to run the test in order to obtain the required data. A variety of factors can go into such a calculation, but for a simple example, we assume an exponential distribution for the software reliability and that we are testing to determine if we can have a 95% confidence that the software has a failure rate of $\lambda = 4 \times 10^{-4}$ or less (corresponding to a value of 2500 hours or higher for the software mean time between failures [MTBF]). We also assume a time-terminated test. Using Table 6.7, we use

$$\lambda = \frac{\chi^2_\alpha(2n + 2)}{2T}$$

where $\alpha = 0.05$ ($0.05 = 1 - 0.95$) and T is the required test time. The variable n is the number of failures experienced during the test. Rearranging, we have

$$T = \frac{\chi^2_\alpha(2n + 2) \, MTBF}{2} = \chi^2_{0.05}(2n + 2) \, 1250$$

Table 6.6 Confidence Interval for Exponentially Distributed Failure-Terminated Data.

	Lower Limit	Upper Limit
One-Sided Limits (Upper)	0	$\frac{\chi^2_\alpha(2n)}{2T}$
One-Sided Limits (Lower)	$\frac{\chi^2_{1-\alpha}(2n)}{2T}$	∞
Two-Sided Limits	$\frac{\chi^2_{1-\alpha/2}(2n)}{2T}$	$\frac{\chi^2_{\alpha/2}(2n)}{2T}$

Table 6.7 Confidence Interval for Exponentially Distributed Time-Terminated Data.

	Lower Limit	Upper Limit
One-Sided Limits (Upper)	0	$\frac{\chi^2_\alpha(2n+2)}{2T}$
One-Sided Limits (Lower)	$\frac{\chi^2_{1-\alpha}(2n)}{2T}$	∞
Two-Sided Limits	$\frac{\chi^2_{1-\alpha/2}(2n)}{2T}$	$\frac{\chi^2_{\alpha/2}(2n+2)}{2T}$

Notice that the required test time is dependent on the number of test failures. Table 6.8 summarizes the results for zero through five test failures. As the table shows, the required test durations for high confidence (95%) can be large. The total testing duration ("calendar time") can be reduced by running multiple independent test sessions at the same time.

Another consideration is that we do not know how long we will be required to run the test without knowing how many failures we will experience, and we will not know that until we run the tests. One approach is to assume some maximum number of failures to plan the testing and if more than this number is experienced, stop testing and correct the failures that were found. Then, retest with the corrected software. Good software reliability predictions, supported by the results of previous testing such as units testing, integration testing, and verification testing, can help determine a value for the assumed maximum number of failures.

As a final note, when performing multiple runs of the software, we often consider each run to be statistically independent of all other runs. In reality, this is often not truly the case. Runs are rarely truly independent, although effort should be made to make the runs as independent as practical.

Example 6.3.2 Sometimes, the result of a test is either "pass" or "fail." For example, we may run a software product either to run completion or the software run is stopped by the occurrence of a downing failure during the run. In this example, we want to estimate the probability p of not having a downing failure during a software run and want to have a 95% confidence in this value. We also want our estimate $\hat{p}$ of the probability to be within ± 0.01 of the true probability value. In symbols, we want to have a 95% confidence that $|p - \hat{p}| \leq 0.01$. As we are performing independent "pass/fail" tests, we can use a binomial distribution with a probability of p of success for the test results. Assuming that we have been running randomly selected runs from a test suite that follows the OP of the software, we can estimate the probability p of not having a downing failure by

$$\hat{p} = \frac{k}{n}$$

Table 6.8 Test Duration Example.

Number of Test Failures, n	Required Duration (hours)
0	7 490
1	11 860
2	15 740
3	19 385
4	22 884
5	26 283

where k is the number of test runs to completion that occur and n is the total number of test runs. It is important, however, to know how much confidence we can have in this value. If we have only made a few runs, we are not justified in having a high level of confidence in this result. Given that the runs can be taken to be independent, that they are taken from the same distribution, and that we have 30 or more runs, we can assume that the samples follow a normal distribution. If we have fewer than 30 samples, we should use a Student's t-distribution.

Assuming a normal distribution, we can find the required number of runs to have at least a 95% confidence that the estimated probability is within ± 0.01 of the true probability value. We want

$$1 - \alpha = P(p - d \le \hat{p} \le p + d)$$

where $\alpha = 0.05$ and $d = 0.01$. As we can assume a normal distribution, this expression is the same as

$$1 - \alpha = P\left(p - z_{\alpha/2} \frac{\sigma}{\sqrt{n}} \le \hat{p} \le p + z_{\alpha/2} \frac{\sigma}{\sqrt{n}}\right)$$

where σ is the population standard deviation, n is the sample size, and $z_{\alpha/2}$ is the critical normal deviate found from

$$\frac{1 - \alpha}{2} = \frac{1}{\sqrt{2\pi}} \int_0^{z_{\alpha/2}} \exp\left(-\frac{x^2}{2}\right) dx$$

(Note that tables and computer programs usually give

$$\frac{1}{\sqrt{2\pi}} \int_{-\infty}^{z_{\alpha/2}} \exp\left(-\frac{x^2}{2}\right) dx$$

rather than

$$\frac{1}{\sqrt{2\pi}} \int_0^{z_{\alpha/2}} \exp\left(-\frac{x^2}{2}\right) dx$$

so we can use

$$\frac{1}{\sqrt{2\pi}} \int_0^{z_{\alpha/2}} \exp\left(-\frac{x^2}{2}\right) dx = \frac{1}{\sqrt{2\pi}} \int_{-\infty}^{z_{\alpha/2}} \exp\left(-\frac{x^2}{2}\right) dx - 0.5$$

when $z_{\alpha/2} \ge 0$). As a result

$$d = z_{\alpha/2} \frac{\sigma}{\sqrt{n}}$$

so that

$$n = \left\lceil \left(\frac{\sigma z_{\alpha/2}}{d}\right)^2 \right\rceil$$

where $\lceil \cdot \rceil$ denotes that we round up to the next integer.

If the standard deviation σ of the population is known or a good estimate from previous tests is available, we can use this formula to find the required number of test runs.

Unfortunately, we often do not have this value and must estimate it. One way to do this is to use the expression for variance of the binomial distribution,

$$\sigma^2 = p\,(1 - p)$$

and assume a worst case of $p = 0.5$ so that $\sigma^2 = 0.25$. Another approach is to run a pilot series of tests to estimate σ^2. This pilot series of runs can also be used to check the test setup and procedures. The pilot run should have more than 30 test runs.

Completing this example, suppose that we perform a pilot series of tests and calculate a value $\sigma = 0.09$ ($\sigma^2 = 0.0081$). For a 95% confidence, $z_{\alpha/2} = 1.96$ and $d = 0.01$, so $n = 312$. As a final note, the results of this example are used in Example 6.4.10, where we consider the capture–recapture method of estimating the remaining defects.

9. Reliability growth testing: Reliability growth testing is a method of measuring and projecting the reliability of a system and using the measurement and projection to determine if the project is on track to meeting its reliability requirements or if additional actions are needed. A reliability growth curve is constructed to describe the planned overall reliability trend of the project. Tests are run on the product based on the OP, and as with software reliability demonstration testing, the software should be run on the target hardware with the same operating system, memory, drivers, load from other users, and other background processes that are expected in real-world operation. The results of reliability measurements from testing are assessed against this reliability growth curve to determine if the measured trend is consistent with the planned trend. It is also important that each defect found during testing is carefully analyzed to find its root cause and that a significant number of these defects are corrected. Otherwise, there is no reliability growth.

   Reliability growth testing can identify potential reliability problems early, although not before operational software has been developed. Reliability growth testing can be used to monitor the reliability performance of the product while defects are being identified and removed. Reliability growth activities are coordinated with a software reliability growth plan (SRGP). See [12, 25–27], and [28] for more details on reliability growth testing.

10. Robustness, or stress testing: Robustness testing of software is analogous to accelerated life testing of hardware. With robustness testing of software, we test the software against a wide variety of test situations in an effort to find software defects.

    For robustness testing, it is important to have a method of determining how long to perform the testing, such as the use of a capture–recapture technique to estimate the remaining defects. (The capture–recapture technique is covered in more detail in the Software Reliability Estimations topic in Section 6.4.) We can then test and fix until the estimated total number of defects is below some number while using code coverage metrics to ensure that a significant portion of the code has been exercised. Design of experiment (DOE) techniques can be used to make the testing more cost-efficient. See Section 6.5 for more details on DOE techniques.

    A related testing technique is often referred to as "bug bounties." With bug bounties, people outside the project, and typically outside the company that is running the project, look for bugs (defects) in the software and are rewarded for each valid bug that they are the first to find and report. The reward is usually financial. Like robustness testing, the main effort is to find defects so that they can be corrected.

Another related type of testing is exploratory testing, which is covered above.

11. Soak testing: When running software over an extended period of time, data corruption issues may occur. Per [17], "Data corruption is the accumulated degradation in data with execution time that results from anomalies in intermediate variables that do not represent failures." For example, software may run as desired over a one-hour period, but because of memory leaks, it may behave incorrectly over a 100-hour period. Soak testing involves running the software under its OP for extended periods of time (called soak times) to determine if data corruption is an issue. Because of the length of time typical for soak tests, test automation is recommended.

12. Software state model testing: Model-based design allows for model-based testing. With this approach, the software specification is expressed as a state machine. A state machine is an abstract machine that takes one state at a time and transitions from one state to another based on external inputs. Expressed as a state machine, the software specification specifies all possible state combinations, all possible transitions, and all required responses to inputs. Through the use of (typically commercial) software tools, the state machine specification can be used to generate tests. Usually, the state machine is used to create a behavior model in the form of a Markov chain. The tool uses this behavior model to generate test cases consistent with the OP.

13. System testing (requirements testing, verification testing): System testing, also called requirements testing or verification testing, is a set of tests used to determine if the system as a whole meets its specified requirements. These tests are performed after the software and hardware have been constructed and integrated and is used to determine if the construction satisfies its requirements. These tests are usually performed by a test organization and are typically "black box" tests, meaning that the internal workings of the software are not used. System testing is performed after unit testing and integration testing but before validation testing. This type of testing is sometimes supplemented with verification by formal methods.

14. Timing and performance testing: Timing tests use timing diagrams, scheduling diagrams, and other inputs to test for race conditions, deadlocks and livelocks, CPU utilization, waiting times and response times, and other timing and performance issues.

15. Unit testing: Unit testing is a type of software test performed on relatively small "units" of code to check its performance and compliance. Unit testing is conducted before the unit is ready for integration into the rest of the code and is typically performed by the code developers during code development. Unit testing is usually "white box" testing, taking advantage of the code configuration.

16. Validation testing: Validation testing is used to determine if the system meets the needs of the user. It usually consists of a series of test scenarios chosen to exercise the critical aspects of the system. Validation testing is performed after system or verification testing.

Software testing requires a test plan to ensure that the objectives of the tests are fulfilled. A test plan requires specifying one or more OPs and a precise statement of test objectives, allowing an unambiguous "pass/fail" decision to be made. It also specifies what data are to be logged and how. The plan should also provide one or more test cases, i.e. a set of test inputs, execution conditions, and expected results. It should, either directly or by reference, detail how the test is to be recorded and reported.

There are many metrics related to software testing. Section 4.5.4 provides a list of some of them. A few of the more common types include:

1. Defect metrics: Collect failure data sufficient to promote reliability growth
2. Coverage metrics: Check that most of the code is covered by the testing
3. Requirements metrics: Ensure that nearly all of the requirements are covered
4. Defect process metrics: Assess how effective the defect discovery and removal process is

The right software test tools can save time and money and significantly improve the test results. For example, software test coverage monitors keep track of the code that has been exercised. Such a tool is particularly useful for tests such as system testing and robustness testing. Test generation tools help create test cases based on information such as timing diagrams or state models, and special tools are used for testing with fault insertion. Also, test execution tools can aid in running tests and collecting data.

Finally, formal methods can be used to make testing more efficient (see for example [29]). Formal methods are covered in more detail in Section 6.4. Software testing is described in more detail in [2, 25, 30, 31], and [11].

Static Analysis of Code
Static analysis tools are used to find defects in software code. These tools do not execute the code but rather the code is an input for the tool. They often calculate various software metrics such as lines of code or cyclometric complexity. They may also run checks for conformance to coding standards or tell us about the code structure. Some static analysis tools perform formal analysis. Static analysis tools are generally most useful during code development.

Static analysis of code has the advantages that the analysis is usually quick and effective for finding the types of defects that they are designed for, the tools can be easy to integrate into the development cycle, and they can help guide the testing activities. Their disadvantages include the fact that they do not detect certain types of defects, and these analysis tools, particularly the more advanced formal methods tools, sometimes give false results, so each tool result must be checked. More details may be found in [32].

Test-Driven Development
TDD is an approach to code development that uses small incremental cycles of testing, coding, and refactoring. It is similar to BDD covered above. Code is developed in small increments, first by developing a test that the small unit of code to be built should be able to pass, then by developing just enough code to pass the test, and once the code passes the test (and all previous tests in the incremental development), the code is refactored. In more detail:

1. Determine the desired feature, behavior, or functionality: TDD develops a test before developing the code that is tested, so the first step is to determine the feature that the small increment of new code (five lines or fewer) is to have. Then, think of a test that the

code will fail unless it has this feature. The test should also be small (about five lines of code).

2. Write the test: Write the code for the test that was determined in the previous step. This test should only test the feature specified in that step and is ideally five or fewer lines of code. After writing the new test code, run the entire set of tests and watch the current code fail. Note that TDD is an incremental process, so unless we are just starting, there will be previous tests and the code should pass them, but fail the new test because the code has not been written for the new feature. If the current code happens to pass this new test, investigate. There is a good chance that the test is incorrect.

3. Write code: Write just enough code to pass the test. Ideally, the new code is no more than five lines. Also, it does not need to be elegant. It just needs to be able to pass the test.

4. Retest the code: Next run the test suite against the code again. Ideally, the code will pass all of the tests in the suite. If it fails a test, continue by either correcting the new code or deleting the new code and starting over writing code for the new test.

5. Refactor: After the code passes the latest test suite, refactor. Refactoring code is the process of making the code clearer, and more clean, efficient, and maintainable without changing the external behavior or functionality of the code. Do the refactoring in small steps and run the test suite after each of the small steps. Refactor as many times as needed to make good code but limit the coding to the features that are currently chosen.

6. Repeat: After successful refactoring, repeat this cycle again until all features have been successfully included and refactored. A key to TDD is keeping the cycles short and quick. Twenty cycles an hour are not unreasonable, but do not focus on performing the cycles quickly. Focus on keeping the increments small and performing the steps correctly.

Advantages of TDD over more "traditional" code development processes include the following:

1. Code developed using TDD is typically much more reliable than code developed using more traditional methods.
2. TDD typically reduces the time required to debug the code.
3. TDD produces checked code rapidly.

Although TDD has some significant advantages, it also has disadvantages, such as

1. TDD does not lend itself well to legacy code.
2. TDD requires the code developers to learn a new develop process. There may be a steep learning curve.
3. Isolating code for unit testing may be more difficult with TDD.
4. TDD works best if the entire coding team uses it and not just one of several coders.

More details may be found in [10].

6.4 Techniques for Reliability Engineering

The techniques covered next are traditionally associated with reliability engineering. These techniques are used to plan for reliability, determine reliability and availability objectives and requirements, predict and estimate the values of reliability and availability measures, determine what might go wrong with the software and when defects are found, and determine why they occurred and what to do about them. Table 6.9 lists 20 such techniques, and like the tables for SE and software techniques, these techniques are divided into techniques that are typically used early in a project, those that are usually used later in a project, and those that are used throughout the project.

Table 6.9 lists seven traditional reliability techniques that are typically used early in a project, although they may be updated, refined, and applied throughout the project. A good example of such a technique is reliability risk assessments. Ideally, this technique is performed early in a project so that risks can be mitigated before they cause serious negative consequences. However, as the project progresses, situations involving these risks may change or new risks may emerge, meaning that the risk assessment needs to be updated. Software reliability and availability predictions is another technique that should be performed early in the project and be updated as the design progresses and more information becomes available.

The table also lists two traditional reliability techniques that are typically used later in a project. These techniques are typically used to estimate the reliability of existing code or to analyze and correct defects in executable code. As with the techniques listed for use early in the project, when they are used depends on the project.

Finally, the list covers 11 traditional reliability techniques that are often used throughout the entire project time frame. These include techniques such as a Defect Reporting and Corrective Action System (DRACAS) and Defect Review Board (DRB) that are used to collect, analyze, and act on errors and defects. Defects affecting software can occur at any time in a project, so these and similar techniques, in some form, should be used throughout the project. Another technique covered below is a software failure modes and effects analysis (FMEA) or failure modes, effects, and criticality analysis (FMECA). Such an analysis may be performed on a functional level before the software design and coding, or later in a project when it is performed at a detailed level on specific code.

Table 6.9 lists 20 software techniques that can improve the reliability of a software product. Table 6.10 lists three other techniques that are related to reliability engineering activities, the organization that the technique is described under, and how the technique is related to reliability engineering.

Next, we describe each technique listed in Table 6.9.

Defect Detection Rate Monitoring

An important part of a well-run project is early detection of defects. By detecting defects early in the project, we can correct them before they become too embedded in the project's products. Early project activities usually produce documents of some type, such as project plans, or a requirements document (which may be documented as a collection of software tests). Defects in these products are often detected early in a project by inspections and reviews, or later in the project through testing and a DRACAS. Ideally, we will know if our rate of detecting defects is high enough to have adequately reliable software. One way of monitoring the defect detection rate is through the use of software defect predictions and a Rayleigh distribution.

Table 6.9 Techniques for Reliability Engineering.

Technique	When Used	Notes
Defect detection rate monitoring	Throughout	This technique compares the expected number of defects with the number found as the project progresses to detect potential problems early when corrections are more cost- and schedule-effective
DRACAS, DRB	Throughout	DRACAS and a DRB constitute an important technique for controlling and reducing defects. This technique is a generalization of the traditional FRACAS technique
Failure definition and scoring criteria	Throughout	This technique supports key reliability activities such as FMEA and reliability testing by defining failure and non-failure events and further dividing failure events by criticality
Formal methods	Throughout	This technique uses mathematical models to clarify requirements and to prove software correctness
FRACAS, FRB	Later	This traditional technique is used to control and reduce failures of existing software
Operational profiles	Throughout	This technique produces a quantitative characterization of how the software will be used
Orthogonal defect classification	Throughout	This technique categorizes defects into classes that point to which project processes need the most attention
Reliability risk assessment	Early	This technique is a process for identifying situations that can lead to reliability issue
Reliability trade studies	Throughout	This technique is a process for making decisions affecting some aspect of the reliability of the system
Software defect root cause analysis	Early	This technique identifies likely causes of defects to enable the project to manage defect detection, analysis, and reduction more effectively
Software fault tree analysis	Throughout	This technique is a top-down analysis used to find sources of defects and unsafe or unreliable system states
Software FMEA/FMECA	Throughout	This technique is a bottom-up analysis used to find failure modes, their effects, and their criticality
Software reliability advocate	Throughout	This technique helps coordinate and manage the project's overall software reliability activities
Software reliability and availability allocations	Early	This technique helps efficiently manage resources to achieve reliability goals and objectives

(continued)

Table 6.9 (Continued)

Technique	When Used	Notes
Software reliability and availability objectives	Early	This technique helps efficiently plan reliability activities
Software reliability and availability predictions	Early	This technique helps efficiently manage reliability activities
Software reliability casebook	Throughout	This technique provides assurance that the overall reliability plan is being successfully followed
Software reliability estimations	Later	This technique uses existing code and software testing to assess reliability measures of the software
Software reliability growth plan	Early	This technique plans for improving software reliability in an efficient and doable manner
Software reliability program plan	Early	This technique systematically details the software reliability approach for the project

Table 6.10 Techniques Related to Reliability Engineering.

Technique	Main Organization	Notes
Fault tolerance	6.3 Software	Fault tolerance is an often-used reliability technique that allows for successful software performance in spite of software defects
Software rejuvenation	6.3 Software	This technique has software maintenance implications by helping to mitigate software aging effects
Software testing	6.3 Software	This technique may be used to gather data for reliability growth and reliability estimates

In the discussion on software reliability and availability predictions, one of the models that can be used to predict the number of defects in software is based on the observation that the number of defects detected throughout the development of the software usually follows a Rayleigh distribution. By tracking defects in the early phases of the project and fitting these to a Rayleigh distribution, we can predict the number of defects that will be detected later in the project. The idea of using a Rayleigh distribution to monitor the project defect detection rate is to use a trusted model to predict the number of defects that can be expected to be in the software over its development lifetime and use this information and information about the duration of the project before software delivery to determine the parameters of the Rayleigh distribution applicable for the project's defects. We compare the number of detected defects with the number of defects that are expected based on the Rayleigh distribution and, if there are significant differences, assess what is likely to be the problem. More on this technique and related techniques may be found in [17, 33], and [34].

Example 6.4.1 Recall that, given defects detected during the development of the software, a Rayleigh distribution can be used to predict the number of defects remaining in the software. Given a prediction that the software has a total of N defects over the life of the software development, we can use a Rayleigh distribution to assess if the project is finding defects at a rate that should be expected given the total number of defects.

Suppose that using a trusted prediction model, we predict that the software will have M defects in it at product release. We also believe that our software reliability practices catch and remove 97% of the product defects before software release. Given these assumptions, the total number of defects placed into the software over its development lifetime is approximately $N = M/(1 - 0.97) = 33.33\ M$. Let t_d be the number of months from the start of the project to the time that the software is expected to be fully ready for operation. Recall that the cumulative probability distribution function for a Rayleigh distribution is given by

$$F(t) = 1 - \exp\left(-\frac{t^2}{2\sigma^2}\right)$$

so

$$N(t) = N\left[1 - \exp\left(-\frac{t^2}{2\sigma^2}\right)\right]$$

gives the total number of defects expected to be found by time t, where in this example, time is in terms of months after project start. The number of defects that are expected to be found in a given month is found from

$$n(t) = \frac{N\,t}{\sigma^2} \exp\left(-\frac{t^2}{2\,\sigma^2}\right)$$

To find the value of σ^2, we use an estimate of the fraction of the total number of defects that will have been found by time t_d. If we assume that by time t_d, 0.97 of the defects will have been found, then

$$0.97N = N(t_d) = N\left[1 - \exp\left(-\frac{t_d^2}{2\sigma^2}\right)\right]$$

Therefore,

$$\exp\left(-\frac{t_d^2}{2\sigma^2}\right) = 0.03$$

and

$$\sigma^2 = -\frac{t_d^2}{2\,\ln(0.03)} \approx \frac{t_d^2}{7.01}$$

As a result, we use

$$n(t) = \frac{7.01\ N\ t}{t_d^2} \exp\left(-\frac{7.01\ t^2}{2\ t_d^2}\right)$$

to gauge if the number of defects detected in a given month t is close to what is expected. The cumulative number of defects detected up to and including month t should roughly follow

$$N(t) = N\left[1 - \exp\left(-\frac{7.01\ t^2}{2t_d^2}\right)\right]$$

It may be decided that 0.97 is not the correct fraction of detected defects at time t_d. Depending on the project, different values may be more appropriate. For example, 0.95 is often used, giving

$$n(t) = \frac{6 \, N \, t}{t_d^2} \exp\left(-\frac{3 \, t^2}{t_d^2} \right)$$

More generally, for a fraction f of defects detected by time t_d, we have

$$n(t) = \frac{-2 \, \ln(1-f) \, N \, t}{t_d^2} \exp\left(\frac{2 \, \ln(1-f) \, t^2}{2 \, t_d^2} \right)$$

where

$$\sigma^2 = -\frac{t_d^2}{2 \, \ln(1-f)}$$

A couple of notes to keep in mind when using this technique are as follows:

1. This assessment provides an indication that defects are, or are not, being found at an expected rate, but the assessment requires interpretation. If the detected rate of defects is less than what the Rayleigh curve indicates that the detection rate should be, it could be due to a lower-than-expected defect detection rate, an incorrect defect prediction N, poor logging of defect detection, or various other reasons. If the defect detection rate exceeds the rate predicted by the Rayleigh curve, it could be that defect detection is better than expected, or that there are more defects than predicted, or again, other reasons.
2. When using this technique, it is useful to predict an upper and a lower bound for the number of defects and use these to calculate an upper and a lower bound to the Rayleigh curve. These can be used to better assess if the observed number of detected defects is in line with expectations or not.

DRACAS, DRB

DRACAS stands for Defect Reporting, Analysis and Corrective Action System. It is a generalization of the traditional FRACAS in that it covers defects across the project and not just failures. By doing this, defects and failures can be handled in a systematic manner, and defects found earlier in a project can be related to failures found later. In the discussion below, we refer to defects; however, the same discussion applies to failures unless otherwise stated. DRACAS is a system that provides a process for the following:

1. Capturing and recording defect data sufficient to perform timely root cause analysis (RCA) and defect removal
2. Prioritizing the defects
3. Ensuring that adequate analysis is performed on each reported defect, ideally analyzing each to its root cause
4. Ensuring adequate communication across the functional areas of defects and their impacts
5. Identifying, implementing, and verifying corrective actions for defects
6. Collecting and storing defect data used to support metrics, assess trends, build reliability models, and develop fault tree analysis (FTA)

7. Aiding the development of test methods and troubleshooting techniques
8. Creating a summary report to keep personnel informed of defect status and trends
9. Using the metrics, trend information, and summary reports to positively affect the project

One key aspect of a DRACAS is that it is a closed-loop system. A defect accepted into the system stays active in the system until a corrective action is implemented and verified to be effective. Information about the defect, its corrective action, and verification of corrective action effectiveness remains in the system so that if there is a similar occurrence, it can be compared with this recorded information and the validity of the verification can be reassessed.

Another key aspect of a DRACAS is ensuring that all appropriate project personnel have easy access to the system so that defects and failures can be reported quickly and accurately. The more that the defect witness relies on memory, the more likely that important defect information is lost. This information is particularly important when trying to find the root cause or causes of a defect. Working with the root cause gives the greatest likelihood of preventing future defect occurrences.

In more detail, a DRACAS process uses the following steps:

1. Enter the defect number, defect description, and other information into the defect reporting and tracking system.
2. Perform defect verification to ensure that the entry is truly a defect.
3. Have someone assigned to perform RCA. Suggestions for RCA for software defects include the following:
   a) Check for error messages. If the error message is from a purchased product, check the appropriate sources for its meaning.
   b) Check all pertinent logging.
   c) Find where the defect manifests itself.
   d) Find the actual line or lines of code that are causing the problem or incorrect actions.
   e) Use the process of elimination. Rule out various possibilities to narrow the search. Fault trees are often used in this step.
   More on RCA is covered in a separate topic in Section 6.5.
4. Have the RCA reviewed by pertinent personnel and approved or corrected.
5. Determine if a corrective action is required, and, if so, determine the corrective action and have someone assigned to correct the defect.
6. After the defect is claimed to be corrected, have the correction verified. This step may use regression testing.
7. Once the defect has been corrected and verified, close the defect report.
8. Ensure that adequate configuration control is maintained.
9. Produce report summaries to keep appropriate project personnel informed.
10. Use the information gathered to add to the statistical information about the current software and other products and processes.

A DRACAS is usually governed by a DRB, which is a group consisting of all affected parties and run by someone with the authority to make decisions for the project. Each defect should have a DRB-assigned defect investigator with clear instructions and a due date for the analysis. Corrective action assignments may go to a different board, but the same process of assigning a responsible person and a due date applies to corrective actions and to verifications. A DRB has the following high-level tasks:

1. Review reported defects for acceptance. Some reported instances may result from misunderstandings and not represent a defect. Multiple reports of the same type of incident should be treated in multiple incident reports, but multiple reports of the same incident should be combined into a single incident report. Also, a reported defect may require more information before acceptance.
2. Assign an investigator to each accepted defect, along with a due date and any investigation instructions.
3. Review the results of the defect analysis for acceptance.
4. Assign corrective actions.
5. Review completed corrective actions and provide approval or additional instructions.
6. Close defect reports when appropriate.
7. Review defect trends and metrics.
8. Review DRACAS and DRB trends and metrics to ensure that the system is effective and efficient, making improvements as necessary.

DRACAS should use a database that allows for easy defect data entry and ensures that each step in the DRACAS process is successfully performed and recorded. Effort should be made to ensure that everyone who might see a defect has access to the database. If there are people without access, paper forms can be used, but the forms should include questions about all of the relevant input information, and as soon as practical, the information on the forms should be entered into the database. It is important to record the defect information soon after it is known to ensure that correct, complete information is obtained. As a minimum, the following information should be recorded:

1. Description of the defect
2. Time and date of occurrence (or when noticed if occurrence information is not available)
3. Name of the person observing the defect
4. Operating environment when the incident occurred
5. Actions taken related to the incident
6. Details of any related or possibly contributing factors

As failure information is used to estimate software reliability, information on the run history of the system is useful. Software reliability estimates often use CPU time rather than calendar time, so collecting this information and relating it to the occurrence of the failure are valuable.

There may be a need for a DRACAS training program. DRACAS requires discipline from nearly all project personnel. A clear understanding of what is required of each person may make compliance easier and more error free, although the decision about the type of training should consider the amount of project disruption that it will involve.

DRACAS and FMECA are related in that when analyzing a failure for its root cause, the FMECA may provide useful information. Also, if a failure occurs that has not been covered in the FMECA, it generally should be added to the FMECA to keep it as complete as possible.

When planning for a DRACAS and DRB, consider the following planning tasks:

1. Determine the objectives for the system. Consider how a DRACAS and DRB can support each phase of the project and also how information gathered during one project can help future projects.
2. Determine how defect and failure data will be collected. This collection may vary based on the project phase. Typically, a project-accessible database is used, but often paper copies are used as a back-up method if defects or failures occur at times when the database is not accessible.
3. Define categories of defects and of failures in ways that support the objectives of the system. It is critical that "fault," "failure," and other terms related to a DRACAS are defined in such a way that is agreed by both the project and the customer, and the sooner that this is done, the better.
4. Determine what analysis is to be performed on the defect and failure information to meet the system objectives and ensure that the data collection system collects this information.
5. Decide what tools to use. Activities that may be supported by tools include but are not limited to collect and log defect and failure information, review the reports for completeness and clarity, assign analysis tasks, perform analysis of the defect or failure, review the analysis, report results to the DRB, assign a status to the defect or failure, store and use information from all reported defects and failures, and write and distribute defect and failure reports to inform appropriate personnel of the most current results and trends.
6. Assign task responsibilities to appropriate groups and personnel.
7. Determine how the results of the DRACAS and DRB will affect the project.
8. Document the DRACAS and DRB in sufficient detail.
9. Determine if DRACAS training will be needed and, if so, plan for it.

The DRACAS process should be documented in a project-wide procedure, and the DRACAS process and its implementation should be monitored and updated as needed. See [12, 25, 26], and [35] for more information.

Failure Definition and Scoring Criteria

A Failure Definition and Scoring Criteria (FDSC) document defines failure and non-failure events and further divides failure events by criticality. Traditionally, the main use of an FDSC is to score test results and support the assessment of whether the item under test meets its reliability-related requirements. However, it is important to have an FDSC early in the project to have a common understanding of what constitutes a failure and to assess its severity so that these can be taken into account throughout the design process. For example, FDSC information is critical for the development of an FMECA. As a result, an initial version of the FDSC document should be prepared in the Concept Development and Planning phase of the project.

Development of an FDSC should involve the customer as it is the customer experience that determines customer satisfaction. Steps for developing an FDSC include the following:

1. Clearly define the system that the FDSC applies to.
2. Define the categories of failures. Typical categories are system aborts, essential function failures, and non-essential function failures, but categories can be project specific.

For example, failures could be categorized as catastrophic, critical, moderate, or negligible, with a detailed definition of each. This detailed definition is important because for a given failure event, there may be different opinions as to whether it is (for example) "catastrophic" or just "critical," and a detailed definition of each will make determinations less contentious.

3. Define the categories of non-chargeable failures. Non-chargeable failures are often partitioned into non-relevant failures and relevant failures caused by a condition previously not specified as being the responsibility of the given project's product. A non-relevant failure is a failure that is either verified to have been caused by a condition not present in the operational environment or a failure that is verified as being peculiar to an item design that will not be a part of the delivered system.

4. Identify system failures for the system. These failures are typically described in terms of functions, such as a function not performed or performed incorrectly.

5. For each system failure, identify if the software can cause the failure or significantly contribute to the failure by improper behavior.

6. Assess the criticality of each software failure by putting it into one of the categories from step 2 above.

7. Consider how to handle real-world situations that are often not thought of until they occur. These include the following:

   a) How to score events that will not be removed or corrected

   b) How to score events that result in multiple failures

   c) How to score events in a fault-tolerant system

   There will likely be other situations that have not been accounted for but cover as many as possible in the FDSC.

See [25] and [12] for more information on this topic.

Example 6.4.2 With some projects, an FDSC document is provided by the customer. For commercial projects, this is not usually the case. When designing the software, assessing its failure modes and reliability, or performing tests executing the software, knowing what constitutes a failure and how critical the failures are can be extremely important. Therefore, even for a commercial project, a usable FDSC is very beneficial.

To produce an FDSC, start by defining the system. This description should include any major functions that the system performs. For this example, we describe producing an FDSC for a software game. The game has two players that compete in an effort to find the most treasure. Players are allowed to create obstacles for the other player and to steal from the other player. In describing the system, develop a description that provides enough understanding of the product that the software functions, failures, and their level of criticality will make sense.

Next, define what a failure is. For example, we might define a failure as an event in which some aspect of the product does not perform as specified or as desired within the specification. Note that this particular definition is intended to get around the possibility that the software does something, or does not do something, that is not covered in the specification but is obviously unintended behavior. Note, however, that not performing a function that is not specified is not a failure as failures are defined for performance "within the specification." Unfortunately, this part of the definition also potentially adds subjectivity to the process.

We also need to define levels of criticality of failures. What constitutes a catastrophic failure for a video game may be different than what constitutes a catastrophic failure for an airplane. A software failure in a video game does not usually result in the life of a player being lost. Definitions that might apply for a video game are as follows:

1. Catastrophic failure: A catastrophic failure is a failure that potentially causes a player to reject playing the game.
2. Critical failure: A critical failure is a failure that prevents the performance of a major function of the game.
3. Moderate failure: A moderate failure is a failure that is a nuisance, but players can still adequately play in spite of the failure.
4. Negligible failure: A negligible failure is a failure that is not noticeable or is easily ignored.

These definitions may need adjusting based on the needs of the project, but they provide an idea of possible failure categories.

Next, we define and list non-chargeable failures. These are failures that are not charged, or counted against, the game software. For example, a hardware failure that shuts the computer down or negatively affects its performance to the degree that the play of the game is impacted is a non-chargeable failure for the game software. If the computer is affected by software that is not a part of the game software and this software overrides all other software and negatively impact the play of the game, this is also non-chargeable to the video game software. Continue making a list of non-chargeable failures. It is not unusual for unexpected examples to occur at a later time, but consider all reasonable cases that the team can come up with.

Typically, the next step is to define the chargeable failures for the software. Usually, this step is performed based on the functions or behaviors that the software is to perform. For example, the software should go to the start-up screen when the software program is started. Software can fail to do this in a variety of ways. It could completely fail to perform this function by never going to the start-up screen. It could usually go as desired but occasionally fail to go. It could go there, but take much longer than specified, or it could go there but take only slightly longer than specified. Each of these behaviors is a failure, but their levels of criticality differ. The list of functions should be detailed enough that failures versus non-failures and chargeable versus non-chargeable failures can be differentiated. It should also allow us to assign a degree of criticalness to each failure.

After defining the failures, we determine their levels of criticality, meaning that for the chargeable failures, we determine if a given failure is catastrophic, critical, moderate, or negligible. The level of criticality of a failure of a function typically depends on how it affects the user, although impacts to administrators, maintenance personnel, and others may need to be considered as well. The failure of never going to the start-up screen is catastrophic. The failure of going to the start-up screen slightly later than specified may be a negligible failure.

Finally, consider how the FDSC will be used and how to handle these various situations. For example, if it is determined that a failure will not be removed or corrected, it may be decided that the failure is to be recorded along with all of its root cause information and the reason that it is not removed. New occurrences of the failure may then be considered non-chargeable. Events that result in multiple failures may be given, as a minimum, the level of criticality of the most critical failure of the multiple failures. It should also be considered if the effect on the user of multiple failure events results in a greater level of criticality than any of the individual failures.

Formal Methods

Formal methods use mathematical models of the software requirements, design, or code to reduce ambiguity and to prove their correctness. They can help specifications, development, and verification activities. Formal specifications can be used to prove a correspondence between the specification and the requirements. Formal methods check software design through model verification and software code through code verification. Using a formal specification followed by formal verification allows for incorrect designs to be found and corrected early. It can prove the absence of certain types of design and runtime errors. Areas where formal methods can be applied include the following:

1. Requirements and specifications: A formal specification for software is a concise description of the behavior and properties of a software system and is written in a mathematics-based language. To use formal methods, we translate requirements into a well-defined language so that its semantics allow formal deductions. This formal language is mathematics based. There are different types of specification languages, although most fall into the categories of model-based or algebraic languages. For example, the model-based specification language Z (pronounced "Zed") uses notation from set theory, lambda calculus, and first-order logic. Model-based specification languages are often used to specify system behavior. Algebraic languages describe the system in terms of operations and their relations. These languages tend to be better for formal interface specifications because they can specify the system in terms of relations between interfaces. Even if we are only using formal methods to find defects in the code, there is an "implicit" specification that describes constraints on the code, such as no memory errors or divide by zero situations.

   A formal specification can be beneficial in several ways. First, the process of producing a formal specification forces the project to understand the system more precisely than it might otherwise, resulting in fewer requirement ambiguities, omissions, and other errors and defects. This benefit is perhaps the most notable positive aspect of formal methods from a cost/benefit point of view. It also improves communication of requirements.

   A second use of a formal specification is to compare the specification with requirements. For example, for software with security implications, there may be certain security policies that must be adhered to. If these are expressed formally, the formal specification can be compared with these requirements.

   Sometimes formal methods are used to support the generation of specifications, but the specification is not fully formal. In these cases, some natural language aspects of the specification are replaced with formal specification notation and methods, while the remaining aspects use the more traditional "natural language" specification approach. This partial approach can reduce ambiguities without the cost of a fully formal specification. This approach may be particularly attractive when first attempting formal methods or if the objective for using formal methods is to analyze some particular aspect of the software, such as a particular algorithm or module.

Other uses of formal specifications are described below, namely, their use in verifying the design and code, generating test cases, and finding defects.

2. Design: Models are often used when designing software to express and develop the design. For example, for object-oriented software, UML may be used. Formal methods can be used to verify these models against overflows, state errors, and other potential problems.

3. Code: The software design is translated into code, and formal methods can be used to verify that the code does not have certain defects such as various types of runtime errors. Code can be checked against the formal specification using mechanized model checking or theorem proving. Although this approach requires expertise and a considerable amount of work, it may be the best approach to use for some critical systems.

4. Test: One of the most expensive aspects of a project is its test program, with one of the major expenses being the creation of test cases. Another use of formal methods is to develop complete test cases from a formal specification. For example, when using a model checking tool (see below), we can specify the negation of a desired property and the tool will produce an event trace that results in the failure of this negated property. This trace can then be used to develop a test case.

Tools to perform analysis using formal methods typically fall into two categories:

1. Theorem proving: Theorem proving tools use a formal specification to construct proofs that show that the item being analyzed (such as the specification, design models, or code) has certain properties, such as the absence of certain types of defects. This type of tool generally can handle large and complex programs but requires more expertise to use than is required for model checking tools.

2. Model checking: Model checking tools usually use a finite state model of the system and desired properties of the system, such as the correct sequence of events and required precedences of activities. These tools have the advantage of being fully automated in their analyses so that less expertise is required, but these tools cannot currently handle as large or complex a system as those that can be handled by theorem proving tools.

Software testing can show the existence of a software defect but not the absence of defects. Formal methods potentially have the ability to make assertions and prove that these assertions are correct. One approach to this is through the use of abstract interpretation. Abstract interpretation is a theory of sound approximations of the semantics of programming languages. Its main application is automated static analysis at compile time to determine runtime properties of computer programs. We can view the behavior of a computer system as trajectories of the state of the system as a function of time. A system specification provides "forbidden zones" that no trajectory should enter. An example of a forbidden zone is a location in state space where the system divides by zero. Note that a specification may be "implicit" in that it checks for memory leaks, infinity loops, and other general runtime errors and requires no user inputs. A formal proof of correctness is proving that the semantics implies that none of the possible trajectories enter a forbidden zone. Software testing attempts to do this but can rarely test all trajectories and therefore "underapproximates" the software program. This underapproximation means that testing can miss defects, resulting in false negatives. Abstract interpretation for static analysis "overapproximates" in that it covers all possible trajectories and possibly others as well. All possible trajectories are covered by the abstraction, and the software program is proven to satisfy the specification if none of the trajectories enter a forbidden zone. The overapproximation means that formal methods with abstract interpretation may produce false alarms.

As with software testing, any defects found with formal methods should be recorded and analyzed for root causes and trends. Trend tests are covered in more detail in a separate topic in Section 6.5. This information may help identify more defects in the current product and can be used to improve processes for future products.

Formal methods can be highly effective and are complementary to code reviews and inspections and software testing. Used properly, formal methods can result in higher quality software at the same or even lower cost, and unlike reviews and testing, they can prove the absence of certain types of defects. Also, they can simplify debugging. Software testing can identify the existence of a defect, but potentially time-consuming debugging is required to find the cause of the problem. Formal methods can directly identify the source of the problem. These methods involve investing more effort early in the software development and so can catch defects early and reduce the cost and schedule of later efforts to find and remove these defects. However, they can have high setup costs and require expertise. Formal methods for specifications can also limit the expressiveness of a specification. These methods check software correctness relative to the specification so they will not catch a defect because of the specification missing requirements. These approaches may only be cost-effective if limited to parts of the software, such as safety-critical, security-critical, or mission-critical software, but should be considered for any software that imparts a high cost if it fails. For more information, see [29].

FRACAS, FRB

FRACAS stands for Failure Reporting, Analysis and Corrective Action System. FRB stands for Failure Review Board. For purposes of this book, we use DRACAS and DRB covered above to emphasize the need to cover defects in general.

Operational Profiles

The probability of a software failure depends on how the software is used. To properly allocate development resources, to make accurate reliability predictions and estimates, and to perform meaningful reliability tests, we need a quantitative characterization of how the software will be used. This characterization is called an OP and consists of a complete set of software operations along with their probabilities of occurrence.

OPs are found hierarchically using the following "levels":

1. Customer profile: A customer is any person or group that acquires the system.
2. User profile: Each customer has one or more users and different user groups may use the system differently.

3. System mode profile: A system mode is a set of functions or operations grouped together for analyzing execution behavior. Each user group uses the system in different ways, resulting in different system mode frequencies.
4. Functional profile: Each system mode uses sets of tasks or operations that we identify as functions.

Each of these profiles consists of several disjoint alternatives called elements. We assign a probability to each element of each level and multiply the probabilities together as we go down the hierarchy to obtain an OP.

To explain further, consider the following example:

Example 6.4.3   As a simple example of creating an OP, we consider a company that produces graphics design software. To create an OP for this software, we go through the following steps:

1. For this example, the company has two customer profiles. It sells 80% of its products to customers in Country $A1$ and 20% to customers in Country $A2$. We can denote the results at the customer profile level by $A1 = 0.8$ and $A2 = 0.2$.
2. Both $A1$ and $A2$ have two user profiles. Under $A1$, user profile $B11$ is for amateur users and $B12$ is for professionals. Similarly for company $A2$, user profile $B21$ is for amateur users and $B22$ is for professionals. For these companies, $B11 = 0.5$, $B12 = 0.5$, $B21 = 0.3$, and $B22 = 0.7$.
3. Under each user profile, there are three system modes, two-dimensional graphics, three-dimensional graphics, and video. Under $B11$, we have two-dimensional graphics $C111 = 0.7$, three-dimensional graphics $C112 = 0.2$, and video $C113 = 0.1$. For $B12$, we have $C121 = 0.2$, $C122 = 0.25$, and $C123 = 0.55$. Continuing with $B21$ and $B22$, $C211 = 0.8$, $C212 = 0.15$, $C213 = 0.05$, $C221 = 0.1$, $C222 = 0.15$, and $C223 = 0.75$.
4. Finally at the function profile level, there are three functions: create a graphic, modify a graphic, and print or play a graphic or video. For these functions, we have the following:
   (a) $C111$: Under $C111$, $D1111 = 0.2$ (create a graphic), $D1112 = 0.60$ (modify a graphic), and $D1113 = 0.2$ (print or play).
   (b) $C112$: Under $C112$, $D1121 = 0.1$, $D1122 = 0.65$, and $D1123 = 0.25$.
   (c) $C113$: Under $C113$, $D1131 = 0.05$, $D1132 = 0.7$, and $D1133 = 0.25$.
   (d) $C121$: Under $C121$, $D1211 = 0.4$, $D1212 = 0.3$, and $D1213 = 0.3$.
   (e) $C122$: Under $C122$, $D1221 = 0.45$, $D1222 = 0.35$, and $D1223 = 0.2$.
   (f) $C123$: Under $C123$, $D1231 = 0.5$, $D1232 = 0.15$, and $D1233 = 0.35$.
   (g) $C211$: Under $C211$, $D2111 = 0.2$, $D2112 = 0.65$, and $D2113 = 0.15$.
   (h) $C212$: Under $C212$, $D2121 = 0.2$, $D2122 = 0.5$, and $D2123 = 0.3$.
   (i) $C213$: Under $C213$, $D2131 = 0.1$, $D2132 = 0.5$, and $D2133 = 0.4$.
   (j) $C221$: Under $C221$, $D2211 = 0.45$, $D2212 = 0.35$, and $D2213 = 0.2$.
   (k) $C222$: Under $C222$, $D2221 = 0.4$, $D2222 = 0.4$, and $D2223 = 0.2$.
   (l) $C223$: Under $C223$, $D2231 = 0.5$, $D2232 = 0.2$, and $D2233 = 0.3$.
   These values are shown in a tabular form in Table 6.11 for $A1$ and Table 6.12 for $A2$.

Table 6.11 Example of Operational Profile, A1.

Customer																	
							A1 = 0.8										
User	B11 = 0.5									B12 = 0.5							
System mode	C111 = 0.7			C112 = 0.2			C113 = 0.1			C121 = 0.2			C122 = 0.25			C123 = 0.55		
Functional	0.2	0.6	0.2	0.1	0.65	0.15	0.05	0.7	0.25	0.4	0.3	0.3	0.45	0.35	0.2	0.5	0.15	0.35

Table 6.12 Example of Operational Profile, A2.

Customer							A2 = 0.2						
User			B21 = 0.3							B22 = 0.7			
System mode	C211 = 0.8		C212 = 0.15		C213 = 0.05		C221 = 0.1		C222 = 0.15		C223 = 0.75		
Functional	0.2	0.65	0.15	0.2	0.5	0.3	0.1	0.5	0.4	0.45	0.35	0.2	0.4	0.4	0.2	0.5	0.2	0.3

5. Each element has a probability of occurrence associated with it, so if we multiply the probability of $A1$ by that of $B11$, followed by that of $C111$ then $D1111$, we have one piece of the OP, in this case, one piece of the probability that the user will use the software for creating two-dimensional graphics.
6. The complete OP is obtained by going through each such path. In this example, users create graphics 33% of the time, modify graphics 41% of the time, and print or play graphics 26% of the time. Of the created graphics, 31% is two dimensional, 19% is three dimensional, and 50% is video.

The best source for the probabilities is usage data from the latest release of the software or from that of a similar system. Sometimes, usage data can be obtained from system logs. These values will need to be adjusted to account for new, removed, or changed features, which require estimates. If the system is completely new with no previous system available as a guide, estimates must be used. These estimates could be inaccurate but are still likely to be the best picture of the customer use of the system. See [25, 36] and Chapter 5 of [17] for more details.

Orthogonal Defect Classification

Orthogonal Defect Classification (ODC) is a means of categorizing defects into classes that point to the part of the project process that needs the most attention. In addition to this "orthogonality" or non-redundancy, the classification should be consistent across project phases and be uniform across products. We also want to keep the number of classes relatively small.

With defects, we want to relate causes with effects. Defect type is an important attribute that helps us narrow down where in the development process more attention is needed. In [17] and [37], the following eight defect types are considered:

1. Function
2. Interface
3. Checking
4. Assignment
5. Timing/serialization
6. Build/package/merge
7. Documentation
8. Algorithm

For example, if a function defect is found, it points to the high-level design phase as the likely source of the defect. A timing error is more likely to be associated with the low-level design phase. The distribution of defect types provides an indication of the current maturity of the software and its readiness for the next phase as (continuing with the example) defects for a system in the design phase should be more predominantly function defects over being timing defects, just as during the system test phase, if we see a lot of function defects (and a smaller percentage of timing defects), we should suspect that the software is not ready for this phase. The defect distribution changes with time and provides an indication of the maturity of the software development. Note that it takes time to calibrate the distributions and their changes, and the calibration is dependent on the development process. However, even before this calibration, trend analysis can provide evidence of whether a process is progressing correctly or not.

Another useful attribute is defect trigger, which is a condition that allows the defect to surface. There may be certain environmental factors, system states, or hardware platforms that are required to trigger the defect. While defect types provide an insight into the development process, triggers provide an insight into the verification process. For example, the defect trigger distribution for fielded software should be similar to the defect trigger distribution for system test. A significant difference indicates a potential problem with the system test environment. Defect trigger information coupled with defect-type information can provide an insight into weaknesses of various verification processes.

Triggers can also aid in assessing the effectiveness of document reviews and code inspections. By considering the activities performed by different reviewers based on their backgrounds and skill levels, we can map defect triggers to required skills for the reviewers. The defect trigger distribution, along with the defect type, can then be used to gain an insight into how effective the review or inspection is and what changes to the reviewer or inspector skill set are needed.

There are other ODC attributes, but the above provides some of the flavor of ODC. See Chapter 9 of [17] and [37] for more details. ODC can be used to improve the effectiveness of RCA while reducing the time and cost of the process. It also helps identify issues early and manage resources more effectively. Finally, ODC aids in tracing a defect to where it was introduced.

Reliability Risk Assessment

A reliability risk assessment is a process for identifying situations that can lead to reliability issues for a project. These situations are identified, prioritized, and analyzed so that their consequences can be estimated. Based on this analysis, actions are taken to eliminate or reduce the likelihood of those considered to be actionable and sufficiently severe. A reliability risk assessment should be in addition to a project-wide risk program, although the two should support each other. Typical steps include the following:

1. Determine the risk areas to assess. These generally include safety and security risks as well as risks that negatively affect the reliability of the software. For example, an issue with the software not being sufficiently reliable when performing a safety feature is likely to be extremely important and the reliability risk assessment should address it.
2. Determine how risks will be identified. Reliability risk assessment inputs are usually from software reliability engineering (SRE) personnel supported by safety and security personnel, although inputs from other project personnel and the customer should also be considered. These inputs are usually event driven, condition driven, or standards driven:
   (a) Event driven: Event-driven risks are risks identified because of the occurrence of some event, such as a failure.
   (b) Condition driven: A condition-driven risk is identified because of the status of some reliability indicator or metric, such as having a reliability prediction below a certain threshold.
   (c) Standards driven: Finally, a standards-driven risk is identified by project standards, such as safety standards. Inputs from the project risk program should also be assessed frequently to see if any of these inputs affect the reliability risks.

When identifying potential reliability risks, consider historical data, not just for the software but also for the project. Schedule risks, the introduction of new processes or tools, inexperienced personnel, and other project factors can increase risk. Also, identify reliability sensitive items, such as estimations of the number of lines of code or the expected reliability growth rate. Pay close attention to these as errors in them can have a significant impact on estimates.

3. Develop measures for assessing risk so that risks can be prioritized. These measures usually provide information on the likelihood of the risk occurring and the severity of the effect on the system if it does occur. An approach similar to that used in creating an FMECA (covered below) can be used to create these measures, with assessments being refined as more information becomes available.
4. Determine how to measure the effectiveness of the risk process and how to improve it.
5. Document the reliability risk assessment.

In addition, make a determination on how to address each risk. Addressing a risk includes the following:

1. Determine who (by project function) will be informed of a risk and its status and when and how they will be informed.
2. Determine what to do for each risk, such as which to address with corrective or mitigative actions, how to perform verification, and how to eventually close the risk.
3. Assess who to assign each risk-related task to and how to schedule the tasks.
4. Determine how the risks and the associated actions will be documented and lessons learned captured.

Often, further actions for the risks identified in a reliability risk assessment are handled in the project risk program. More on the topic of reliability risk assessment may be found in [25] and [38].

Reliability Trade Studies

A trade study is a decision-making analysis using technical criteria to objectively choose the "best" decision from a collection of alternative decisions. Trade studies are used when there is more than one course of action, it is not obvious which course of action is best, and the decision is important enough to merit the time and cost of performing a trade study. A reliability trade study is a trade study used to make a decision concerning some aspect of the reliability of the system. Typical steps in a trade study are as follows:

1. Identify the decision to be made.
2. Identify any constraints and assumptions.
3. Determine the decision criterion or criteria to be used, and if multiple criteria are used, determine how to arrive at a final decision if the multiple criteria do not agree. For example, criteria may be weighted by priority.
4. Identify the alternatives to choose from. Care should be taken that all viable alternatives are included in this list.
5. Choose a decision-making approach, perform the required analysis, and choose one of the alternatives. (See the topic on decision-making techniques in Section 6.5 for several techniques that can be used for this decision-making analysis.)

6. Perform a sensitivity analysis to determine if the decision is repeatable. Sensitivity analysis typically involves making small changes to weightings and scores and seeing if the results change, or changing the way that multiple criteria are combined to see if reasonable but different approaches give different results. If the results do not change, we gain more confidence in the final results of the study.
7. Document the results and how the results are derived. This documentation should as a minimum include the requirements, assumptions, decision criteria with their priorities, and enough study details to reproduce the study results. Also ensure that the appropriate personnel receive a copy of these documented results. The results of a trade study often constitute an important milestone for the project and may determine its future course.

See [2] for more details on this topic.

Software Defect Root Cause Analysis

A software defect RCA is an analysis to identify the most likely causes of software defects for a project. It is typically performed early in the project and updated as the project progresses. Early analysis can take advantage of defect information from earlier releases of the product or similar projects and products, provided that the products, processes, and personnel are similar enough. The purpose of the analysis is to enable the project to manage defect detection, analysis, and reduction more effectively. It also provides a good base of knowledge for software failure modes and effects analysis (SFMEA) and software fault tree analysis (SFTA).

Steps for producing a software defect RCA include the following:

1. Collect relevant information, such as the project FDSC or other definitions of failures, a list of the project processes that could result in product defects and pertinent information about these processes, and defects and failure information from the current and related projects.
2. Collect defect information from similar projects. Early in the project, use defect information from similar projects with similar products, processes, and personnel. This information is not as reliable as the defect information on the target project that will be collected as the project matures but can serve as a starting point to ensure that "obvious" defects are not missed.
3. Make a list of defect root cause categories. This list should be project specific, but Table 12 of [25] gives a representative such list with keywords from defect reports associated with each category member.
4. Collect defect information as the project progresses through its phases, adding defect categories that are not included in the earlier list as needed. Assign each defect to a category and keep a list of how many defects correspond to each category. Also record other information about each defect, such as who reported it, how it was found, and what activities were in process when the defect occurred and when it was found.
5. Produce a Pareto chart of the defects. This type of chart is a (vertical) bar chart that lists the defect categories on the abscissa (x-axis) and the number of such defects on the ordinate (y-axis). The defects are displayed from most prevalent to least. Pareto analysis is covered as a separate topic in Section 6.5.

The Pareto chart produced by this analysis provides an easy-to-understand visual of the most common causes of defects. If needed, further analysis can be performed to determine what activities produce the most defects or are responsible for finding the most. See [25] for more information. Note that this analysis fits in well with the activities of Section 3.3.

Software Fault Tree Analysis

An SFTA is a systematic, top-down approach for analyzing a fault or undesirable software event. The approach is hierarchical, with the event under investigation being the root of the tree. This root is fed from the next level, which consists of events that can cause the root event. Lower levels support the preceding level. A supporting level typically uses AND, OR, and other logical operators to combine potential causes. Most fault trees are deterministic, but a fault tree can be probabilistic. A probabilistic fault tree adds a conditional probability to each node of the tree to indicate the probability of the associated state, given that the events in paths higher up in the hierarchy occur.

Using fault trees can be time-consuming, so they are often only used for high-priority concerns and troubleshooting, such as safety-critical or mission-critical events. However, SFTA is sometimes used earlier in a project to systematically identify high-level faults and defects and decompose them into lower-level causes so that mitigations can be designed for them. They are often used in conjunction with FMECA and to serve as a guide to simplify maintenance activities. See [25, 26, 31, 39], and Chapter 15 of [17], for more details. Also, see Chapter 30 of [19] for probabilistic risk assessment, a technique for tree analysis that assigns likelihoods to events.

Software FMEA/FMECA

An SFMEA is a bottom-up analysis of potential failure modes and their effects at higher levels. An FMECA is an FMEA plus a critical items list (CIL), which is a prioritized list of the most critical items. We will use the acronym SFMEA whether a CIL is created or not. An SFMEA is a useful tool for the following:

1. Finding single points of failure
2. Identifying critical faults
3. Identifying faults that do not have adequate mitigation
4. Providing data for SFTA and troubleshooting, including RCA
5. Aiding in system design, such as the investigation of design alternatives
6. Supporting reliability, safety, security, vulnerability, logistics, and maintenance analysis
7. Supporting test planning and development
8. Aiding in the development of maintenance procedures
9. Developing a deeper understanding of the software design and its implications

For years, the standard reference for an FMEA or FMECA has been MIL-STD-1629A (see [40]), although this reference is for hardware rather than software. For many practitioners, SFMEA is not as well understood as FMEA or FMECA for hardware. SFMEAs are newer than the hardware versions of the analysis. Also, hardware failures tend to be independent of each other, but software failures often are not. Furthermore, software is often a "one-of-a-kind" hand-made product, and so software failures often do not repeat in as predictable a pattern as is observed with hardware failures. Finally, with hardware FMEA, hardware redundancy is often a mitigating factor, but with SFMEA, redundancy may not be as attractive a mitigation technique.

An SFMEA can be a labor-intensive analysis and often most of the results have no impact on the project. However, it is frequently the case that a few of the results have a significant impact. The analysis considers requirements, design, and code to assess what defects may be in the code, what effects the defects could have at various levels, and how they can be detected, isolated, and mitigated. An SFMEA is usually constructed in a tabular form with a row for each failure mode. There are various commercial tools that may aid the production of an SFMEA, although often they are produced using spreadsheets. Typical columns for a row include the following:

1. Row number: An identifying number
2. Unit: The name of the module, function, class, or other appropriate level
3. Function: A brief description of the function of the unit
4. Failure mode: The manner in which the failure occurs
5. Probable cause of failure: A description of the defect that causes the failure
6. Effect of the failure on
   (a) Software unit
   (b) Subsystem
   (c) System
7. Failure detection method: The means by which the failure can be detected by an operator or a user under normal conditions
8. Failure criticality or severity: The potential severity of an occurrence of the failure mode
9. Failure likelihood or probability of occurrence: The potential likelihood of an occurrence of the failure mode
10. Risk priority number (RPN): A number used to prioritize risks that combines the probability of occurrence, severity of occurrence, and possibly the likelihood of failure detection
11. Failure mitigation: What can be done to prevent or reduce the likelihood or severity of the failure mode
12. Remarks: Comments that add clarity to the row

For example, in MIL-STD-1629A, failure criticality is measured as follows:

1. Category I: Catastrophic
2. Category II: Critical
3. Category III: Marginal
4. Category IV: Minor.

Other sources define different categories, often resulting in 5 or 10 such categories. As criticality, or severity, is a factor used to calculate the RPN, criticality categories are often assigned a value of one for a very low severity and higher numbers for greater severity. Ultimately, the number and definitions of these categories should be based on the needs of the product being analyzed.

For failure likelihood, MIL-STD-1629A uses either a quantitative method or a qualitative method. The quantitative method uses the failure rates of the applicable parts. The qualitative method uses the following levels:

1. Level A – Frequent: Level A is defined as a high probability of occurrence with a single failure mode probability of >0.20 of the overall probability of failure during the item operating time
2. Level B – Reasonably probable: This level is defined as a moderate probability of occurrence with a single failure mode probability of >0.10 but <0.20 of the overall probability of failure during the item operating time
3. Level C – Occasional: Level C is defined as a single failure mode probability of >0.01 but <0.10 of the overall probability of failure during the item operating time
4. Level D – Remote: Level D is defined as a single failure mode probability of >0.001 but <0.01 of the overall probability of failure during the item operating time
5. Level E – Extremely unlikely: This level is defined as a single failure mode probability of <0.001 of the overall probability of failure during the item operating time

Again, other sources use different categories. Another example is as follows:

1. Level 1: Less than 1% likelihood of the failure mode occurrence
2. Level 2: Between 1% and 5% likelihood of the failure mode occurrence
3. Level 3: Between 5% and 25% likelihood of the failure mode occurrence
4. Level 4: Between 25% and 50% likelihood of the failure mode occurrence
5. Level 5: Greater than 50% likelihood of the failure mode occurrence

As with severity, choose levels that make sense for the product that the FMEA is analyzing.

Detection is a measure of likelihood that the failure will be detected in a reasonable amount of time. An example list of levels is as follows:

1. Level 1: Certain to be detected
2. Level 2: Almost certain to be detected
3. Level 3: There is a high probability of failure detection
4. Level 4: There is a moderate probability of failure detection
5. Level 5: There is a low probability of failure detection
6. Level 6: The failure is undetectable

A detection value is not always used, but when it is, it is almost always used as a factor when calculating the RPN.

The risk priority number (RPN) is an overall measure of risk and is found either by multiplying the failure severity number by the failure likelihood number or by multiplying both of these by the detection number. These factors are expressed as integers starting at 1 and usually going to 5 or 10. Higher numbers cause more concern, so an RPN of 1 is of minimal concern, but a high number indicates that mitigation is needed.

Reference [25] lists eight SFMEA viewpoints, such as functional, interface, and detailed, and has a template for each. Each template has column headings and a description of the contents of the column. These templates can be very useful and save considerable work when organizing an SFMEA.

The first part of producing an SFMEA is preparation. Issues such as whether the SFMEA will cover all of the software or only specific parts must be addressed. Ground rules should be made explicit. Levels of severity or criticality need to be defined and agreed to. If using [25] as a guide, choose the viewpoints to use for the SFMEA and ensure that the right resources and experts are available for those viewpoints. Choose the team of people to perform the analysis carefully, and make sure that each is qualified and properly instructed and trained for the assigned tasks. The SFMEA should be considered a primary task for each team member and not additional work to be done quickly so that they can return to their "main" tasks. If these and other up-front issues are not addressed early, the analysis will be inconsistent and will need excessive rework.

For an SFMEA to be useful, it is important to have a good list of applicable failure modes. The point-of-view templates in [25] each have a list of typical failure modes associated with the given point of view. These examples are a good starting point, but each project should tailor their failure modes to their experience and applications.

If a software FMECA is produced, a CIL is added to the software FMEA. This list contains the most critical RPN items listed in order of criticality. The list is usually combined with the hardware CIL if it exists.

An SFMEA is an important tool for preventing defects and designing mitigations. It is also useful in DRACAS RCA, where the failure symptoms may match those found in the SFMEA. Conversely, defect reports should be used to update the SFMEA if a defect or failure is reported that is not covered in the SFMEA.

As an SFMEA is a bottom-up analysis and an SFTA is a top-down analysis, they can be used to support each other. Although an SFTA is often used to troubleshoot a failure after it has occurred, when used to systematically identify faults and defects and decompose these to causes, SFTA and SFMEA can be complementary analyses that "meet in the middle." Reference [25] provides more information on SFMEA and also provides guidance on when to perform them. Other references are [12, 31], and [40].

Example 6.4.4   For this example of a software FMEA, we consider a project that writes software for a telescope that finds and tracks asteroids. The software receives data from various sources, including other telescopes, and calculates where the subject telescope needs to point to find and track a specified asteroid or potential asteroid. The software uses a numerical integration routine as a part of these calculations. This SFMEA example considers a small part of this integration routine.

More specifically, this example considers one piece of the error-handling aspect of the numerical integration code. It considers an I/O error failure mode and finds three potential failure causes: faulty data, a data format error, and data that are outside of the expected data bounds. For example, the data may be faulty because of excessive noise or outliers. The faulty data could result in unrealistic velocities or accelerations and can potentially be partially mitigated by feasibility checks for these quantities. Table 6.13 shows how the SFMEA can document these, along with the effects on the software unit (the integration routine), the subsystem, and the system. For this example, we assign row numbers 232–235 to these items.

Table 6.14 continues the example by considering the detection methods, criticality, likelihoods, and RPN values, along with mitigations and remarks. For this SFMEA, the following failure criticality levels are used:

1. Level 1: Negligible, no noticeable impact
2. Level 2: Minor, noticeable but does not impact the mission

Table 6.13   Software FMEA.

Row Number	Unit	Function	Failure Mode	Failure Cause	Effect on SW Unit	Effect on Subsystem	Effect on System
232	Numerical integration	Error handling	I/O error	Faulty data	Produces incorrect result	Points telescope incorrectly	Misses asteroid observation
233				Format error	Potential crash	Potential crash	Potential crash
234				Data bounds	Produces incorrect results	Points telescope incorrectly	Misses asteroid observation
235					Potential crash	Potential crash	Potential crash

3. Level 3: Moderate, does impact the mission but there is a work-around that prevents mission degradation
4. Level 4: Serious, it impacts the mission and at least some of the mission objectives are negatively impacted
5. Level 5: Critical, mission failure

Similarly, the following failure likelihoods are used:

1. Level 1: Less than 1% likelihood of the failure mode occurrence
2. Level 2: Between 1% and 5% likelihood of the failure mode occurrence
3. Level 3: Between 5% and 25% likelihood of the failure mode occurrence
4. Level 4: Between 25% and 50% likelihood of the failure mode occurrence
5. Level 5: Greater than 50% likelihood of the failure mode occurrence

Using these values, an RPN of 1 is likely to be acceptable, whereas an RPN of 25 is a serious matter that should not be allowed.

Software Reliability Advocate

Reliability is built into software from early in the project. For example, people working on the conceptual design are impacting software reliability, but these people are probably focused on traditional aspects of their job and not reliability. Different organizations and disciplines on a project often have different objectives and approaches, making it less likely that there is a coherent cross-discipline plan for creating and maintaining reliable software. A software reliability advocate works with people in each project phase and across disciplines to ensure that impacts to software reliability are understood and considered. The advocate also serves as a liaison between groups. This person "parallels" the person in the traditional role, at least with respect to software reliability. The advocate performs reliability audits, handles the reliability casebook, and provides advice. In some organization, the role of the advocate may already be performed by a reliability engineer, but it is important to emphasize the pro-active, up-front nature of the advocate's job. Although assessing

Table 6.14 Software FMEA (cont.).

Detection Method	Criticality	Likelihood	RPN	Mitigation	Remarks
Multiple sources are used when available	3	3	9	Stop processing when detected	Additional data quality checks can reduce this risk
Error handling for format	5	1	5	Stop processing when detected	Different data sources provide different formats
Minimal error handling for bounds	3	3	9	Stop processing when detected	Different data sources provide different bounds
Minimal error handling for bounds	5	2	10	Stop processing when detected	Different data sources provide different bounds

reliability may be a part of the advocate's job, the main effort is to prevent defects. It is also important to note that because of the breadth of knowledge and abilities required (and to be sufficiently respected by the people in the various disciplines and across the various phases), this role may require a senior technical person. The advocate must also have sufficient "political" skills to work with various people and point out potential reliability weaknesses of their work without offending them and making a "toxic" work environment. Finally, as the advocate's tasks span several groups, this position should report to someone in a high-level position.

Software Reliability and Availability Allocations
Once the overall project reliability and availability objectives have been determined, they need to be allocated down to the appropriate level. Often, this level is the software LRU level, particularly if different organizations are responsible for different software LRUs. Also, new releases of the product may update some software LRUs and not others, so tracking reliability at the LRU level is again appropriate. Correctly performed allocations ensure that if the individual allocations are met, then the system objectives are met. They should also be done in such a way that the difficulties associated with meeting the allocations for each of the LRUs are manageable.

There are various approaches for allocating reliability and availability objectives. Reference [12] details four software allocation methods:

1. Allocation using equal apportionment, sequential operation: With this method, we have an allocation to the overall software and reason to believe that the software LRUs are operated sequentially, meaning that software LRU number 1 is used, then we cease to use this LRU and instead use software LRU number 2, and so on. With this method, if the reliability allocation for the software is R, then each software LRU is also allocated a reliability of R. If an overall failure rate allocation of λ is used, then each LRU is allocated a failure rate of λ as well.

2. Allocations using equal apportionment, concurrent operation: With this method, we have an overall software allocation and reason to believe that the software has N LRUs that are operated concurrently. With this method, if the reliability allocation for the software is R, then each software LRU is allocated a reliability of

$$R^{\frac{1}{N}}$$

In terms of failure rate, if λ is used for the overall software allocation, then λ/N is allocated to each software LRU.

3. Allocation using operational criticality factors: Allocations based on operational criticality factors consider the impact that a failure will have on the system and allocate failure rates to LRUs based on these factors. The criticality of a software LRU is the degree to which the reliability of the system depends on the proper functioning of the LRU. The greater the criticality, the lower the failure rate should be. Let the total software allocation, in terms of failure rate, be λ, and let the system have N software LRUs. Also, let the criticality of the ith software LRU be c_i where a smaller value of c_i denotes greater criticality. Finally, let τ_i be the total active time for the ith LRU and let T be total mission or operating time. Then, a total adjustment factor K is found from

$$K = \frac{\sum_{i=1}^{N} c_i \, \tau_i}{T}$$

and the allocated failure rate for the ith software LRU is

$$\lambda_i = \frac{\lambda \, c_i}{K}$$

4. Allocation based on complexity factors: This allocation method follows the same pattern as the previous one except that rather than using criticality factors, it uses complexity factors (McCabe complexity or some other such measure). Let the total software allocation, in terms of failure rate, be λ, and let the system have N LRUs. Now, let the complexity of the ith software LRU be w_i, where a smaller value of w_i denotes greater complexity. Finally, let τ_i be the total active time for the ith LRU and let T be total mission or operating time. Then, the total adjustment factor K is found from

$$K = \frac{\sum_{i=1}^{N} w_i \, \tau_i}{T}$$

and the allocated failure rate for the ith software LRU is

$$\lambda_i = \frac{\lambda \, w_i}{K}$$

Table 28 of reference [25] lists seven allocation methods. These seven include several of the methods listed above plus some other methods. Three of these are as follows:

1. Allocation using cost: This method uses estimated cost in dollars to fund the research and development required to produce a given software LRU.

2. Allocations based on predictions: For this method, predictions for the failure rate of each software LRU are made and these values are used to weight the failure rate allocations. If the total allocated software failure rate is λ, there are N software LRUs, and the predicted

failure rate for the ith software LRU is μ_i, then the allocated failure rate for the ith LRU is

$$\lambda_i = \lambda \, \frac{\mu_i}{\sum_{i=1}^{N} \mu_i}$$

3. Allocation based on previous failure rates: If a similar system has been produced in the past and the failure rates of its software LRUs are known, this method uses these historical failure rates to weight the allocation process.

These methods can be used to find allocated failure rates and therefore allocated values for MTBF, mean time between critical failures (MTBCF), and other related quantities. To find allocated reliability, a mission or operating time t_m is needed. We then use

$$R(t_m) = \exp\left(-t_m/\text{MTBF}\right)$$

Note that reliability values using MTBCF or mean time between essential function failures (MTBEFF) may also be used, either instead of or in addition to the above MTBF value.

Allocating software availability is handled similarly, using mean time to software restore (MTSWR) or a related downtime value. Reference [25] provides a set of steps for taking a system-level availability requirement and allocating down to software LRU MTBF values.

After performing the initial allocation, compare the results with the corresponding predictions. If there is a significant mismatch, try reallocating. If no reallocation enables the software to meet it overall reliability objectives, these objectives may need to be changed or significant changes to the project may be required.

Example 6.4.5 In this example of allocating down to the LRU level, we use allocation by prediction for a design that has the reliability block diagram (RBD) shown in Figure 6.1. The design has six hardware LRUs, labeled H1 through H6, and three software LRUs, labeled S1, S2, and S3. As can be seen in the RBD, hardware LRUs H2, H3, and H4 are in parallel as are hardware LRUs H5 and H6. These are in series with H1 and S1, S2, and S3.

The overall system has a reliability requirement of 0.935 over a 24-hour time period. We start the allocation by making a prediction for the system. The first column of Table 6.15 lists the LRUs and the second column gives the predicted critical failure rates for each of them. The third column converts the critical failure rate to 24-hour reliability $R(24)$ where for critical failure rate λ_c,

$$R(24) = \exp(-\lambda_c \times 24)$$

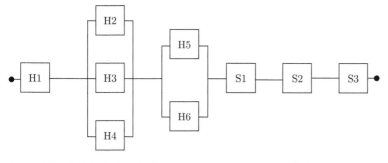

Figure 6.1 Reliability Block Diagram.

Table 6.15 Predicted Critical Failure Rate Values.

LRU	Predicted λ_c	Predicted R(24)	Allocated λ_c
H1	0.001	0.976	0.00093
H2	0.002	0.953	0.00186
H3	0.002	0.953	0.00186
H4	0.002	0.953	0.00186
H5	0.0005	0.988	0.000465
H6	0.0005	0.988	0.000465
S1	0.0005	0.988	0.000465
S2	0.0005	0.988	0.000465
S3	0.001	0.976	0.00093

The RBD is then used to make a reliability prediction for the design, $R_D(24)$. Suppressing the time dependency part of the notation, let

$$R_3 = 1 - (1 - R_{H2})(1 - R_{H3})(1 - R_{H4})$$

and

$$R_2 = 1 - (1 - R_{H5})(1 - R_{H6})$$

then per the RBD

$$R_D = R_{H1} \ R_3 \ R_2 \ R_{S1} \ R_{S2} \ R_{S3}$$

With the critical failure rates in the second column of Table 6.15, we have a reliability $R_D = 0.93$, although we have a requirement for a value of 0.935.

To raise the reliability to the desired value, we decrease the required critical failure rate values of the LRUs, each by the same factor. By multiplying each critical failure rate value in the second column by 0.93, we obtain the allocated values shown in the fourth column of Table 6.15. The factor 0.93 is chosen so that using the same RBD and LRU reliability values found using the critical failure rates in the fourth column, we have $R_D = 0.935$. (Note that although the value of the factor and the value of the predicted reliability R_D are the same, this is only a coincidence.) In this example, we have allocated to where the system just meets its reliability requirement. Often, we allocate to a slightly higher reliability to provide some margin, although care needs to be taken to prevent an expensive over design of the system that is based on margin and not need.

This process is a modification to the "Allocations based on predictions" process described above. In this example, we adjusted the critical failure rate of each LRU based on our overall prediction and by the same factor, but we used the RBD to account for the effects of parallel LRUs. Other modifications are also possible. For example, if we have good historical data for the critical failure rate for a particular LRU and believe that improving that rate would be difficult or impractical, it can be used for the subject LRU's known critical failure rate and use the technique shown in this example to find allocations for the remaining LRUs.

Software Reliability and Availability Objectives

Sometimes, the reliability and availability objectives for a product are given to the project by a customer. In other cases, the project must determine reliability and availability objectives based on what will satisfy the customer along with what will be achievable. In either case, these objectives should be obtained or determined in the Concept Development and Planning (CDP) phase and reviewed for reasonableness as the project progresses.

Information required to have useful and understandable reliability objectives includes the following:

1. A definition of the system that the objectives apply to: Products can be complex with many options and potential added features, with various operating modes, all running on various platforms. The objective or objectives should specify which are applicable to a given objective.
2. A definition of what constitutes a failure: Ideally, an FDSC is included, but as a minimum, what constitutes a failure with respect to the objective or objectives should be defined.
3. The time in the life of the product that the system is to meet the objectives: The reliability and availability of a product at delivery is often different from that after "customer shakedown" or at a year after delivery. The point in time that the objectives are intended to apply should be specified.

It is also important to consider how the objectives will be verified. For example, if an objective is to be verified by software testing, there needs to be a determination of the required level of confidence. Also, the cost and schedule impacts of the testing should be considered.

There are several different figures of merit for reliability and availability objectives. Sometimes, some combination of MTBF, MTBCF, mean time between system aborts (MTBSA), and MTBEFF is used, often along with a mean down time (MDT) or MTSWR. Other common objectives are reliability and availability. Sometimes, combinations of both are used, but care should be taken to ensure that the objectives are compatible and that the system is not overspecified.

To derive the reliability and availability objectives for a project, consider the following three situations:

1. If there is a previous product that is sufficiently similar to the product in question, use it as a guide. Adjust the values based on differences in the two products. See Figure 35 of [25] for more on this topic.
2. If the system is a new product and not particularly similar to other systems, use accepted techniques to predict the hardware and software reliability (or other figures of merit) for the product. Check the values for reasonableness to the customer and feasibility for the project. See Figure 36 of [25] for more on this topic.
3. If the product is mass-deployed, objectives can be determined based on information such as the maximum acceptable number of maintenance actions in a given time period. See Figure 37 of [25] for more on this topic.

Ideally, once these objectives have been identified, they remain stable throughout the life of the project. Sometimes, as the project matures, it becomes apparent that one or more of the objectives are unrealistic either because of unforeseen changes to the project, aspects of the project or product that were unknown at the time that the objectives were set, or mistakes that were made.

Finally, for some (typically commercial) products, reliability objectives are met by successful completion of one or more processes, such as a predefined "pass/fail" performance on a series of software reliability tests. Such an approach may be acceptable, but before adopting it, the project should consider that the quantitative objectives form the basis of quantitative goals used in earlier phases of the project. Without quantitative reliability objectives, other means need to be considered to ensure that the project is producing sufficiently reliable products in the early project phases.

Example 6.4.6 For this example, we consider a situation where we plan to update an existing product and want to determine a new reliability objective for the new product. The reliability is over a time period of 168 hours (1 week), so our remaining task is to find the objective critical failure rate. The product has both hardware and software components. Other related information includes the following:

1. The previous product has a critical failure rate objective of 2.0×10^{-4} critical failures per hour, with a hardware value of 1.4×10^{-4} and a software value of 6.0×10^{-5}.
2. Based on field data, the actual critical failure rate for the existing product is 2.8×10^{-4} critical failures per hour. The hardware failure rate is 1.3×10^{-4}, and the software failure rate is 1.5×10^{-4}.
3. For the existing product, 50% of the software is autogenerated code and 50% is "hand-generated" code. Per the fielded data, the autogenerated code has a critical failure rate of 4.5×10^{-6}, and the hand-generated code has a critical failure rate of 1.455×10^{-4}.
4. For the new product, there is little change to the hardware. A hardware switch that contributed to some of the hardware failures is being replaced with what should be a more reliable switch, but it is decided to set the hardware objective to the actual fielded critical failure rate of the current product.
5. The software for the new product will have new functions. Because the new functions and code size increase due to code age (there is an average increase in code size of 10% per year), the new code is expected to be larger than the existing product by a factor of 1.75. It is expected to be 70% autogenerated code and 30% hand-generated code. The same processes and tools and a similarly experienced software team will produce and test the software.

Based on this information, the objective hardware critical failure rate is set to 1.3×10^{-4} critical failure per hour, the same value as found from the field data of the existing hardware. As for the software critical failure rate, we assume the field data critical failure rate per line of code for both the autogenerated code and for the hand-generated code (as the same processes, tools, and level of expertise are being used). Let μ_A denote the critical failure rate per line of autogenerated code and similarly let μ_H denote the critical failure rate per line of code for hand-generated code, both according to the field data. Also, let L be the total lines of code in the existing product and recall that half of the total lines of code are autogenerated. Also, recall that the autogenerated code has a critical failure rate of 4.5×10^{-6}. Then,

$$4.5 \times 10^{-6} = (\mu_A)(0.5)(L)$$

so that

$$\mu_A = \frac{4.5 \times 10^{-6}}{(0.5)(L)} \tag{6.1}$$

and using the same calculation with the hand-generated code,

$$1.455 \times 10^{-4} = (\mu_H)(0.5)(L)$$

giving

$$\mu_H = \frac{1.455 \times 10^{-4}}{(0.5)(L)} \tag{6.2}$$

The number of lines of code for the new product, L_N, is

$$L_N = (L)(1.75) = (0.7)(L)(1.75) + (0.3)(L)(1.75)$$

where the first term on the right side is the number of autogenerated lines of code and the second term is the number of hand-generated lines of code. Therefore, the critical failure rates for the autogenerated code and for the hand-generated code, λ_A and λ_H, respectively, are

$$\lambda_A = (\mu_A)(0.7)(L)(1.75) \tag{6.3}$$

$$\lambda_H = (\mu_H)(0.3)(L)(1.75). \tag{6.4}$$

Substituting Eq. (6.1) into Eq. (6.3) and Eq. (6.2) into Eq. (6.4) gives

$$\lambda_A = \left(\frac{4.5 \times 10^{-6}}{(0.5)(L)}\right)(0.7)(L)(1.75) = 1.10250 \times 10^{-5}$$

and

$$\lambda_H = \left(\frac{1.455 \times 10^{-4}}{(0.5)(L)}\right)(0.3)(L)(1.75) = 1.52775 \times 10^{-4}$$

Therefore, the total critical failure rate λ for the new product is found by adding the critical failure rate for the hardware, for the autogenerated code, and for the hand-generated code, giving

$$\lambda = \left(1.3 \times 10^{-4}\right) + \left(1.10250 \times 10^{-5}\right) + \left(1.52775 \times 10^{-4}\right) = 2.938 \times 10^{-4}$$

With the 168-hour reliability time period, the reliability objective R (to three digits) is

$$R = \exp(-\lambda \times 168) = 0.952$$

Software Reliability and Availability Predictions

When referring to software reliability predictions, we are referring to the use of models to predict software reliability based on the characteristics of the software and the processes used to create it. These models can make predictions before the creation of the software. The models typically use historical data from similar systems and processes. Explicitly or implicitly, software reliability predictions usually use three broad steps:

1. Predict the number of defects in the software.
2. Use the number of defects to predict the failure rate for the software.
3. Use the failure rate to predict the MTBF, reliability, availability, and other figures-of-merit for the software.

Predictions of software defects, failure rates, and reliability can be important for ensuring that a product is reliable. It helps management assess the progress of the project and plan activities. Project personnel can use the predictions to develop achievable requirements and to assess whether the project is ready to progress to the next project phase. For example, if the software is not predicted to be sufficiently reliable, it may be determined that it is not ready for the IV&V phase. Predictions can also support sensitivity analysis to determine the best areas to improve. By performing these predictions early and updating them throughout the progress of the project, they provide early guidance for concept feasibility and for requirements as well as providing early indications of potential issues.

There are many models that can be used to predict the number of defects in software and a model that is appropriate for one project may not be the right model for another. Techniques that do not require running the software can be used early in the project but generally do not provide as accurate an assessment as a technique that has the advantage of test and operational data. Even if these early predictions are of limited accuracy relative to the needs of the project, trends in the predictions may still provide useful information. If predictions are trending away from expectations (in an unsatisfactory manner), then further investigation is in order. As an additional note, changes in project processes, personnel, platforms, or product type may require changes to aspects of how defects are predicted. For example, the historical data may not be applicable any more.

The following steps are useful for software reliability predictions:

1. Collect information about the software and the project.
2. Choose a model.
3. Apply the model to predict the total number of software defects.
4. Predict defects over time.
5. Predict the failure rate and MTBF for the software.
6. Predict the reliability, availability, and other measures for the software.
7. Check the results.

These are covered in more detail below.

Step 1: Collect information
Collect information typically relevant to software prediction models. Examples of potentially relevant information include what software development processes are used, staff size, start and stop times for activities, what language or languages are used, a list of software LRUs, estimated lines of code for each software LRU, and the type of code for each (new code, slightly modified code, heavily modified code, autogenerated code, and other types).

Step 2: Choose a model
Several software prediction models are listed below. Choosing a model typically starts with determining which models can be used based on the information available. If a model requires information that cannot reasonably be obtained, it is removed from consideration. The remaining models are then assessed based on factors such as expected accuracy of the results, effort required to use the model, which models are more typical for the industry, which are most up to date, and which have the most applicable sets of features and options relative to the project needs. Reference [25] lists and explains the following models:

1. Shortcut model: This model uses a list of 22 questions to assess strengths and risks to predict a defect density value.

2. Application-type model: Software for different industries and different applications tend to have different defect densities. This model predicts defect density based on this information.

3. Capability maturity index model: The capability maturity index model uses the Software Engineering Institute (SEI) Capability Maturity Model Integration (CMMI) level evaluation of the software to provide a defect density estimate, with error bounds, at testing and at three or more years of fielding.

4. Detailed survey assessment models: Similar in concept to the shortcut model, the detailed survey assessment models are three different models, each with their own more detailed set of survey questions.

5. Company-developed model based on historical defect density data: If a company has carefully collected defect data on their software and is developing similar software using similar processes and with a similarly skilled team, then defect predictions can be made based on this historical data set. This method can be accurate if the data used in the model are both good and kept current.

6. Neufelder prediction model: The Neufelder prediction model, incorporated into the Quanterion Solutions 217Plus^{TM} model, considers a collection of software development practices and correlates these with software defect densities.

7. Rayleigh models: Unlike the other models on this list, Rayleigh models use the number of defects found on the project. It uses defects found throughout the development process, not just those found in the software. As a result, predictions can be made before having functioning software. This approach assumes that defect detection occurs and is recorded throughout the software development process. For example, inspections or peer reviews may be used for requirements, software designs, and other documents. The cumulative distribution function for a Rayleigh distribution is

$$F(t) = 1 - \exp\left(-\frac{t^2}{2\sigma^2}\right)$$

and the probability density function is

$$f(t) = \frac{t}{\sigma^2} \exp\left(-\frac{t^2}{2\sigma^2}\right)$$

The parameter σ is called the scale parameter. For defect modeling, let the independent variable be time t and multiply $F(t)$ by K, the total number of defects. Therefore, the cumulative function (thus modified) gives the total number of defects found to date. The maximum of the probability density function is found where $t = \sigma$, so at the maximum, we have found $1 - \exp(-0.5) = 0.3935 \approx 0.39$ of the total number of defects. With this information, we assume a Rayleigh distribution for the defects, fit the defects found as a function of time to a Rayleigh curve, and when we have reached the maximum of the probability density function, we have found approximately 39% of the total number of defects. Therefore, we can predict the total number of defects (the number at the peak value times $1/0.39 = 2.56$) and also how many defects we can expect to find in any time interval. Note that these results are dependent on how well the curve fits the defects, but when performed properly, the results are generally good.

Rayleigh distribution defect prediction techniques make predictions based on discovered defects per project phase. For this application, project phases often include the requirement phase, design, implementation, unit testing, integration testing, and possibly others. These models typically fit the data to a Rayleigh distribution where the abscissa is project phase, starting with the earliest and going to the latest, and the ordinate axis is defect density, such as defects per thousand lines of code. These models predict defect density by phase, so decisions concerning readiness to proceed to the next phase can be made. To be accurate, the models require consistently tracked defect counts. Examples of this type of model are the Software Error Estimation Program (SWEEP) model and the Software Error Estimation Reporter (STEER) model. For more information, see [12, 33], and [34].

In Chapter 3 of [17], a Rayleigh model is described that models the number of faults per thousand lines of code detected from time $t - 1$ to time t, $\Delta V(t)$, by

$$\Delta V(t) = V\left[\exp(-B\,(t-1)^2) - \exp(-B\,t^2)\right]$$

where V is the lifetime fault rate in faults per thousand lines of code and t is the fault discovery index given by

(a) Requirements analysis: $t = 1$
(b) Software design: $t = 2$
(c) Implementation: $t = 3$
(d) Unit test: $t = 4$
(e) Software integration: $t = 5$
(f) System test: $t = 6$
(g) Acceptance test: $t = 7$

For B, we have

$$B = \frac{1}{2\tau_p^2}$$

where τ_p is the defect discovery phase constant value of t at the peak of the curve of faults, i.e. where we have approximately 39% of the total number of defects. Note that the total number of defects per thousand lines of code discovered through fault discovery index t is

$$V(t) = V\left[1 - \exp(-B\,t^2)\right]$$

Two more models that can be used for defect prediction are

1. COnstructive QUALity MOdel (COQUALMO): The COQUALMO is a part of the COnstructive COst MOdel (COCOMO) II suite of models. This suite of models predicts software cost and schedule. COQUALMO predicts software defects based on sources of defect introduction and defect discovery techniques. It considers the requirements phase, design phase, and coding phase and uses the size of the product and 21 quality adjustment factors that characterize the people, product, platform, and project. It uses a defect introduction submodel that considers defect source, defect introduction rate (by source), and quality adjustment factors. It also has a defect removal submodel. This submodel considers defect removal activities, such as peer reviews, automated analysis, and testing and removal fractions for each of these. Additionally, as the model incorporates a cost and schedule model to tie cost, schedule, and quality together, it

makes resource allocations more practical. There is also a version of COQUALMO that predicts software defects that are introduced or removed and classifies them based on their ODC type. The model can provide an estimate of the number of defects remaining in the software and can be used to do "what if" exercises to determine what process to change to "best" improve the software. See [41] for more information.

2. Orthogonal Defect Classification (ODC): ODC is a means of categorizing defects into classes. When used with historical information, ODC can be used to predict software defects. ODC is covered in more detail under the ODC topic above in this section.

Step 3: Apply the model to predict the total number of software defects

Software defect prediction techniques usually predict a defect density, such as defect per normalized effective thousand source lines of code (EKSLOC) or defects per function point. Function points are a unit of size based on the amount of functionality provided. To predict the number of defects, we need to assess the size of the software. We then determine the total number of defects by multiplying the defect density by the effective size of the code. Reference [25] explains how to find normalized EKSLOC. Also, see Example 6.4.7.

When the project uses an incremental software development process or an evolutionary process such as agile processes, the same basic approach for predicting defects can be used, but predictions should be made for each increment. If the applicable requirements for an increment are the same as for the previous increment, update the size estimate and the prediction. If the applicable requirements have changed in the new increment, do a separate prediction on the new increment.

Another potential issue is predicting defects for subcontractor-generated software, Commercial-Off-The-Shelf (COTS) software, and Free Open Source Software (FOSS). Suggestions for these situations may be found in [25].

At the end of this step, we have a prediction of the total number of defects in the software.

Step 4: Predict defects over time

The models outlined above typically predict the number of software defects at the time that the software is deployed, although some of the models also predict the number of software defects at the beginning of software testing. To convert this value into a failure rate, we need to predict defect discovery as a function of usage time. Models for this analysis listed in [25] include the following:

1. Exponential model: Perhaps the most simple and commonly used model for predicting when defects in software will be experienced is the exponential model. (See for example [42].) This model is based on a function of the form

$$N(t) = N \left[1 - \exp(-\lambda t)\right]$$

where $N(t)$ is the number of defects detected by time t, N is the total number of defects, and λ is the defect detection rate. t is typically software run time or CPU time and here is in terms of months. The number of defects detected from time t_{i-1} to t_i is

$$N(t_i) - N(t_{i-1}) = N \left[\exp\left(-\lambda t_{i-1}\right) - \exp\left(-\lambda t_i\right)\right]$$

In [25], λt_i is expressed as

$$\frac{Qi}{TF}$$

Here, Q is the growth rate in defects per month, or how fast defects are predicted to become known, after deployment. The larger the growth rate Q, the faster the software reliability can grow. i is the index indicating the month, and TF is the growth time period in months. It is typically considered to be the time from deployment to the time that no more defects are found. TF is usually three times the average time between releases and has a default value of 48 months. The range of values for Q is from 3 to 10 with a default value of 4.5. Reference [25] provides further guidance for finding appropriate values of Q and TF.

Using these values, the number of defects N_i predicted to be found in month i is

$$N_i = N \left[\exp \left(-\frac{Q\,(i-1)}{TF} \right) - \exp \left(-\frac{Q\,i}{TF} \right) \right]$$

The result is a profile of new detected defects per month, starting with deployment and continuing through the time period TF. Example 6.4.8 uses the exponential model.

2. Duane models: The Duane model is based on an observation (see [43]) that, taking into account reliability growth, the number of failures versus operating hours is a straight line when plotted on log–log paper. This observation is often referred to as the Duane postulate. (Note that the observation was based on hardware failures, not software failures.) More specifically, the model uses the following values:

   (a) t: Test time
   (b) ω: Scale constant
   (c) α: Shape constant
   (d) $N(t)$: Expected cumulative number of failures observed by test time t

   Duane observed that

   $$\ln \left(\frac{N(t)}{t} \right) = \ln \omega - \alpha \ln t$$

   so that

   $$N(t) = \omega\, t^{1-\alpha}$$

   In terms of failure rate $\lambda(t)$ as a function of time, we have

   $$\lambda(t) = \omega\,(1-\alpha)\,t^{-\alpha}$$

   and in terms of MTBF,

   $$\text{MTBF}(t) = \frac{t^{\alpha}}{\omega\,(1-\alpha)}$$

   If we use the notation

   $$M_I = \frac{1}{\omega\,(1-\alpha)}$$

   we have

   $$\text{MTBF}(t) = (M_I)\,(t^{\alpha}) \tag{6.5}$$

   Note: When using this model for reliability testing, it is often the case that testing is run for a time period t_I, resulting in an empirical initial MTBF of M_I so that for time greater than t_I,

   $$\text{MTBF}(t) = \frac{M_I}{1-\alpha} \left(\frac{t}{t_I} \right)^{\alpha}$$

If historical values for ω and α are available and these are based on similar software, software development processes, and teams, then these are the best values to use for this model. For hardware, typical values for α are $0.25 \leq \alpha \leq 0.4$ and 0.6 is often considered to be an upper limit. Values above 0.5 are rare and indicate very rapid reliability growth. For software, particularly commercial software sold to many users, α may be larger. For example, in [44], the authors consider reported failures as a function of time for a widely distributed software product. Even without product changes, their data indicates a significant reduction in the failure rate of the software. This reduction closely matches $\alpha = 0.77$. There are probably several reasons for this failure rate reduction, such as the users becoming more familiar with how to use the software and the user updating drives and other supporting software. The point is not to explore this interesting aspect of software reliability but rather to note that for software, values of α may be larger than for hardware.

As for ω, we sometimes let $\omega = N(1)$ and assume that at $t = 1$, the expected cumulative number of failures is the predicted number for the beginning of deployment. If we have an initial MTBF M_I and an MTBF goal of M_G, the time T required to grow the MTBF from M_I to M_G is

$$T = \left(\frac{M_G}{M_I} \right)^{1/\alpha}$$

Similarly, if we have T, M_I, and M_G, then

$$\alpha = \frac{\ln M_G - \ln M_I}{\ln T} \tag{6.6}$$

Note however that values of α found in this way may not be realistically achievable. See [12] for more on this approach. Also, see Example 6.4.9.

3. Army Materiel Systems Analysis Activity (AMSAA) PM2 model: The AMSAA Planning Model based on Projection Methodology (PM2) model is described in [45]. It assumes that there are a large number of independent failure modes, that each failure mode is exponentially distributed, that the system fails when a failure mode occurs, and that new failures are not introduced by corrective actions. It also assumes that the software operation is representative of the OP. The model is designed for planning and tracking reliability growth activities, but we can use its growth curves to estimate the software failure rate as a function of time after deployment as defects are found and corrected. Quantities used for the model are:

(a) λ_A: Initial failure rate for failures that are not addressed with corrective actions
(b) λ_B: Initial failure rate for failures that are addressed with corrective actions
(c) μ_d: Planned average fix effectiveness factor (FEF), where the FEF is the reduction in an individual initial mode failure rate because of the implementation of a corrective action
(d) *MS*: Management strategy (MS), the fraction of the initial failure rate that is addressed with corrective actions
(e) T: Duration of reliability growth activities
(f) M_I: Planned initial mean time between failures (MTBF)
(g) M_G: Goal mean time between failures (MTBF)

The model assumes that some failure modes will not be addressed for corrective actions and groups the failure rates for these into λ_A. (Corrective actions are design or procedural changes that prevent or reduce the likelihood or impact of the failure mode in future operation of the software.) The failure rates for failure modes that are addressed with corrective actions are grouped into λ_B. The planned average FEF μ_d accounts for the fact that fixes do not always eliminate the failure mode but may simply reduce its likelihood. The management strategy MS is defined by

$$MS = \frac{\lambda_B}{\lambda_A + \lambda_B}$$

and is the fraction of the initial failure rate that the project will address with corrective actions. Note that $\lambda_i = \lambda_A + \lambda_B$ is the initial failure rate, which in this case is the failure rate predicted for deployment, and $M_I = 1/\lambda_i$. Often, M_G is chosen based on the required MTBF, M_R. For example, we can use

$$M_G = \frac{M_R}{0.9}$$

meaning that our goal is to achieve an MTBF that is approximately 11% higher than the requirement. A similar factor is typically used when working with empirical results obtained in the development test (DT) environment. The DT environment usually cannot anticipate all of the things that will occur in an operational test (OT) environment. Because of this, there is also often a DT/OT degradation factor to account for differences in going from the DT environment to the OT environment. For example, if an MTBF of 1000 hours is desired in the operational environment and a 10% degradation factor is assumed, then a DT goal of $1000/(1 - 0.1) = 1111$ is used.

The expected failure rate for this model is given by

$$\lambda(t) = \lambda_A + (1 - \mu_d)(\lambda_B - h(t)) + h(t)$$

where

$$h(t) = \frac{\lambda_B}{1 + \beta t}$$

and

$$\beta = \left(\frac{1}{T}\right)\left(\frac{1 - \frac{M_I}{M_G}}{MS\,\mu_d - \left(1 - \frac{M_I}{M_G}\right)}\right)$$

The planned MTBF value at time t is $\text{MTBF}_{PL}(t) = 1/\lambda(t)$. The growth potential M_{GP} is given by

$$\lim_{t \to \infty} \text{MTBF}_{PL}(t) = M_{GP} = \frac{M_I}{1 - MS\,\mu_d}$$

An example use of this approach is found in Example 6.4.9.

These values are used to predict failure rates at different times in the project, such as when testing begins, at deployment, or at a certain number of years after deployment. These predicted values are in turn used to predict the reliability and availability of the software at these times. Techniques for these calculations are covered below.

At the end of this step, we have a prediction of the number of software defects found as a function of usage time.

Step 5: Predict the failure rate and MTBF for the software
After step 4, we have a prediction of the occurrence of defects over usage time, but this prediction does not fully take into account how often the software will be used. To do this, we need to calculate the monthly duty cycle (in hours per month) for each software LRU and divide the number of faults predicted for that month by the duty cycle.

The duty cycle for a software LRU is specific to the project and to the end user of the software. Some software is used nearly constantly. Other software is only used during normal work hours. Still, other software is only used if specific events occur. Example 6.4.8 provides more on how to work with duty cycles. See [25] for more on duty cycle estimation. Once we have the duty cycle for a particular software LRU, we find the failure rate (as a function of month but expressed in failures per hour) by dividing the predicted number of faults for that month (from step 4) by the monthly duty cycle.

Given the failure rate λ of a software LRU, we find its MTBF by MTBF $= 1/\lambda$. To find the MTBCF, divide the MTBF by the fraction of failures that are expected to be critical. Ideally, this factor is based on historical information from similar software, although sometimes industry averages are used. MTBSA and MTBEFF are found in the same way. If historical data sets are not available, an SFMEA can be helpful in determining these factors. More details about SFMEA can be found in the Software FMEA/FMECA topic in this section.

Step 6: Predict the reliability and availability of the software
The reliability of a software LRU is the probability that the LRU will perform its intended functions for a specified time interval under stated conditions. For a given software LRU, let

$$\lambda_c = \frac{1}{\text{MTBCF}}$$

then the reliability of the LRU is given by

$$R(t) = \exp\left(-\lambda_c\, t\right)$$

where t is the operation time. If λ_c is critical failures per hour, then t is in hours. As the predicted number of defects may be changing each month, the MTBCF may also be a function of month. Note that this definition considers the software to be reliable even if there are failures provided that these failures are not critical.

To predict the availability of the software LRU, we need to find the MTSWR. The value to use for MTSWR is dependent on the LRU but should consider restart and reboots times, work around times, and other times needed to end the "unacceptable" behavior and return to proper behavior. Reference [25] contains more information on calculating MTSWR.

Once we have a value for MTSWR, we calculate the availability of the software LRU using

$$A = \frac{\text{MTBCF}}{\text{MTBCF} + \text{MTSWR}}$$

This calculation assumes that only critical failures become availability issues. If other types of failures cause availability issues, then MTBCF should be replaced by a value that includes these failures.

To predict the reliability or availability of the software system, it is usually sufficient to use RBDs and combine the LRU values in series, in parallel, or a combination of these.

Most of the time, software LRUs are in series. If the final product has both hardware and software, consider what hardware will host the software, and if that software fails, how it will affect the entire system. For example, the system may have three computer hardware systems running in parallel. From a hardware prospective, the three computers are in parallel. However, if the three computers are all running the same software essentially at the same time, it is unlikely that we can justify assuming that the three instances of software are also in parallel.

See [25] for more on software reliability and availability predictions. Software reliability can also be measured for commercial mass market software products based on user experiences. For more details, see [46]. Bayesian techniques have become more accepted and can provide useful additional information to predictions. See [26] for an overview of Bayesian techniques for reliability predictions.

Step 7: Check the results
There are several ways to check the results. First, go through the work performed and look for defects in the work. Peer reviews are a good technique for this effort. Take particular note of software size estimates, reliability growth estimates, reliability risks, and parameters used to estimate the software defect density. (Table 29 of [25] provides more details on these considerations.) Consider applying a different but equally applicable model and see if the results are similar. Drastically different results (relative to the accuracy required by the project) indicate that a deeper investigation is needed. Also check to see if the predictions are in line with any applicable historical data. Finally, [25] provides a useful table (Table 25) of typical MTBF values based on the size of the team writing the code. Additionally, Table 39 in the same reference provides average defect density values per normalized EKSLOC for three time periods: at three years of fielded operation, at system testing, and at integration testing.

Example 6.4.7 In this example, we want to predict the number of defects in a given piece of software. This software has six software LRUs with characteristics as listed in Table 6.16.

To predict the total number of defects for the software, we perform the following steps:

1. Choose a software prediction model for each of the software LRUs
2. Calculate a defect density for each LRU

Table 6.16 Characteristics of the Software LRUs.

LRU Number	KSLOC	Software Type	Application	Language
LRU1	71.0	New	Commercial	C
LRU2	527.0	Autogenerated	Commercial	C
LRU3	6.5	Moderate modification	Commercial	C
LRU4	28.0	New	Commercial	C
LRU5	5.1	New	Device	C++
LRU6	7.0	New	Device	C++

3. Calculate a normalized EKSLOC for each LRU
4. For each LRU, multiply the defect density by the normalized EKSLOC to give the predicted number of defects in the LRU
5. Add the number of defects for the LRUs, resulting in the total number of defects in the software

In this example, the company has a software reliability prediction model (based on historical data) that is applicable for the first four LRUs. The company software reliability prediction model predicts a defect density at a time of delivery of 0.251 defects per normalized EKSLOC for newly developed code.

For the fifth and sixth LRUs, they use the shortcut model and the application-type model, both from [25]. With the shortcut model, there are 15 questions on strengths and 7 on risks. These questions are listed in Table 47 of [25]. For each question, answer either "yes" or "no." Questions 5, 12, and 13 on strengths allow an answer of "somewhat," as do questions 4 and 5 on risks. Score a 1 for each "yes" answer, and a 0.5 for each "somewhat." Add the number of points from the strength questions and likewise for the risk questions and subtract the total risk score from the total strengths score. If the resulting score is greater than or equal to 4.0, the predicted defect density per normalized EKSLOC is 0.110. If the score is less than or equal to 0.5, the predicted defect density per normalized EKSLOC is 0.647. Otherwise, the predicted value is 0.239. Both LRUs 5 and 6 have a combined score of 2.0 and so have a predicted defect density of 0.239 per normalized EKSLOC from this prediction method.

For the application-type model, the software in question is classified into one of the application types listed in Table 48 or Table 49 of [25]. Both of these tables predict defect density based on what is typical for software of that application type. Table 48 of [25] gives predicted operational defect density and predicted testing defect density per normalized EKSLOC. Table 49 of [25] gives the predicted defect density per function point at delivery. Both LRUs 5 and 6 are of application-type "device" and so per Table 48 of [25], they have an operational defect density of 0.338 ± 0.259 per normalized EKSLOC. The shortcut method gives a prediction that is well within the range of the application-type method. The company uses a value of 0.338 defects per normalized EKSLOC to be slightly conservative. Table 6.17 shows the predicted defect density values per normalized EKSLOC for each of the software LRUs.

Table 6.17 Predicted Defect Density Values for Software LRUs.

LRU Number	Defect Density
LRU1	0.251
LRU2	0.251
LRU3	0.251
LRU4	0.251
LRU5	0.338
LRU6	0.338

Next, we find the normalized EKSLOC for each LRU. The company has found that auto-generated code has a defect density of <0.02 of that of new code. Based on the size and complexity of the modifications made to LRU 3, it is estimated that its defect density will be 0.35 of that of new code. The EKSLOC for the first four LRUs is therefore

$$EKSLOC = (1 \times 71.0) + (0.02 \times 527.0) + (0.35 \times 6.5) + (1 \times 28.0) = 111.8$$

For LRUs 5 and 6,

$$EKSLOC = (1 \times 5.1) + (1 \times 7.0) = 12.1$$

Finally, we must normalize the EKSLOC by taking into account the language used. For LRUs 1–4, the language C is used. Per Table B.2 of [25], this language has a normalization conversion of 3.0. LRUs 5 and 6 are in C++, which by the same table has a language conversion of 6.0. Therefore, the normalized EKSLOC for the first four LRUs is $3.0 \times 111.8 = 335.4$ and for the fifth and sixth LRUs, $6.0 \times 12.1 = 72.6$.

Multiplying the defect density of the first four LRUs by the normalized EKSLOC for these four LRUs gives $0.251 \times 335.4 = 84.2$ defects. For the fifth and sixth LRUs, $0.338 \times 72.6 = 24.5$ defects. Adding these together (and rounding up) results in a prediction of 109 defects in the operational software at the time of delivery.

Example 6.4.8   This example extends the previous one by taking the prediction of the number of defects in the software at the time of delivery and finding the reliability of the software. We do this by first predicting the defect discovery rate for these defects as a function of time. We can use this information to predict a software failure rate and then the MTBCF and finally the reliability. Recall that we described three ways of performing this task. We can use the exponential method, the Duane method, or the PM2 method. All of these methods assume that most defects are corrected as they are found and so reliability is improving. In this example, we use the exponential method.

This example starts with step 4, the Predict defects over time step, and uses the exponential method of predicting defects as a function of time. The exponential method uses an exponential function of the form

$$N(t) = N \left[1 - \exp(-\lambda\, t)\right]$$

where N is the initial number of defects, λ is the defect detection rate, and $N(t)$ is the number of defects at time t. To use this, we predict the number of defects detected per month using

$$N_i = N \left[\exp\left(-\frac{Q\,(i-1)}{TF}\right) - \exp\left(-\frac{Q\,i}{TF}\right)\right]$$

where

1. i is the index denoting the month after software delivery.
2. TF is the number of months that the software in under operation before defect plateau. i ranges for $i = 1$ to $i = TF$. The default value for TF is 48 months.
3. N is the predicted number of defects in the software at delivery.
4. N_i is the number of detected defects for the month i.
5. Q is the growth rate. Q typically ranges from a value of 3 to a value of 10 and has a default value of 4.5. Table 51 of [25] provides more guidance.

Table 6.18 Predicted Defects, Failure Rate, MTBF, MTBCF, and Reliability.

Month	Defects	Failure Rate	MTBF	MTBCF	Reliability
1	9.75	0.0448	22.31	223.1	0.9648
10	4.20	0.0193	51.79	517.9	0.9847
20	1.64	0.0075	132.62	1326.2	0.9940
30	0.64	0.0029	339.84	3398.4	0.9976

Continuing with our example, $N = 109$ defects and we used the default values $TF = 48$ and $Q = 4.5$. As a result,

$$N_i = 109 \left[\exp(-0.09375\,(i-1)) - \exp(-0.09375\,i) \right]$$

Using this approach, we expect to find 9.75 defects in the first month, 4.20 in the tenth month, 1.64 in month 20, and 0.64 in month 30 as shown in Table 6.18.

Next, use step 5 to find the software duty cycle and use it to find the software failure rate and MTBF. For the software in this example, all LRUs are used every time that the software as a whole is used. The software is used by the office of a company, and in this office, people use the software from 7:00 a.m. to 5:00 p.m. Monday through Friday for a total of 50 hours a week. There is an average of 365.25 days per year, $52.18 = 365.25/7$ weeks per year, and $4.35 = 52.18/12$ weeks per month. At 50 hours per week, the average number of hours of software use per month is therefore $50 \times 4.35 = 217.5$ hours per month.

Using this duty cycle, we find the predicted failure rate by dividing the number of defects that are predicted to occur in a given month by the duty cycle. The predicted failure rate for the first month is $9.75/217.5 = 0.0448$ failures per hour and the MTBF is $217.5/9.75 = 22.31$ hours per failure. Table 6.18 shows the other failure rate results.

Not all failures are critical enough to affect the reliability of the software. We account for this in step 6. If we have historical information on failures for similar software, or have a software FMEA with criticality information, we can estimate the fraction of defects that result in a critical failure. For this example, we use a factor of 0.1. Using this value, $MTBCF = MTBF/0.1$. Finally, we can find the reliability for a given time period. For this example, we use an 8-hour time period for reliability. To find the reliability, we use

$$R(8) = \exp\left(-\frac{8}{MTBCF}\right)$$

The results are shown in Table 6.18.

Example 6.4.9 In the previous example, we used the exponential method to predict the discovery of defects in time and used this information to predict the failure rate, MTBF, MTBCF, and reliability of the software. Although the exponential method is probably the most commonly used method, other methods can be used, such as the Duane method or the PM2 method.

In this example, we again use the results of Example 6.4.7, and we also use the initial failure rate found in Example 6.4.8. We then use both the Duane method and the PM2 method to predict the failure rate as a function of time.

Recall for the Duane method,

$$\lambda(t) = \omega \, (1 - \alpha) \, t^{-\alpha}$$

where $\lambda(t)$ is the failure rate as a function of time, ω is the scale constant, and α is the shape constant. For this example, we assume that we do not have historical values for ω and α that are applicable. One possibility around this is to use the growth rate from Example 6.4.8. In that example, at month 1, we have an MTBF of 22.31, and at month 30, it has grown to 339.84. Using Eq. 6.6,

$$\alpha = \frac{\ln M_G - \ln M_I}{\ln T}$$

we find that for this growth rate, we need $\alpha = 0.8$, which represents rapid growth but may not be unreasonable for widely distributed commercial software. However, for this example, we assume that an MTBF goal at 48 months is 66.67 hours. This MTBF goal is found from an assumed requirement of 60 hours, which is then divided by 0.9 to produce a goal value. Then, $\alpha = 0.2828$ and the MTBF at 30 hours is 58.38 hours. Here, we have used Eq. 6.5 evaluated at $T = 30$ hours. Again using a ratio of critical failures to total failures of 0.1, the MTBCF value is 583.8 hours and the 8-hour reliability is $R(8) = 0.9864$.

We can use the PM2 method in a similar way. We start with the initial failure rate found in Example 6.4.8 and use the following parameters for the PM2 model:

1. $\mu_d = 0.9$: μ_d is the average FEF and accounts for the fact that not all attempts to fix a defect will completely remove the associated failure rate. Some fixes only remove some of the failure rate. Also with software, removing a defect may add another defect and μ_d is also a way to account for this condition.
2. $MS = 0.95$: MS is the fraction of discovered defects that the company will attempt to remove. Some defects may be too difficult or too low a priority to remove, or perhaps there is not enough schedule and the defect will be removed later.
3. $T = 48$: T is the duration of reliability growth activities.
4. $\lambda_I = 0.0448$: λ_I is the initial failure rate, and we use the value found for $\lambda(1)$ in Example 6.4.8.
5. $M_I = 22.31$: $M_I = 1/\lambda_I$.
6. $M_R = 60$: M_R is the required MTBF.
7. $M_G = M_R/0.9 = 66.67$: M_G is the goal MTBF.

With these values, $\lambda_A = (1 - MS)\,(\lambda_I) = 0.0022$ and $\lambda_B = (MS)\,(\lambda_I) = 0.0426$. Using

$$\beta = \left(\frac{1}{T}\right) \left(\frac{1 - \frac{M_I}{M_G}}{MS\,\mu_d - \left(1 - \frac{M_I}{M_G}\right)} \right)$$

we have $\beta = 0.0731$ and from

$$h(t) = \frac{\lambda_B}{1 + \beta t}$$

at $t = 30$ months, $h(30) = 0.0133$, so using

$$\lambda(t) = \lambda_A + (1 - \mu_d)(\lambda_B - h(t)) + h(t)$$

we have at $t = 30$ months, MTBF $= 54.35$, MTBCF $= 54.35/0.1 = 543.5$, and $R(8) = 0.9854$.

Software Reliability Casebook

Whereas a Reliability Program Plan (RPP) details the reliability activities to be undertaken by the project and success criteria for these activities, a Reliability Casebook documents evidence of how well the reliability activities have been performed and how well the reliability goals and requirements have been met at each phase of the project. A Software Reliability Casebook provides a continuous assessment, documenting the project's understanding of the software reliability requirements and providing a well-reasoned and auditable argument that the product meets, or does not meet, these requirements. This argument is incremental, building with each project phase. A Reliability Casebook typically has the following sections:

1. Product description: This section includes narratives, product quantities and identifiers, block diagrams, and other descriptions.
2. Reliability requirements: This section includes a statement of the reliability requirements, their rationales, progress made toward achieving the requirements, and current predictions and estimates of them.
3. Risk areas: Current risks to meeting the reliability requirements are listed here, along with their severities and probabilities. In addition, steps to alleviate the risks are addressed.
4. Planned reliability activities: Provide a brief description of planned reliability activities.
5. Performed reliability activities: Give a list of reliability activities that have been performed and an assessment of how adequate the results are.
6. Mitigations: Of those activities planned to have been performed but as yet have not been, or performed activities that show indications of inadequate results, state if mitigations have been performed and, if so, if they are adequate.
7. Results of analyses: Results of reliability and reliability-related analysis, as well as interpretations and implications of the results, are included here.
8. Results of tests: Results of reliability and reliability-related tests, as well as interpretation and implications of the results, are included here.

The RPP and Reliability Casebook work together with the RPP telling what is to be done and what success is and the Casebook explaining and providing evidence for what actually happened, why it happened, and how it relates to the reliability requirements. See [47] and [26] for more details on Reliability Casebooks. Software reliability program plans (SRPPs) are covered in more detail later in this section.

Software Reliability Estimations

Software reliability estimations require the existence of executable software. Good software reliability estimation models use numerous test runs of production-representative software to produce statistically significant estimations of the reliability of the software. Here, "production-representative" means that the software product used in the test runs is essentially the same software that will be delivered to the customer and that the tests are run on the same type of hardware and in the same environment that the customer is expected to have and it is run with the same OP that the customer will use. The greater the deviation from this ideal, the less confidence that we can have in the results of the testing and estimations. Such an estimate, properly produced, can be used to estimate the reliability of the software and help determine if it is ready for release. Software reliability estimation models can be more accurate than those that predict software reliability without executable code, but as these models require mature operating software, the results are only available in later project phases.

For reliability testing, we have the following general steps:

1. Establish software reliability goals or requirements: Reliability testing is expensive and can be time consuming. Without reliability goals or requirements, it may be unclear how much reliability testing is required. For example, a reliability estimate must have a confidence level associated with it, and obtaining a result with a 99% confidence requires significantly more test runs than obtaining one with an 80% confidence.
2. Develop operational profiles: The results of the reliability testing should enable us to estimate the reliability that the user will experience with the product. As a result, reliability tests should be conducted as though the product is being operated by the user in the user's environment. To do this, OPs are developed giving a set of relative frequencies (or probabilities) of occurrences of disjoint software operations during operational use. These, along with the target hardware using the same operating system, memory, drivers, load from other users, and other background processes that are expected in real-world operations, should be used for the reliability tests. Again, the further the tests deviate from these ideals, the less confidence we can have in the results of the testing.
3. Plan and execute the tests: Reliability tests should be planned based on the requirements and OPs. The number of runs and durations of these runs should be determined to ensure statistically accurate results. Sometimes, data from multiple tests can be combined to give a larger data set; however, care must be taken to do this productively. See [26] for more on this task.
4. Analyze the test results: After a sufficient test duration, reliability parameters and their associated confidence bounds can be estimated. If the bounds are not sufficient, a longer duration may be called for or perhaps correction of defects and retesting is in order.

   Note that reliability testing is often combined with reliability growth testing per the SRGP, meaning that a round of testing is performed and defects are found. After analysis, some or all of these defects are addressed for corrections and another round of testing is performed. This process continues until a pre-determined end point is reached.

More on software testing is covered in the software testing topic in Section 6.3.

The typical process for software reliability estimation is to collect fault information from suitable software runs, fit the data to a software reliability model curve, and extrapolate to a future time. The following steps are useful in this effort:

1. Develop reliability objectives and operational profiles
2. Create reliability tests
3. Measure test coverage
4. Collect fault data
5. Choose a software reliability model
6. Apply the model to obtain results
7. Assess the accuracy of the results
8. Validate the reliability in the field

These are covered in more detail below.

Step 1: Develop reliability objectives and operational profiles
Software testing can be expensive and time-consuming and should be well planned. Testing to estimate software faults or software reliability requires knowing the reliability objectives to determine how much testing is required. It also requires an approved OP as the number of faults uncovered while running the software is dependent on how the software is run. Accurate results require that the software be run per the OP or profiles. More on these subject matters may be found in the OPs and Software Reliability and Availability Allocations topics in this section and the software testing topic in Section 6.3.

Step 2: Create reliability tests
As noted in the previous step, reliability tests need to be carefully planned. The software testing topic found in Section 6.3 provides information to help plan the duration of such a test. Ideally, software tests used for assessing software reliability are performed based on the software OP. (The OP is covered in more detail in its own topic in this section.) Unfortunately, this is not always feasible. One approach to accommodating this issue is to use an empirically derived test compression factor. A test compression factor is the ratio of the MTBF during operations to the MTBF during testing. More on software reliability testing may be found in [12, 17, 25, 28], and the software testing topic.

Step 3: Measure test coverage
In addition to testing per the OP, test coverage is important. To be comfortable with the results of our fault estimation, we should ensure that we have tested a significant portion of the code. There are different ways of measuring test coverage, such as statement coverage, decision coverage, and condition coverage. Also, 100% coverage does not ensure defect-free software. Some faults only occur after an unusual combination of conditions, and this combination may not be realized in a given test suite. Although high test coverage does not ensure high reliability, it is an important part of it. It is recommended to set some minimum acceptable test coverage value and to measure test coverage to ensure that this minimum is met before software release. More on software testing can be found in the software testing topic in Section 6.3.

Step 4: Collect and analyze fault data
Accurate collection of fault data is critical for accurate reliability estimations and extrapolations. Unfortunately, collection of fault data is too often performed by untrained personnel or not given the priority that it deserves, and the resultant collection may make any accurate quantification of software faults impossible. Software reliability estimations generally use the following:

1. The number of non-cumulative faults occurring in a given time period
2. The number of non-cumulative operating or CPU hours occurring in a given time period
3. The number of cumulative faults occurring in a given time period
4. The number of cumulative operating or CPU hours occurring in a given time period

Note that the cumulative values may be calculated from the non-cumulative values. Fault data should be collected and processed for correctness and accuracy in the DRACAS and DRB. More on this step is covered under the DRACAS, DRB topic in this section.

Step 5: Choose a software reliability model

After collecting fault data, we can choose a software reliability model. There are many models to choose from. A few examples are listed below. For each of these models, $m(t)$ is the expected number of failures detected by time t. The expected number of failures per unit time at time t, $\lambda(t)$, is called the failure intensity and may be found from

$$\lambda(t) = \frac{d}{dt}\, m(t)$$

1. Jelinski–Moranda model: The Jelinski–Moranda software reliability growth model is a simple and frequently used model. Some of the assumptions for this model are as follows:
   (a) At the start of testing, the software has an unknown but fixed number of defects in it.
   (b) Failures are not correlated.
   (c) Each defect is equally likely to cause a failure.
   (d) The times between failures are independent and exponentially distributed.
   (e) The software is being used in a manner similar to the way that the end user will use it.
   (f) The rate of failure detection is decreasing or stabilized.
   (g) There are no defects masking other defects.
   (h) Faults are corrected immediately and fault correction does not introduce any new faults.

   The Jelinski–Moranda model uses failure times or times between failures. The formula for $m(t)$ for this model is as follows:

   $$m(t) = a\left(1 - e^{-b\,t}\right)$$

   where a, $b > 0$. The parameter a is the initial number of faults in the code before testing and b is the failure detection rate. The curve is concave and exponential and does not generally provide a good fit to S-shaped growth situations. See [48] for more details.

2. Goel–Okumoto model: This model results in the same formula for $m(t)$ as the Jelinski–Moranda model but under different assumptions and with a different interpretation. The Goel–Okumoto model is a non-homogeneous Poisson process (NHPP) model that assumes the following:
   (a) The number of failures experienced by time t follows an NHPP.
   (b) The number of failures that occur in the time interval $(t,\ t + \Delta t]$ as $\Delta t \to 0$ is proportional to $a - m(t)$, where a is the initial number of faults in the software.
   (c) The numbers of failures collected in any of the disjoint collection periods $(0,\ t_1)$, $(t_1,\ t_2)$, ..., $(t_{n-1},\ t_n)$ are independent.
   (d) The software is being used in a manner similar to the way that the end user will use it.
   (e) The rate of failure detection is decreasing or stabilized.
   (f) There are no defects masking other defects.
   (g) When a fault is found, it is fixed instantaneously and without introducing any new faults.

The Goel–Okumoto model uses the number of failures observed in a time interval. The formula for $m(t)$ for this model is

$$m(t) = a\left(1 - e^{-b\,t}\right)$$

where a, $b > 0$. The parameter a is the expected initial number of faults in the code prior to testing and b is the fault detection rate. There is a version of this model that has been modified to account for imperfect debugging. For this modified model, $m(t)$ is given by

$$m(t) = \frac{a}{p}\left(1 - e^{-b\,p\,t}\right)$$

where p is the probability that a detected fault is successfully removed. See [49] for more details.

3. Duane model: The Duane reliability growth curve has been used for hardware reliability since the practice of reliability growth modeling began. The model has also been used for software reliability, typically when the failure rate is decreasing linearly. Some of the assumptions for the model are as follows:

   (a) The cumulative number of defects found by time t follows an NHPP with mean function $m(t)$.
   (b) When a fault is found, it is fixed instantaneously and without introducing any new faults.
   (c) There are an infinite number of defects in the software.
   (d) The software is being used in a manner similar to the way that the end user will use it.

   The Duane model uses failure times or times between failures. The expected number of failures detected by time t, $m(t)$, is given by

   $$m(t) = a\,t^b$$

   where a, $b > 0$. The parameter a is the expected total number of faults at $t = 1$ and b is the rate of reliability growth. See [43] for more details.

4. Musa basic model: The Musa basic model is another frequently used software reliability growth model. Some of the assumptions for the Musa basic model are as follows:

   (a) The cumulative number of failures by time t follows an NHPP with a mean function $m(t)$.
   (b) There are a finite number of defects in the software.
   (c) The faults in the software are independent of each other.
   (d) The execution times between failures are piecewise exponentially distributed.
   (e) The software is being used in a manner similar to the way that the end user will use it.
   (f) Resources such as fault identification and fault correction personnel are constant over the software observation time.
   (g) Computer utilization is constant.
   (h) The rate of failure detection is decreasing or stabilized.
   (i) There are no defects masking other defects.

   The Musa basic model uses either the actual times of software failures or the elapsed times between failures. The expected number of failures detected by time t, $m(t)$, is given by

$$m(t) = a \left(1 - e^{-b\,t}\right)$$

where $a, b > 0$. Parameter a is the limiting number of faults as t approaches infinity and b relates to the fault detection rate. See [50] for more details.

5. Log-logistics model: The log-logistics model was developed to capture failure trends that many models do not handle well. For example, this model can be used when the failure rate is decreasing nonlinearly. The expected number of failures detected by time t, $m(t)$, is given by

$$m(t) = a \frac{(\lambda\,t)^c}{1 + (\lambda\,t)^c}$$

where $a, \lambda, c > 0$. The parameter a is the expected initial number of defects in the software, λ is related to the failure rate, and c is a shape parameter. See [51] for more details.

6. Weibull model: The Weibull model, also called the generalized Goel Non-Homogeneous Poisson Point process model, can be used regardless of the failure rate trend. Several assumptions for this model are as follows:
   (a) There is an unknown but fixed expected number of defects initially in the software.
   (b) The failure rate function has the shape of a Weibull probability density function.
   (c) Defects do not necessarily cause failures with the same detection rate.
   (d) Software is tested in a manner similar to how the software will be used operationally.
   (e) Defects are fixed immediately without causing other defects.
   The expected number of failures detected by time t, $m(t)$, is given by

$$m(t) = a \left(1 - e^{-bt^c}\right)$$

where $a, b, c > 0$. The parameter a is the estimated total number of defects in the software and b and c are shape parameters that reflect the severity of the testing of the software. See [52] for more details.

7. Musa–Okumoto model: The Musa–Okumoto model considers the fact that the reduction of the failure rate due to the correction of a defect may be greater earlier in the testing than later because these earlier defects tend to occur more than once. This model can be used when the failure rate is decreasing nonlinearly. Some of the assumptions made by the Musa–Okumoto model are as follows:
   (a) Failures are independent of each other.
   (b) The software is tested in a manner that is similar to how the software will be used operationally.
   (c) The failure rate decreases exponentially with the expected number of failures experienced. This model should not be used if the failure rate is increasing.
   (d) The cumulative number of failures found by time t follows an NHPP.
   (e) Defects are fixed immediately without causing other defects.
   The model requires either actual failure times or elapsed times between failures. The expected number of failures detected by time t, $m(t)$, is given by

$$m(t) = a \, \ln(1 + b\,t)$$

where a, $b > 0$. Parameter a is one over the initial failure rate and b is the fault rate decay parameter. See [53] for more details.

8. Yamada delayed S-shaped model: The Yamada delayed S-shaped model is one of the first models designed to fit S-shaped fault detection processes. The model takes into consideration that programmers may improve their debugging as development progresses so that at the beginning of the testing, the fault detection is fairly flat but then rapidly increases. Later, it levels off as defects become more difficult to find. The model can be used on data where the fault rate is increasing and then later starts to decrease. Some of the assumptions of this model are as follows:
   (a) Failures occur randomly and are independent.
   (b) The software is tested in a manner similar to how the software will be used operationally.
   (c) The initial number of defects in the software is a random variable.
   (d) Defects are fixed immediately without causing other defects.
   (e) The time between the $(i-1)$th failure and the ith failure depends on the time to the $(i-1)$th failure.

   The expected number of failures detected by time t, $m(t)$, is given by

   $$m(t) = a \left[1 - (1 + bt)\, e^{-bt}\right]$$

   where a, $b > 0$. The parameter a is expected total number of faults in the software and b is the fault detection rate. This model uses times of occurrences of failures. See [54] for more details.

9. Gompertz growth curve model: The Gompertz model is a simple model with an S-shaped curve. The model assumes perfect debugging. The mean value function for this model is given by

   $$m(t) = a\, b^{c^t}$$

   where $a > 0$, $0 < b < 1$, and $0 < c < 1$. The parameter a is the expected total number of faults, b is the rate at which the failures detection rate decreases, and c produces the growth pattern. A small value of c results in rapid reliability growth early in the project, whereas a larger value of c produces later reliability growth. See [55] for more details.

10. Inflected S-shaped model: The inflected S-shaped model considers the observation that software reliability growth tends to be S-shaped when faults in the software are mutually dependent, meaning that some faults are not detected until some others are removed. The model also assumes that defects are fixed immediately without causing other defects. The mean value function for this model is given by

   $$m(t) = a\, \frac{1 - e^{-bt}}{1 + \beta\, e^{-bt}}$$

   where a, b, $\beta > 0$. The parameter a is the expected total number of defects and b is related to the fault detection rate. The fault detection rate function is given by

   $$f(t) = \frac{b}{1 + \beta\, e^{-bt}}$$

   Sometimes, β is replaced by the function

   $$g(r) = \frac{1 - r}{r}$$

where $r > 0$. In this case, r is the inflection rate and indicates the ratio of detectable faults to the total number of faults in the software. The model assumes that the defect discovery rate increases throughout the test period. See [56] for more details.

11. Logistics growth curve model: The logistics growth curve model uses an *S*-shaped curve and assumes perfect debugging. Its mean value function $m(t)$ is given by

$$m(t) = \frac{a}{1 + k \, e^{-b \, t}}$$

with $a, \ b, \ k > 0$. The parameter a is the expected total number of faults in the software and k and b are fitting parameters. The inflection point for the curve occurs when

$$t = \frac{\ln(k)}{b}$$

and if $k > 1$, then this inflection time is greater than zero. See [57] and [58] for more details.

12. Modified Duane model: The Duane reliability growth model is mentioned above. One of the issues with the original Duane model is that, given continued effort, reliability growth continues indefinitely. In reality, reliability growth will level off. The modified Duane model addresses this issue. Like the original Duane model, the modified model assumes that defects are fixed immediately without causing other defects. For this model, the expected number of failures detected by time t, $m(t)$, is given by

$$m(t) = a \left[1 - \left(\frac{b}{b+t} \right)^c \right]$$

where $a, \ b, \ c > 0$. The parameter a is the expected total number of faults in the software and c is the rate of reliability growth. The parameter b is a shape parameter. See [59] for more details.

13. Yamada exponential imperfect debugging model: Many software reliability growth models assume that software defects, when found, are corrected immediately and that each correction is perfect, meaning that a correction completely corrects the defect and does not introduce new defects. This assumption may be adequate for some projects, but for many, it is not. The Yamada exponential imperfect debugging model assumes that the fault detection rate is proportional to the total number of faults in the software plus the faults introduced because of the imperfect fault correction. For this model, the expected number of failures detected by time t, $m(t)$, is given by

$$m(t) = ab \left(\frac{e^{ct} - e^{-bt}}{c + b} \right)$$

where $a, \ b, \ c > 0$. The parameter a is the total expected number of faults to be detected and b is the fault detection rate. The parameter c is the constant fault introduction rate because of fault corrections. Notice that if $c = 0$, then

$$m(t) = a \left(1 - e^{-bt} \right)$$

as expected. See [60] for more details.

14. Yamada linear imperfect debugging model: As with the Yamada exponential imperfect debugging model, the Yamada linear imperfect debugging model assumes that correcting faults can introduce new faults. The expected number of failures detected by time

t, $m(t)$, is given by

$$m(t) = a \left(1 - e^{-b\,t}\right)\left(1 - \frac{c}{b}\right) + a\,c\,t$$

where a, b, $c > 0$. Parameter a is the total expected number of faults to be detected and b is the fault detection rate. Parameter c is the fault introduction rate. Again if $c = 0$, then

$$m(t) = a\,\left(1 - e^{-bt}\right)$$

as expected. See [60] for more details.

15. Yamada Rayleigh model: Like the Yamada imperfect debugging models, the Yamada Rayleigh model assumes imperfect debugging. In addition, it attempts to account for testing effort, where the testing effort is modeled with a Rayleigh distribution curve. The overall model follows an S-shaped curve. The expected number of failures detected by time t, $m(t)$, is given by

$$m(t) = a\left[1 - \exp\left(-r\,\alpha\,\left[1 - \exp\left(-\frac{\beta\,t^2}{2}\right)\right]\right)\right]$$

where a, r, α, $\beta > 0$. The parameter a is the total expected number of faults to be detected. The fault detection rate is given by

$$f(t) = r\alpha\beta t\ \exp\left[-\frac{\beta\,t^2}{2}\right]$$

See [54] for more details.

16. Yamada exponential model: The Yamada exponential model, like the Yamada Rayleigh model, attempts to account for testing effort but assumes an exponential distribution rather than a Rayleigh distribution curve for the testing effort. The model assumes imperfect debugging. The expected number of failures detected by time t, $m(t)$, is given by

$$m(t) = a\,(1 - \exp[-r\alpha(1 - \exp[-\beta\,t])])$$

where a, r, α, $\beta > 0$. The parameter a is the total expected number of faults to be detected. The fault detection rate is given by

$$f(t) = r\,\alpha\,\beta\ \exp(-\beta\,t)$$

See [54] for more details.

17. Pham–Nordmann–Zhang model: The Pham–Nordmann–Zhang model assumes that the fault introduction due to debugging is a linear function of testing time and that the fault detection rate is a nondecreasing S-shaped function. More specifically, fault introduction is given by

$$f_1(t) = a(1 - \alpha t)$$

and the fault detection rate is given by

$$f_2(t) = \frac{b}{1 + \beta\,e^{-b\,t}}$$

The expected number of failures detected by time t, $m(t)$, is given by

$$m(t) = \frac{a\,(1 - e^{-b\,t})\left(1 - \frac{\alpha}{b}\right) + a\,\alpha\,t}{1 + \beta\,e^{-b\,t}}$$

where a, b, α, $\beta > 0$. Notice that if both α and β are equal to zero, then $f_1(t) = a, f_2(t) = b$, and

$$m(t) = a\left(1 - e^{-bt}\right)$$

as expected. See [61] for more details.

18. Pham–Zhang model: For the Pham–Zhang model, we assume that the fault introduction due to debugging is an exponential function of testing time and as with the Pham Nordmann–Zhang model, that the fault detection rate is a nondecreasing S-shaped function. For this model, fault introduction is given by

$$f_1(t) = c + a(1 - e^{-\alpha t})$$

and the fault detection rate is again given by

$$f_2(t) = \frac{b}{1 + \beta e^{-b t}}$$

The expected number of failures detected by time t, $m(t)$, is given by

$$m(t) = \frac{1}{\left(1 + \beta e^{-b t}\right)} \left((c + a)(1 - e^{-b t}) - \frac{ab}{b - \alpha}(e^{-\alpha t} - e^{-b t}) \right)$$

where a, b, c, α, $\beta > 0$. See [62] for more details.

19. Zhang–Teng–Pham model: In the Zhang–Teng–Pham model, like the Pham–Nordmann–Zhang and Pham–Zhang models, there is fault introduction due to debugging and there is a function for the fault detection rate. Some of the assumptions for this model are as follows:

(a) The occurrence of software failures follows an NHPP.
(b) At any given time, the failure rate is a function of both the fault detection rate and the current number of faults remaining in the software.
(c) Upon the occurrence of a failure, debugging begins immediately with probability p.
(d) There is a probability of β, $\beta \leq p$, that for any given debugging effort (successful or not), new faults will be introduced into the software.

Like the previous two models, the Zhang–Teng–Pham model assumes that the fault detection rate is a nondecreasing S-shaped function. In this case, we use

$$f(t) = \frac{c}{1 + \alpha e^{-b t}}$$

Note that in the previous two models, the numerator is b, whereas here, we use a potentially different value c. This change is so that the upper bound of the fault detection rate can be a different value than the learning curve increasing rate. The expected number of failures detected by time t, $m(t)$, is given by

$$m(t) = \frac{a}{p - \beta} \left(1 - \frac{(1 + \alpha) e^{-b t}}{1 + \alpha e^{-b t}} \right)^{\frac{c}{b} (p - \beta)}$$

where a, b, c, p, $\alpha > 0$, $\beta \geq 0$ The parameter a is the expected number of initial faults in the software, p is the fault removal efficiency, and β is the fault introduction probability. See [63] for more details.

More on these and other models may be found in references [16, 17, 19, 25], and [64].

Unfortunately, no model is best for all data sets, so care must be taken when choosing a model. The best model choice may be different even across different releases of the same software product. Several considerations when choosing which software reliability growth model to use are as follows:

1. The model should be suitable for the needs of the project.
2. The model should be applicable to the fault rate trend.
3. The model should fit the data well.
4. The model distribution should closely match the distribution of the data.
5. The model should show good predictive ability.
6. The confidence bounds of model should be acceptable.

We discuss each of these in more detail below.

1. Suitable for the Needs of the Project
First, make sure that the model provides the outputs that the projects needs and in general can support the project. Important information to know when choosing and using a model include:

1. Outputs for the model: Examples of outputs include estimated current and future failure rates, MTBF, and reliability values.
2. The form of the model: Each model uses an equation that describes the general shape that the failure occurrences (as a function of time) are expected to follow.
3. The assumptions for each model: Some of the assumptions may be violated without undue negative impacts, but the closer the assumptions are to being met, the better.
4. Know how to estimate the parameters for the model.
5. Know how to develop confidence bounds or goodness-of-fit estimates for the model.

As mentioned above, when choosing a software reliability model, pay careful attention to the assumptions made by the model. The assumptions may not need to be rigorously held, but significant deviations may make a given model too inaccurate for the current application. Several examples of software reliability growth model assumptions include the following:

1. The times between failures are independent: This assumption may approximately hold well enough for the model to be accurate, but it depends on the testing. For example, if several failures are due to a particular software module, there may be a decision to intensify testing of this module, resulting in shorter times between failures.
2. When detected, a fault is corrected immediately: Strictly speaking, this assumption rarely holds, but it is often the case that after a sequence of software tests, defects are identified and corrected, and the model estimations reflect this post-correction state of the software. However, sometimes, a fault is detected and the decision is made to correct it later. It is also possible that such an uncorrected fault prevents some software paths from being tested until the fault is corrected. In such cases, this assumption may not be acceptable.
3. Fault corrections are perfect: Some models assume that the number of defects never increases during the testing period. This assumption may be reasonable if no or few new features are being added and the defect correction process is rigorously controlled, but in the real world, fixes can introduce new defects. If historically, similar projects have

introduced a high number of new defects during defect correction activities, then this assumption may not be appropriate.

4. The failure rate is proportional to the number of defects remaining in the software: This assumption may be reasonable if the tests execute all portions of the code with equal probability. If, instead, some software paths are executed more frequently, defects in these paths are more likely to be found and defects in less frequently executed path may not be found.

5. The software is being tested in a manner similar to the way that the end user will use the software: This assumption is very common with software reliability growth models and often does not hold. Because of the expense of collecting reliability data, these models are sometimes applied to data collected from the wrong type of test. Also, a good OP may be difficult to develop. Sometimes, we simply do not know how the user will use the software.

Keep in mind that we are going to use the model to extrapolate the reliability of the software to sometime in the future. To be useful, a model should have a good fit to the currently available data, but a good fit does not guarantee that the model will successfully extrapolate the reliability to our desired time in the future. To increase confidence that the model will extrapolate accurately, we must make sure that the model is designed with assumptions suitable to our situation. These assumptions affect the shape of the extrapolated curve and therefore the accuracy of the extrapolation.

Another difference in software reliability models is the type of failure data that they use. Some models use failures per time period and others use the time between failures:

1. Failure times, failures per time period: With this type of failure data, the time of the occurrence of the failure is recorded. Sometimes, failures are recorded at the end of a time period, such as a work shift.

2. Time between failures: The time between failures, or inter-failure time, is the time between each recorded software failure.

Note that some models can use either type of data. Finally, there are also discrete time models that use test cases as the unit of fault detection and removal.

2. Applicable to the Fault Rate Trend

Different models fit the data to different curves, so the choice of a software reliability estimation model should depend on the data collected. Failure data from different projects may follow differently shaped curves, and so to accurately model the failures, plot the failure data to determine trends. Trend tests are covered in more detail in a separate topic in Section 6.5. There are several ways to plot the failure data. Reference [25] recommends calculating the cumulative number of faults observed through each point in time and dividing this value by the cumulative number of hours where CPU or operational time may be used. This calculated time quantity is plotted on the abscissa, or "*x*-axis," and the cumulative number of faults observed through each point in time is on the ordinate, or "*y*-axis." With such a plot, we can observe any of several cumulative fault rate trends:

1. Increasing fault rate (This is typical early in the testing phase.)

2. Peaking fault rate (About 40% of the faults should have been observed at this point, assuming that additional defects are not added through means such as software changes or faulty defect corrections.)

3. Increasing, then decreasing fault rate
4. Decreasing fault rate (Assuming that more defects are not introduced, expect the fault rate to decrease as the testing gets up to speed and defects are discovered and corrected.)

Some models assume a decreasing fault rate. Examples of these are the Musa basic, Goel–Okumoto, Jelinski–Moranda, Duane, Musa-Okumoto logarithmic, and log-logistic models. The Yamada S-shaped model allows for increasing and then decreasing fault rates, and the Weibull model can be used regardless of the fault rate trend. It is not unusual to have an increasing cumulative fault rate early in the testing, but if it is still increasing as the project approaches deployment, there may be issues.

It is also useful to plot the non-cumulative faults as a function of time (operating hours or CPU hours). This type of plot provides a different view of the data and may more readily show the typical Rayleigh curve shape of the fault occurrences with time.

Finally, a trend test, such as the Laplace trend test, can be used to statistically determine if there is a trend and what confidence we can assign to that determination.

3. Fit the Data Well
A software reliability growth model has parameters that must be determined. This determination is made by choosing parameters that produce the best fit of the model to the data. There are various ways to fit such a model to data. For example, the Duane model is often fit graphically by fitting a straight line to a plot of the logarithm of the test time versus the logarithm of the MTBF. However, the two most common ways for fitting software reliability growth models to data are the least mean squares method and the maximum likelihood estimate (MLA) method.

1. Least mean squares (LMS) method: The LMS method requires less data than the maximum likelihood estimate (MLE) method. It is computationally simple and provides a good starting point for other methods. The basic method is simple. Let $\{x_1, x_2, \ldots, x_n\}$ be the data set that we want to fit a model to and let $\{\hat{x}_1, \hat{x}_2, \ldots, \hat{x}_n\}$ be the estimated values from the model after using a particular set of model parameters. The indices $\{1, 2, \ldots, n\}$ typically indicate the times that the samples are taken. The LMS method minimizes the quantity

$$E = \sum_{i=1}^{n} \left(x_i - \hat{x}_i \right)^2$$

over a given set of parameter choices. The LMS method uses an algorithm to iteratively change the parameters of the model in such a way so as to find progressively smaller values of E as we iterate. This process is repeated until some criterion is met for being "good enough." There are numerous variations on this technique, but they all require minimizing E or a related quantity.
2. MLE method: The MLE method is probably the "method of choice" for fitting data to software reliability growth models. Two important reasons for this are that it provides more statistical information about the fit than the LMS method, and it also usually provides more precise parameter estimates. In the past, the MLE method had the disadvantage of being more computationally difficult but with modern computers, this is rarely an issue, at least for software reliability growth models.

The MLE method finds values for model parameters by maximizing the likelihood function. Suppose that the probability density function describing the distribution that the software reliability growth model assumes is $f(x_1, x_2, \ldots, x_n; \theta_1, \theta_2, \ldots, \theta_m)$, where there are n data samples x_1 through x_n and the model has m parameters θ_1 through θ_m. The likelihood function $L(\theta_1, \theta_2, \ldots, \theta_m; x_1, x_2, \ldots, x_n)$ is equal to this probability density function. In other words,

$$L(\theta_1, \theta_2, \ldots, \theta_m; x_1, x_2, \ldots, x_n) = f(x_1, x_2, \ldots, x_n; \theta_1, \theta_2, \ldots, \theta_m)$$

We can often assume that the data samples are independent so that

$$L(\theta_1, \theta_2, \ldots, \theta_m; x_1, x_2, \ldots, x_n) = f(x_1; \theta_1, \theta_2, \ldots, \theta_m)$$
$$\times f(x_2; \theta_1, \theta_2, \ldots, \theta_m)$$
$$\ldots \times f(x_n; \theta_1, \theta_2, \ldots, \theta_m).$$

The idea behind the MLE method is to find parameters θ_1 through θ_m that maximize this likelihood function. We want to find the parameters that make the data samples that we have most likely to occur for our model. It is typical to work with the natural logarithm of the likelihood function, ln L, rather than L. As the logarithm function is monotonically increasing, maximizing the logarithm of the likelihood function gives the same results as maximizing the likelihood function itself, but with the logarithm of the function,

$$\prod_{i=1}^{n} f(x_i; \theta_1, \theta_2, \ldots, \theta_m)$$

becomes

$$\sum_{i=1}^{n} \ln f(x_i; \theta_1, \theta_2, \ldots, \theta_m)$$

which is easier to work with.

To find the maximum, we simultaneously solve the following m equations:

$$\frac{\partial}{\partial \theta_1} \sum_{i=1}^{n} \ln f(x_i; \theta_1, \theta_2, \ldots, \theta_m) = 0,$$

$$\frac{\partial}{\partial \theta_2} \sum_{i=1}^{n} \ln f(x_i; \theta_1, \theta_2, \ldots, \theta_m) = 0$$

up to

$$\frac{\partial}{\partial \theta_m} \sum_{i=1}^{n} \ln f(x_i; \theta_1, \theta_2, \ldots, \theta_m) = 0$$

We find the values of $\theta_1, \ldots, \theta_m$ that simultaneously solve these equations and ensure that the solution is a maximum. For some systems, we can find explicit values for the parameters, but for others, we must use numerical techniques. See [17] for some examples.

Once a model has been fit to test data, we want to know how good the fit is. There are several measures of goodness for curve fitting. In [65], the following measures are considered:

1. Mean squared error (MSE): MSE is defined by

$$MSE = \frac{1}{n-N} \sum_{i=1}^{n} \left(y_i - \hat{m}(t_i)\right)^2$$

   where there are n observations and N parameters or coefficients. y_i is the total number of faults in testing data detected up to time t_i, and $\hat{m}(t_i)$ is the number of cumulative faults up to time t_i that the model estimates from $m(t)$. A smaller value of MSE indicates a better fit. Note that a model is penalized for requiring more parameters or coefficients.

2. Regression curve correlation index (R^2): Another popular measure of fit is the R^2 value. R^2 is found with the formula

$$R^2 = 1 - \frac{\sum_{i=1}^{n} \left(y_i - \hat{m}(t_i)\right)^2}{\sum_{i=1}^{n} \left(y_i - \bar{y}\right)^2}$$

   where

$$\bar{y} = \frac{1}{n} \sum_{i=1}^{n} y_i$$

   The closer R^2 is to a value of one, the better, but at least choose models that have $R^2 > 0.95$.

3. Adjusted R^2: The adjusted R^2 value is found using

$$\text{Adjusted } R^2 = 1 - \frac{(1 - R^2)\,(n - 1)}{n - k - 1}$$

   where R^2 is the value of R^2 described above and k is the number of predictors (independent variables or coefficients, excluding the constant) that we fit to for the model. The adjusted R^2 is similar to R^2 but penalizes for increasing the number of coefficients. A larger adjusted R^2 value means a better fit.

4. Predictive power (PP): The PP measure is defined as

$$PP = \sum_{i=1}^{n} \left(\frac{\hat{m}(t_i) - y_i}{y_i}\right)^2$$

   Smaller values of PP mean better fits.

5. Akaike's information criterion (AIC): Another fit metric is the AIC, found using

$$AIC = -2\ln[L] + 2k$$

   where L is the maximum value of the likelihood function and k is the number of parameters used in the model. Recall that we discussed the likelihood function above when discussing the MLE method for fitting the data to a curve. The AIC measures the ability of the model to maximize the likelihood function used for model fitting. To reduce the chances of overfitting, this criterion also penalizes the use of more parameters. A smaller value of AIC means a better fit.

There are many approaches for determining the goodness of a fit to data. Other choices may be found in [58].

4. Model Distribution Should Closely Match the Distribution of the Data
If applicable, apply multiple software reliability growth models to the test data to help assess
the accuracy of the results. Often when we have two (or more) candidate models, we want to
determine which of the two is most likely to be the best for our purposes. Three approaches
for this determination are as follows:

1. Sequential probability ratio test: One approach for choosing between two candidate
   models is to use a sequential probability ratio test (see [17] and [66]). The technique
   is referred to as the prequential likelihood test in [17]. Let f_0 and f_1 be the probability
   density functions of the inter-failure times $\{t_1, t_2, \ldots, t_n\}$ for the two software reliability
   models, Model 0 and Model 1, respectively. To perform the test, we define

   $$R_n = \prod_{j=1}^{n} \frac{f_1(t_j)}{f_0(t_j)}$$

   and choose constants $0 < A \leq 1 \leq B < \infty$, $A < B$, as described below. Let $I \in \{1, 2, \ldots\}$
   be the first number n such that $R_n \geq B$ or $R_n \leq A$. Then,
   (a) If $R_I \leq A$, accept hypothesis H_0, i.e. the sequence $\{t_j\}$ is governed by probability den-
       sity function f_0.
   (b) If $R_I \geq B$, accept hypothesis H_1 that the sequence is governed by probability density
       function f_1.
   If $A < R_I < B$, then no conclusion is reached, and the testing continues with the next
   datum.
   The above description leaves open the question of what to choose for A and B. Let
   hypothesis H_0 be that probability density function f_0 is our best choice and let hypothesis
   H_1 be that f_1 is best. Let α be the required probability of a type I error, i.e. the probability
   that the test will conclude that the result is hypothesis H_1 when it is really H_0. Also, let
   β be the probability of a type II error, in which we decide on hypothesis H_0 when the
   correct choice is really H_1. It can be shown that

   $$\alpha \leq \frac{1 - \beta}{B}$$

   and

   $$\beta \leq A(1 - \alpha)$$

   With this testing technique, we frequently assume that the inequalities are approximate
   equalities, resulting in

   $$\alpha \approx \frac{1 - A}{B - A}$$

   and

   $$\beta \approx \frac{A(B - 1)}{B - A}$$

   Equivalently,

   $$A \approx \frac{\beta}{1 - \alpha}$$

   and

   $$B \approx \frac{1 - \beta}{\alpha}$$

   These approximations are called *Wald's approximations*.

2. u-Plots: u-plots are another approach for choosing which of two candidate models best fits the distribution of the data. For a u-plot, we start with the fitted cumulative distribution function $\hat{F}_j(t_j)$, where t_j is the observed time to the next failure. Let the random variable T_j represent the time between the $(j-1)$st and jth failures. We would like to compare the distribution function $\hat{F}_j(t_j)$ with the true distribution function $F_j(t) = P(T_j \leq t)$, but this true distribution function F_j is unknown. However, we know that by applying the true distribution F_j to the random variables T_j, we have $U_j = F_j(T_j)$, where the random variable U_j is (by the probability integral transform) uniformly distributed over the interval $(0, 1)$. Therefore, we can create an empirical cumulative distribution function from the $u_j = \hat{F}_j(t_j)$ and the observed times t_j to the next failure. The plot of u_j is called a u-plot. We then compare this empirical distribution with a uniform distribution. The Kolmogorov–Smirnov test (see below) is often used to test the significance of deviations from uniform. See [17] for more on this technique.

3. Kolmogorov–Smirnov test: The Kolmogorov–Smirnov test can also be used to more directly test how good a probability distribution fits the data. The one-sample Kolmogorov–Smirnov test is used to compare the observed data with some specific continuous distribution. Suppose that the sample size is n and the cumulative distribution function that we wish to compare these observed data with is $F(x)$. If our observations truly do have a distribution of $F(x)$, we can approximate $F(x)$ by an empirical cumulative distribution function $F_n(x)$. To do this, order the data samples $\{x_1, x_2, \ldots, x_n\}$ from smallest to largest. Let $\{x_{(i)}\}$ with i from 1 to n be the order statistics for the observed data so that $x_{(1)}$ is the smallest value from $\{x_1, x_2, \ldots, x_n\}$, $x_{(2)}$ is the second smallest, and so forth. Then, $F_n(x)$ may be calculated by

$$F_n(x) = \frac{k}{n}$$

for $x_{(i)} \leq x < x_{(i+1)}$, where k is the number of observations less than or equal to x. To make this test useful, we need a means of determining how close $F_n(x)$ is to $F(x)$ and relate this "closeness" to a probability that the observed data have the cumulative distribution function $F(x)$. For this test, our measure of closeness is

$$D = \max |F_n(x) - F(x)|$$

where the maximum is taken over all of the observations. The value of D is compared with a critical value, which is a function of the number of samples n and the desired confidence level α. The null hypothesis H_0 is that $F_n(x)$ has the cumulative distribution $F(x)$. This hypothesis is rejected if D is greater than the critical value. Critical values are usually found with a computer program or with tables, but a few special cases are listed in Table 6.19.

See [19, 25] and [17] for other techniques for choosing between models.

Table 6.19 Approximate Kolmogorov–Smirnov Statistics Critical Values – One-Sample Tests.

	$\alpha = 0.20$	$\alpha = 0.10$	$\alpha = 0.05$	$\alpha = 0.02$	$\alpha = 0.01$
$n > 40$	$\dfrac{1.07}{\sqrt{n}}$	$\dfrac{1.22}{\sqrt{n}}$	$\dfrac{1.36}{\sqrt{n}}$	$\dfrac{1.52}{\sqrt{n}}$	$\dfrac{1.63}{\sqrt{n}}$

5. Show Good Predictive Ability

Note that ultimately, we want models that do a good job of estimating the rate of occurrence of future defects. A good fit to the current data may not mean a good fit to future data. Three measures of this desirable predictive ability are described below, along with a measure of prediction stability:

1. Predictive validity, prediction relative error: The predictive validity, also sometimes called predictive relative error, is found in [67] and is calculated using:

$$RE = \frac{\text{Predicted defects} - \text{actual defects}}{\text{actual defects}}$$
$$= \frac{m(t_q) - q}{q}$$

where we have observed q faults by the end of time t_q and have used failure data up to some time $t_c < t_q$ to estimate the parameters for $m(t)$. We then use $m(t_q)$ to estimate the number of faults by time t_q. Typically, we use something around the first 75% of the test data to fit to the model and use this to make predictions for the remaining 25% of the data. We then compare the predicted faults with the faults found in the last 25% of the test data. Smaller values of RE indicate better prediction ability. Positive values indicate overestimation and negative values show underestimation.

2. Bias: The bias is defined to be

$$\text{Bias} = \frac{1}{n} \sum_{i=1}^{n} (q_i - m(t_i))$$

where q_i is the number of observed faults up to time t_i and $m(t_i)$ is the number of faults that the model has predicted would be observed up to the same time t_i.

3. Prediction *MSE*: The prediction *MSE*, $MSE_{\text{prediction}}$, is defined in [65] to be

$$MSE_{\text{prediction}} = \frac{1}{n - m + 1 - N} \sum_{i=m}^{n} (y_i - m(t_i))^2$$

where the data from t_1 to t_{m-1} is used to fit the data to the model and the data from t_m to t_n is used to estimate the predictive ability. N is the number of parameters in the model and y_i is the number of faults detected up to time t_i. A smaller value of $MSE_{\text{prediction}}$ indicates a better predictive ability.

4. Prediction stability: Although not directly a measure of predictive ability, we also want models that have stable predictions. One criterion for prediction stability is if the predicted number of faults for the ith week is within 10% of the predicted value of the $(i - 1)$th week.

With respect to predictive capability, we typically find that a model is either optimistic or pessimistic. It is generally better to choose a pessimistic model over an optimistic model so that we err on the conservative side.

Note that some authors have found improved predictive ability by using a linear combination of models. Suppose that we have two models that have passed all of the model-choosing criteria, fit the observed data well, and do well with the predictive criteria. Further suppose that one of these models tends to be optimistic and the other tends to be pessimistic.

Let $m_1(t_i)$ be the output of the first model and $m_2(t_i)$ the output for the second. For a suitably chosen value of w, $0 < w < 1$, we may have an improved predictive ability by using $m(t_i) = w\, m_1(t_i) + (1 - w)\, m_2(t_i)$. This approach can be extended to more than two models. Keep in mind, however, that we should try to err on the conservative side.

6. Acceptable Confidence Bounds

The confidence bounds of the model give us an indication of how much trust we can have in the model results. For example, if the 95% confidence bounds for a given parameter are widely separated, then we cannot justifiably have high confidence in the value of that parameter (and hence in the model results) because the true value of the parameter is likely to be anywhere inside a wide interval.

Assuming that there are thirty or more independent samples, all with the same distribution, we can use a normal distribution to estimate the confidence bounds. Suppose that the parameter we are estimating is N and the estimated value is $\hat{N}$. To find the lower and upper $1 - \alpha$ confidence bounds for $\hat{N}$, LB, and UB, respectively, find the variance of the estimates of N, $Var(\hat{N})$, and use

$$LB = \hat{N} - z_{1-\alpha/2}\sqrt{Var(\hat{N})}$$

and

$$UB = \hat{N} + z_{1-\alpha/2}\sqrt{Var(\hat{N})}$$

$z_{1-\alpha/2}$ is the two-sided α critical value for a standard normal distribution. If X is a standard normal random variable, $P(-z_{1-\alpha/2} \le X \le z_{1-\alpha/2}) = 1 - \alpha$. So, for example, for the two-sided 95% bounds, we have that

$$P(X \le 1.96) = \frac{1}{\sqrt{2\pi}} \int_{-\infty}^{1.96} \exp\left(-\frac{x^2}{2}\right)\, dx = 1 - 0.025 = 0.975$$

We also have that

$$P(X \le -1.96) = \frac{1}{\sqrt{2\pi}} \int_{-\infty}^{-1.96} \exp\left(-\frac{x^2}{2}\right)\, dx = 1 - 0.975 = 0.025$$

so that for our two-sided limits, we therefore have that $P(-1.96 \le X \le 1.96) = 0.975 - 0.025 = 0.95$, or in short,

$$P(-1.96 \le X \le 1.96) = \frac{1}{\sqrt{2\pi}} \int_{-z_{0.975}}^{z_{0.975}} \exp\left(-\frac{x^2}{2}\right)\, dx = 0.95$$

so we use $\alpha = 0.05$ and $z_{1-\alpha/2} = z_{0.975}$

There are various ways to find $Var(\hat{N})$. One simple way is to estimate $\hat{N}$ each time that a new fault occurs and use the variance of this collection of estimates. Note that the more sample data that we have, the tighter our confidence bounds will be. Sometimes, formulas can be found for the variance of parameters for a given model, and software reliability growth model software packages usually have a feature that calculates confidence bounds.

The material above on choosing a software reliability growth model may appear daunting; however, it is often the case that a software package is used that performs many of these tasks. Two commonly used packages are Computer-Aided Software Reliability Estimation (CASRE) and Software Failure and Reliability Assessment Tool (SFRAT).

At this point, we mention three other modeling techniques that can be useful:

1. Capture–recapture defect estimation: For the capture–recapture technique, we detect defects of a product multiple times. The idea is that if the overlap in detected defects is large, then there are few defects remaining to be found. Capture–recapture is typically used both to estimate the number of defects remaining and to assess if the product is ready to go to the next phase of the project.

   Assume that we use capture–recapture with two defect detection efforts. For example, if we are running reliability tests on the software, we make a suitable number of runs of the software per the OP. We then run another set of runs, different from the first set but still using the OP. The first time that we detect defects, we label each defect so that we can identify it if we see it again. We then independently detect defects in the same product again and compare the defects found in the second detection with those of the first. If we detect m defects in our first detection effort and n in the second, with k defects found twice, then the most likely number of defects in the product is

   $$N = \frac{m\,n}{k}$$

   Therefore, the estimated number of remaining defects in the product given that we have found N_0 unique defects is $N - N_0$.

   Example 6.4.10 As an example of the capture–recapture defect estimation technique, suppose that a software project uses software testing to estimate the number of downing failures in a given software product. The tests either run to completion or are stopped because of a downing failure. A large test suite of scenarios, designed to cover the OP of the software, is run on the software product. The project selects a random sampling of runs from this test suite that matches the OP and that also provides sufficient statistical confidence in the reliability of the software. See Example 6.3.2 for more on determining the number of test runs required to have sufficient statistical confidence. Using data from Example 6.3.2, suppose that we run 312 test runs and have $m = 29$ downing failures. We run another set of 312 randomly selected test runs from the same test suite and find $n = 33$ downing failures. Of these two sets of downing failures, $k = 19$ are repeated. Therefore, the capture–recapture technique estimates that the most likely number of software defects that will produce the observed downing failures is

   $$N = \frac{m\,n}{k} = \frac{(29)\,(33)}{19} = 50.4$$

   or rounding up, $N = 51$. As we have found $N_0 = m + n - k = 43$ unique downing defects, we estimate that there are $N - N_0 = 8$ downing defects waiting to be found.

   See [68] and Chapter 27 of [19] for more details on capture–recapture techniques. These techniques have also been used to estimate defects in software documentation through multiple inspections. For examples, see Appendix C.5 of [25] and [69].

2. Bayesian models: Bayesian models consider the number of failures observed over time but also incorporate past information via the prior distribution. The use of a prior distribution is both a strength and a weakness. It can be an important source of information, but there can be uncertainty as to how to choose the right prior distribution. See [17, 26] and the Bayesian techniques topic in Section 6.5 for more details.

3. Model recalibration: Recall that one of the approaches for choosing the best software reliability model (among the subset of "good" models) is to use an empirical cumulative distribution function $\hat{F}$ and use $u_j = \hat{F}_j(t_j)$ to create a u-plot. This empirical distribution $\hat{F}$ is a step function, but we can also "smooth" it by using a stepwise linear approximation, spline curves, or other techniques. The resultant curve G_j^* is based on previous data (the set $\{t_1, t_2, \ldots, t_{j-1}\}$) and can be used to potentially improve our estimate of F_j by using

$$\hat{F}_j^*(t) = G_j^*\left[\hat{F}_j(t)\right]$$

Technically, this approach assumes that the error in the reliability model is stationary, but in practice, the technique may give good results even if this assumption does not hold. Reference [17] covers this technique in more detail.

Step 6: Apply the model to obtain results
After using the model on test data, we will have any combination of the following results:

1. An estimation of the reliability of the software at release
2. The amount of additional test time required to reach the reliability objectives for the software
3. The reliability growth resulting from the testing and defect corrections
4. Predictions of the reliability of the software at various times after software release
5. Information required to make a determination of whether the software is ready to be deployed

There are limitations to extrapolating predictions from test or operational data. First, attempting to predict too far into the future is likely to result in inaccurate predictions. Second, the data should be representative of the way that the system will be used during the future period of time. Changes to the software environment may make predictions inaccurate, and the software should be stable enough that extensive changes are no longer expected.

Step 7: Assess the accuracy of the results
As with any analysis results, the sooner we know if the results are accurate, the sooner we know how to proceed and the more confidence we can have in moving forward. After obtaining results in step 6, we can use the following steps to check for accuracy:

1. Go through the work performed and look for errors in the tasks performed. Perform peer reviews to try to detect and reduce errors.
2. Review the OP and ensure that the tests were performed accordingly. Also, make sure that the hardware that the software is run on is representative of that to be used in the field and that supporting software, such as operating systems and libraries, are the same as is intended for the fielded system and its operation.
3. Calculate the confidence bounds for the estimations and ensure that they are tight enough. For example, the specification may state that the MTBCF estimate shall have an 80% confidence. Even if there is no specified confidence bound, determine bounds that are acceptable to the project and make sure that the estimations meet them. If the confidence bounds are too wide, the estimation should not be accepted, and it is likely that more testing is needed.

4. Compare the results from step 6 with the predicted results (see the Software Reliability and Availability Predictions topic in this section). We can obtain partial estimation results before testing is complete by considering the percentage of code that has been tested relative to the total number of normalized EKSLOC. See [25] for more details. If the predictions are within the confidence bounds for the estimations, then we have greater confidence in both the predictions and the estimations. If they do not line up very well, we need to understand why and make appropriate adjustments.

5. If possible, use several (applicable) software reliability models and see if different models give significantly different results. Note that if the models make significantly different assumptions, we can expect different results, but if several models should be applicable, the different models should give comparable results.

6. Compare the most current reliability estimate for time t with the actual (measured) value at the same time. For example, if the model estimates that the failure rate at time t will be λ_e and the measured failure rate at time t is λ_m, then the relative error RE is

$$\text{RE} = \frac{\lambda_m - \lambda_e}{\lambda_e}$$

Again [25] contains a process (Figure 91) for making this comparison.

Step 8: Validate the reliability in the field
After deploying the software, continue to collect fault information. This information can be used to continue to correct defects and to continue assessing the reliability of the software to validate the pre-release software reliability estimation. We should also use this feedback to improve the processes.

See [17, 19, 25], and [12] for more on software reliability estimations. Software reliability can also be measured for commercial mass market software products based on user experiences. For more details, see [46].

Example 6.4.11 For this example of a software reliability estimation, we consider a company that has developed a software product and is testing it to determine if it is expected to meet its MTBF requirement at the end of testing. Testing is per the OP and is expected to last for 30 weeks at 80 hours of software execution per week. Tests are largely automated, and each test has expected results, so deviations can be checked for potential software defects. The process uses DRACAS for assessing test failures, for removing duplicate failure reports, and uses DRACAS and other activities to make the list of failures as appropriate for reliability estimation as possible. Test coverage is checked, and defects are corrected as soon as possible after discovery. The number of failures is recorded at the end of each week of testing.

Table 6.20 shows the results of the testing. The number of detected faults per week and the test time per week are listed in the second and third columns, while the cumulative number of detected faults and cumulative test times are in the fourth and fifth columns. The sixth column shows the cumulative number of detected faults divided by the cumulative test time.

A plot of n/t (x-axis) versus n (y-axis) fits a straight line well, and the least squares fit for this line has a slope of -2145.86 and a y-intercept of 105.73. The rate of detection of faults

Table 6.20 Software Fault Data.

Week	Faults	Time	Cumulative Faults n	Cumulative Time t	n/t
1	4	80	4	80	0.050
2	3	80	7	160	0.044
3	4	80	11	240	0.046
4	3	80	14	320	0.044
5	3	80	17	400	0.043
6	3	80	20	480	0.042
7	2	80	22	560	0.039
8	1	80	23	640	0.036
9	2	80	25	720	0.035
10	2	80	27	800	0.034
11	3	80	30	880	0.034
12	1	80	31	960	0.032
13	2	80	33	1040	0.032
14	2	80	35	1120	0.031
15	3	80	38	1200	0.032
16	1	80	39	1280	0.030
17	1	80	40	1360	0.029
18	3	80	43	1440	0.030
19	1	80	44	1520	0.029
20	1	80	45	1600	0.028
21	3	80	48	1680	0.029
22	1	80	49	1760	0.028
23	1	80	50	1840	0.027
24	1	80	51	1920	0.027
25	1	80	52	2000	0.026
26	1	80	53	2080	0.025
27	1	80	54	2160	0.025
28	1	80	55	2240	0.025
29	1	80	56	2320	0.024
30	1	80	57	2400	0.024

is decreasing and so there are several software reliability models that can be used. For this example, we use the Jelinski–Moranda model.

For the Jelinski–Moranda model, the failure rate $\lambda(t_{n+1})$ is given by

$$\lambda(t_{n+1}) = k\,(N_0 - n)$$

where n is the cumulative number of detected faults and t_{n+1} is the time period between faults n and $n + 1$. N_0 is the number of defects in the software and k is the rate of change in the fault rate. N_0 can be estimated by the y-intercept of the least squares fit line for the data in Table 6.20, which in this case means that we use $N_0 = 105.73$ as our estimate. k is estimated as the reciprocal of the absolute value of the slope of that same line. For this example, the slope is $m = -2145.86$, so $k = 1/|m| = 0.000466$. At the end of 30 weeks of testing, $n = 57$ per Table 6.20. The initial failure rate (when $n = 0$) is $\lambda(0) = k\ N_0 = 0.049271$ and the failure rate at $t = 2400$ hours (at the end of 30 weeks of testing) is

$$\lambda(2400) = k\ (N_0 - n)$$
$$= (0.000466)\ (105.73 - 57)$$
$$= 0.022709.$$

The resulting MTBF value after 2400 hours of testing (and defect correcting) is the reciprocal of the failure rate, so the MTBF is 44.04 hours. Note that we estimate that the total number of defects in the software is 106 (105.73 rounded up to 106), and we have found and corrected 57 of them, meaning that after 30 weeks of testing, there is expected to be $106 - 57 = 49$ defects remaining. We have removed approximately 54% ($57/106 \times 100\%$) of the defects.

What if the requirement for the MTBF is 100 hours? How much more defect removal is needed? As the required failure rate is 0.01, and letting $0.01 = k\ (N_0 - n)$, we can find the total number of defects that will need to be removed using

$$n = N_0 - \frac{0.01}{k}$$
$$= 105.73 - \frac{0.01}{0.000466}$$
$$= 84.27 \quad \text{defects},$$

which rounds up to 85 defects. As 57 defects have been found, we need to find at least 28 of the estimated 49 remaining defects.

Another consideration is how long we must continue testing to find these additional defects. Recall that the Jelinski–Moranda model calculates the expected number of faults by

$$m(t) = a\left(1 - e^{-b\,t}\right)$$

where a is the initial number of faults in the code and b is the fault detection rate. We have already estimated a to be 105.73. Using a least mean squares (LMS) approach, we estimate b to be 0.00033. (Note that in actual practice, we should apply more steps before we are comfortable with this model fit, but to keep the example simple, we omit these steps here.) Therefore, the total time T of testing that we can expect to be required until we have found 85 defects is

$$T = \left(-\frac{1}{b}\right)\ \ln\left(1 - \frac{85}{a}\right) = 4937.343 \quad \text{hours}$$

Example 6.4.12 In Example 6.4.11, we used the Jelinski–Moranda software reliability model, although several other models could have also been used. In this example, we consider the Goel–Okumoto software reliability model and assess which of these two models, the Jelinski–Moranda model or the Goel–Okumoto model, best fits the data.

The Goel–Okumoto software reliability model finds the software failure rate $\lambda(t)$ at time t using

$$\lambda(t) = N_0\, k\, \exp(-k\, t)$$

where N_0 is the number of defects in the software and k is rate of change in the fault rate. As we are using the data in Table 6.20, we use the calculations from the previous example to obtain that $N_0 = 105.73$ and $k = 0.000466$. At the end of 30 weeks of testing, $n = 57$ per Table 6.20. Therefore, at the end of 30 weeks, the Goel–Okumoto software reliability model gives $\lambda = 0.016101$ and an MTBF of 62.11 hours.

In Example 6.4.11, we found a value of MTBF of 44.04 hours, which is significantly different than the 62.11 hours of this example. To determine which model best fits the data, we use the sequential probability ratio test. For this test, we let f_0 be the probability density function for the Jelinski–Moranda software reliability model and let f_1 be the probability density function for the Goel–Okumoto model, each to be evaluated at inter-failure times (the durations between failures) $\tau_j, j = 1, 2, \ldots, m$. To perform the test, we define

$$R_m = \prod_{j=1}^{m} \frac{f_1(\tau_j)}{f_0(\tau_j)}$$

Let α be the required probability of a type I error, *i.e.* the probability that the test will conclude that the best probability density function to use is f_1 when in reality f_0 is best. Let β be the probability of a type II error, in which we decide on f_0 when we should use f_1. Define

$$A = \frac{\beta}{1 - \alpha}$$

and

$$B = \frac{1 - \beta}{\alpha}$$

and let $I \in \{1, 2, \ldots\}$ be the first number such that $R_I \geq B$ or $R_I \leq A$. Then:

1. If $R_I \leq A$, choose f_0.
2. If $R_I \geq B$, choose f_1.

If $A < R_I < B$, then no conclusion is reached and the testing continues with the next datum.

The data in Table 6.20 are coarser than we would like. Ideally, we would have durations between each discovered fault rather than the number of faults found in each 80 hours of execution. As a rough estimate of the time between fault discovery, we let τ_j be the execution time, 80 hours, divided by the number of discovered faults. So, for the first 80-hour execution time period, we have $\tau_1 = 20$, $\tau_2 = 20$, $\tau_3 = 20$, and $\tau_4 = 20$. The other time periods with multiple detected faults are handled similarly, resulting in τ_1 through τ_{57}. The first two columns of Table 6.21 show the results for faults 1 through 16.

For both the Jelinski–Moranda (J-M) model and the Goel–Okumoto (G-O) model, the probability density function is given by

$$f_i(\tau_j) = \lambda_i\, \exp(-\lambda_i\, \tau_j)$$

where $i = 0$ for the J-M model and $i = 1$ for the G-O model. We set $\lambda_0 = 0.022709$, which is $\lambda(2400)$ for the J-M model. Similarly, we set $\lambda_1 = 0.016101$ for the G-O model. With both $\alpha =$

Table 6.21 Sequential Probability Ratio Test.

Fault	τ_j	f_0 (for J-M)	f_1 (for G-O)	R_j
1	20	0.014419	0.011668	0.809215
2	20	0.014419	0.011668	0.654828
3	20	0.014419	0.011668	0.529896
4	20	0.014419	0.011668	0.428800
5	26.6667	0.012394	0.010481	0.362617
6	26.6667	0.012394	0.010481	0.306649
7	26.6667	0.012394	0.010481	0.259319
8	20	0.014419	0.011668	0.209845
9	20	0.014419	0.011668	0.169810
10	20	0.014419	0.011668	0.137412
11	20	0.014419	0.011668	0.111196
12	26.6667	0.012394	0.010481	0.094034
13	26.6667	0.012394	0.010481	0.079520
14	26.6667	0.012394	0.010481	0.067247
15	26.6667	0.012394	0.010481	0.056867
16	26.6667	0.012394	0.010481	0.048090

0.05 and $\beta = 0.05, A = 0.0526$ and $B = 19$. Table 6.21 shows the progress with the sequential probability ratio test.

Notice that in the last row of the table, the product of ratios $R_{16} < 0.0526$, so we choose f_0 and the J-M model. This result is not unexpected. Note that in the last row of Table 6.20, we have 57 discovered faults in 2400 execution hours, resulting in an empirical failure rate of 57 / 2400 = 0.023750. The J-M model gives $\lambda_0(2400) = 0.022709$, which is closer to this empirical failure rate than the value $\lambda_1(2400) = 0.016101$ from the G-O model.

Software Reliability Growth Plan

In this context, software reliability growth occurs when the software is executed on the target hardware, and with the target supporting software (such as operating systems and libraries), the execution is consistent with the OP, and when failures are detected, a significant number of the associated software defects are mitigated. We also assume that no new features are introduced to the software in this time period.

When we say that a defect is "mitigated" or "corrected," we mean that some aspect of the software system design has been changed to remove the defect, reduce the likelihood of the occurrence of the defect, or reduce the impact of the defect. Simply rebooting the system and returning the system to operation does not constitute a correction. It can be argued that reliability growth occurs whenever defects that could result in a software failure are removed. By this definition, removing defects in a requirements specification in the Requirements and Interfaces Phase constitutes reliability growth. However, a traditional SRGP addresses the reliability growth of the existing software by comparing its current reliability with what can be expected for its future reliability and seeing if it is projected to meet its reliability goals at the required time in the future.

If a project produces a product consisting of both hardware and software, there may be a Reliability Growth Plan for the combined product; however, it is often useful to have a separate SRGP to focus on the software. An SRGP typically includes the following:

1. A brief description of the project and product
2. A list of reference documentation
3. A description of the project management structure as it relates to software reliability and testing
4. A description of reliability risks
5. A project schedule, particularly as it relates to reliability growth activities
6. An outline of the reliability objectives and requirements with any necessary descriptions and schedules
7. An outline of the current reliability predictions, allocations, and modeling
8. A description of the OP and systems environment applicable to reliability objectives
9. A description of the reliability growth model used, why it is applicable, and how the model inputs and parameters are determined
10. An outline of the idealized growth curves and planned growth curves
    (a) Idealized curves: An idealized reliability growth curve is a smooth curve depicting the expected reliability growth of the software based on the growth model equation.
    (b) Planned curves: A planned reliability growth curve is a growth curve that is tied to the project schedule. It is typically discontinuous because of the different project increments, test periods, and corrective action periods (CAPs).
11. A description of the outputs to the reliability growth model and how these outputs will be used to affect the product
12. A description of the level of statistical confidence required of the results and how this level of confidence will be achieved
13. A description of the testing to be used for the reliability growth activities
14. A summary and conclusions, particularly as it relates to the attainment of the reliability objectives with the current schedule, and recommended actions if the tests indicate that the product is not on track to meet these objectives

As an overview of the software reliability growth process, the following steps should be considered (partially repeated from the Software Reliability Estimation topic in this section):

1. Determine the reliability objectives: Reliability objectives are necessary to provide target quantities for the growth.
2. Determine the target hardware and operating environment: Running the software on different hardware or with a different operating environment (operating system, libraries, and other supporting software) may result in different reliability estimates.
3. Create a realistic operational profile (OP): As the reliability objectives are to be met when the system is operating per the OP, the software reliability growth testing should be consistent with the OP.

4. Obtain the project schedules: These schedules should indicate when new features are added to the software. The reliability growth period is then taken to be from the start of the test period to when a new feature is added. Adding a new feature is expected to lower the reliability of the software and so a new reliability growth period is started.

5. Produce reliability predictions for the software to be tested: By knowing the reliability predictions for the reliability objective quantities at the start of a reliability growth test period, we can predict the impact that new features have on the updated software. This prediction is used to develop the reliability objective quantities at the start of the next test period.

6. Perform software testing per the OP: In this step, we collect failure and runtime data.

7. Choose or create a model for the software reliability: There are many models to choose from. Choosing the right software reliability model is usually based on guidelines, statistical methods, and experience. See the topic on Software Reliability Estimation in this section for more details.

8. Use the model to estimate the reliability of the software and validate the estimation: As with the previous step, see the Software Reliability Estimation topic in this section for more details.

9. Use the current and previous test results to estimate reliability growth: Fit the failure and runtime information to the model curve and extrapolate to the desired time period. If the model indicates that one or more of the required reliability figures-of-merit will not be met, determine what parameter adjustments are necessary to meet the requirements and what the project will need to do to achieve these adjustments. If these adjustments are unreasonable for the project, significant project changes may be needed.

As noted in the last item listed above, an important part of a software reliability growth program is assessing whether the software can reasonably be expected to achieve its reliability objectives. Factors influencing whether the reliability objectives will be met include the following:

1. The starting value of the reliability objective (M_I): The closer that the starting value is to the objective, the less growth that is required. Typically, M_I is calculated to be the average MTBF over the first test period. The growth potential MTBF is defined to be the theoretical upper limit of the MTBF. If M_I is the initial MTBF for the software, MS is the management strategy, and μ_d is the average FEF, the growth potential for MTBF, M_{GP}, is

$$M_{GP} = \frac{M_I}{1 - (MS)\,(\mu_d)}$$

(MS and μ_d are addressed below.) The growth potential is generally not obtainable. Reference [27] recommends a more attainable goal MTBF, M_G, of

$$M_G = K\,M_{GP}$$

where $0.6 \leq K \leq 0.8$. Note that [27] is primarily considering hardware reliability growth.

2. The management strategy (MS): Some failure modes (called A-mode failures) will not be addressed with corrective action. They may be too expensive, take too long, their consequences may not be severe enough, or mitigating them may be too difficult. The fraction

of the total failure modes that will be addressed with corrective action (called B-mode failures) is referred to as the management strategy (MS). Per [27], the MS usually needs to be at least 0.9 in order to have a successful reliability growth program.

3. The FEF and average FEF (μ_d): Even for the failure modes that are addressed, some will not be completely eliminated. The fraction of decrease in a failure mode's failure rate after corrective actions have been implemented is called the FEF for the failure mode, and the average FEF value over all addressed failure modes is μ_d, the average FEF. Reference [27] notes that typical FEF values are from 0.55 to 0.85 with a mean value of 0.7 and a median value of 0.71. Note that [27] is primarily considering hardware reliability growth.

4. The duration of the reliability testing: The longer that the reliability testing continues (more CPU hours), the more defects that can be found. This statement holds if the testing continues to exercise different parts of the software.

5. The rate of defect encounters: If each test run roughly duplicates the previous runs, it is likely that the same software paths are being exercised in close to the same ways and so the likelihood of encountering a new defect is low. Without encountering and correcting defects, there is no reliability growth. The reliability test runs need to be planned to exercise as much of the software as practical consistent with the OP. Note that test coverage metrics can be important aids with this factor.

Another important consideration is how rapidly defects will be addressed. There may be a temptation to collect the defects and correct them at the end of the overall test period, but this approach does not allow the project to assess how well the corrective action system is working. Care should also be taken to avoid "defect pileup," where a large number of defects must be addressed in a short amount of time. See [17, 25, 70], and [27] for more on SRGPs.

Software Reliability Program Plan

An SRPP details the activities used on the project to ensure that the product will meet its software reliability requirements. The SRPP may be a standalone document or be a part of an overall project RPP. The plan should cover each phase of the project, describing the activities that impact reliability, metrics employed, and expected results. It should also detail what is to be done if the expected results are not obtained. The SRPP should be developed early in the project, preferably in the CDP phase but no later than during the Requirements and Interfaces phase. Section 3.3 above covers steps for planning how to produce reliable software. An SRPP is used to document these activities and relate them to the rest of the project. Here, we consider such a plan at a higher level.

An SRPP typically includes, as a minimum, the following:

1. An overall description of the project and its software deliverables
2. A description of the project's understanding of the customer needs and of the reliability requirements and objectives of the project
3. Visibility into the project's organizational structure to the degree that responsibility for reliability and reliability-related activities are clearly defined
4. A description of the reliability activities for each phase of the project. Section 3.3 details important aspects of this part of a plan. For each phase, the SRPP:

    (a) Describes how the project will design and produce for reliability, including tasks and activities, products, documentation, tools, processes, and metrics

    (b) Describes who is responsible for each task, product, process, and metric

    (c) Describes resources required to perform these reliability activities

    (d) Provides a schedule for reliability activities and how this schedule relates to the overall project schedule

    (e) Describes how the project will monitor reliability progress, including verification and validation, what is expected, and how adjustments will be made when needed

    (f) Lists the most significant reliability risks and how each is mitigated and monitored

5. A description of how subcontractors are chosen and monitored
6. A description of how commercial or freeware software is chosen and monitored

The SRPP provides a roadmap for the software reliability program. It considers resource constraints, and as it is developed with the software development process in mind, it ensures that reliability activities are integrated into the overall project. The SRPP should be coordinated with the systems engineering master plan (SEMP), the software development plan (SDP), the test and evaluation master plan (TEMP), the SRGP, and other governing plans, and it should be reviewed periodically and potentially updated if there are project changes or if project reliability progress is inadequate. More details on an SRPP may be found in [26, 47], and [25].

6.5 Project-Wide Techniques and Techniques for Quality Assurance

The techniques covered in Table 6.22 are used for general defect control, including anticipating, preventing, finding, and correcting them. Although they are not typically thought of as aspects of reliability engineering, they can have a significant impact on the reliability of the software. Many of these techniques are associated with QA activities, but they can be productively used by any organization to produce better products. We list twenty-four such techniques in the table, with four techniques usually used early in a project and twenty that are typically used throughout the project.

For example, determining project best practices is most useful early in a project so that these practices can be applied to all phases of the project. However, if later in the project people become aware of new best practices, these practices may still be worth adopting. Another example of a quality and defect control technique that is ideally performed early in a project is a project premortem. By its very definition, it is designed to be performed before the activities that it addresses. However, it may still be applicable in later phases of a project, such as before the start of formal software testing.

Other program-wide techniques are often used throughout the entire project time frame. These are techniques that either have broad-enough application that they can be used on products or processes in almost any phase of a project or techniques that have multiple forms, some of which can be used early in a project and others later. For example, creating a lessons learned list can benefit any process, no matter when in a project the process is used. Prototypes can be useful early in a project or later, but the type of prototype used early in a

Table 6.22 Project-Wide Techniques and Techniques for Quality Assurance.

Technique	When used	Notes
Affinity diagrams	Throughout	Affinity diagrams are a team technique that helps organize ideas to better use them
Bayesian techniques	Throughout	These types of statistical techniques enable the user to productively use all of the relevant information in the statistical analysis
Benchmarking	Early	Benchmarking is a technique for identifying world-class standards for a product, service, or process and then modifying the team's product, service, or process to meet or exceed those standards
Brainstorming	Throughout	This technique is used to generate new ideas and uses a group of people rather than a single individual
Checklists	Throughout	This technique is useful for reminding and tracking small steps in a task or process
Data collection for metrics	Throughout	These techniques can help a project improve its metrics by collecting the right data efficiently
Decision-making techniques	Throughout	These techniques can be used to aid decision-making for decisions that are both important and complicated
Defect education	Early	Defect education is a simple but powerful technique for reducing the number of errors and defects in almost any human task or process
Design of experiments (DOE)	Throughout	These statistical techniques can be used to efficiently collect the data required for a statistical analysis
Inspections and reviews	Throughout	These types of techniques are highly effective ways of detecting defects in nearly any kind of document, such as requirement specifications, plans, processes and procedures, narrative documents, software code, test cases, and various other document products
Lessons learned	Throughout	This technique helps projects to anticipate and avoid negative experiences and to take advantage of positive experiences in the future

(continued)

Table 6.22 (Continued)

Technique	When used	Notes
Models and simulations	Throughout	These techniques enable us to describe a system, explain some result, predict what will happen in a given situation prior to having a finished product
Pareto analysis	Throughout	This technique is a means of prioritizing actions when it is not feasible to address all of the actions immediately
Process FMEA/FMECA	Throughout	This technique is a bottom-up analysis of a process to detect potential process failure modes and their effects at higher levels
Process improvement techniques	Throughout	These techniques enable projects to understand, communicate, analyze, and improve the various project processes
Project best practices	Early	This technique applies approaches and methods for performing a task or set of tasks that have been shown to be optimal or to work as well or better than other approaches
Project premortem	Early	This technique is useful for anticipating potential issues with a project so that actions can be taken to prevent them
Prototypes	Throughout	This techniques provides a means of creating "shortcuts" to demonstrate certain aspects of a product
Quality circles	Throughout	Quality circles is a group technique to identify and solve work-related problems
Rationale documentation	Throughout	This technique helps clarify and communicate the intent of decisions and preserve this information for future use
Root cause analysis	Throughout	These techniques are used to find the primary, or root, cause or causes of a defect or problem
Statistical process control	Throughout	These statistical techniques are effective ways of determining if a process is producing the results expected of it
Style guides	Throughout	This technique can be used to promote uniformity in a product
Trend tests	Throughout	These statistical techniques can be used to determine if there is a statistical significant trend in a set of data

project may be significantly different than one used later. Again, when a technique is used should be determined by the needs of the project.

These twenty-four techniques include four statistical techniques of particular interest for software reliability. There are many statistical techniques that are useful when producing and supporting reliable software. While this book is not intended to be a statistics tutorial, the following techniques are not seen in many introductory statistics books and tutorials but are useful enough to deserve inclusion here. They are as follows:

1. Bayesian techniques
2. Design of experiments (DOE)
3. Statistical process control
4. Trend tests

Next, we describe each technique listed in Table 6.22.

Affinity Diagrams

An affinity diagram is a useful technique for organizing a lot of ideas or facts, resulting in better decisions. It is often used after brainstorming, when a lot of ideas have been generated, but they are unorganized and the team needs to organize them and determine how to use them. Like brainstorming, this is a team technique. The basic steps for creating such a diagram are as follows:

1. Collect the ideas that are to be organized.
2. Write each idea on its own card or "sticky note."
3. Randomly place the collection of these cards or notes on a table or board.
4. Let people silently start grouping the cards into groups that make sense to them. If there is disagreement as to which group a particular card belongs, make a duplicate and put it in both. Note however that it is important for this step to be done silently.
5. After the grouping is done, create a header for each grouping. Ask the team participants what they see with a particular grouping of ideas. Ask for a label or theme for them.
6. As an optional next step, create a header for groups of headers.
7. After the diagram is complete, discuss the results. Determine what patterns have been found and how they can be used.

See [71] for more details on this technique.

Example 6.5.1   As an example of the use of an affinity diagram, suppose that a project is having difficulty delivering its software on time. An inter-disciplinary team of project personnel is set up to brainstorm for causes and comes up with the following list of ideas for causes of late deliveries:

1. New, inexperienced personnel
2. Unrealistic schedule
3. Too many changes to the schedule
4. Not enough test stations
5. Too many interruptions
6. Uncertainty as to what constitutes successful verification for many of the requirements
7. Unnecessary requirements

8. Poor training
9. Processes are too complex
10. Unclear procedures
11. Too many special cases in the processes
12. Uncertainty as to who has authority in many situations
13. Plans, processes, and procedures do not match
14. Workload is uneven
15. Too few people working on the critical path processes
16. Cannot locate the appropriate procedures
17. Did not know that a procedure existed
18. Schedule not coordinated with the customer
19. Stress causing low employee morale
20. Confusing requirements

After silently arranging the items, the team came up with five groups and the following labels:
Schedules:

1. Unrealistic schedule
2. Too many changes to the schedule
3. Schedule not coordinated with the customer

Plans, processes, and procedures:

1. Processes are too complex
2. Unclear procedures
3. Too many special cases for the processes
4. Uncertainty as to who has authority in many situations
5. Plans, processes, and procedures do not match
6. Cannot locate the appropriate procedures
7. Did not know that a procedure existed

Personnel:

1. New, inexperienced personnel
2. Poor training
3. Stress causing low employee morale

Requirements:

1. Uncertainty as to what constitutes successful verification for many requirements
2. Unnecessary requirements
3. Confusing requirements

Work environment

1. Not enough test stations
2. Too many interruptions
3. Workload is uneven
4. Too few people working on the critical path processes

After considering all of the ideas and the resulting categories, the team notices that most of the problems stem from early project planning. Schedules were not well done. The choice and development of plans, processes, and procedures was poorly done. Staffing was not adequate, and there are workplace environment issues. The only functional area identified as contributing to the problem is the requirements organization. Using these insights, the project can focus on what are more likely to be the key issues.

Bayesian Techniques

Most of the time, when we consider statistical techniques, we think of data samples treated as instances of an underlying random variable. We then try to determine the (fixed) parameter or parameters that describe the probability distribution for this random variable. This determination is done by taking averages, standard deviations, or other statistical quantities. We may also perform more detailed statistical analyses such as analysis of variance (ANOVA) or cluster analysis. This approach to statistical analysis is often called the "frequentist" or "classical" approach. An alternate approach is to treat anything with uncertainty as a random variable, including the governing parameters. This second approach is usually referred to as the Bayesian approach, and the resulting techniques are known as "Bayesian statistics." Some differences in these two approaches include the following:

1. Interpretation of probability: The differences in the classical and Bayesian approaches to statistics start with the definition of probability. The classical approach considers the probability of an event to be the long-term limiting frequency of the event (hence the name "frequentist" approach). The Bayesian approach considers the probability of an event to be a measure of the degree of personal belief that a person has about the event.
2. Samples and parameters: Because of the difference in the interpretation of the concept of probability, the classical and Bayesian approaches have different views concerning the sampled data and the parameters associated with the probability distribution that describes the stochastic nature of the data. The classical approach considers the collection of samples to be determined by an underlying random variable and takes the probability distribution of the random variable to be fixed. As a result, the parameters associated with this probability distribution are also considered to be fixed. The Bayesian approach treats any quantity with uncertainty, including the parameters, as a random variable. For example, the unknown parameter (or parameters) of a probability distribution is treated as a random variable in Bayesian analysis.
3. Prior information: The classical and Bayesian approaches also differ in what information should be used in statistical analysis. The classical approach to statistics only uses information from the collected data samples. The Bayesian approach uses the collected data samples but also uses prior information, which is typically a probability or probability density function for the parameter or parameters of interest. This probability is obtained only with the information available before making observations (or at least observations from the current experiment).

Bayesian techniques provide a rigorous framework that can be used to incorporate all relevant information into the statistical process, including information that is not contained in the data samples. This added information may come from simulations, different types of tests, historical data, or even engineering judgment. However, care must be taken to ensure that the information not contained in the data samples is truly relevant and accurate.

Key to most calculations in Bayesian statistics is the posteriori density

$$p(\theta \mid x) = \frac{p(x \mid \theta) \, \pi(\theta)}{\int p(x \mid \theta) \, \pi(\theta) \, d\theta} = \frac{p(x \mid \theta) \, \pi(\theta)}{p(x)}$$

where $p(x \mid \theta)$ (sometimes denoted by $L(\theta \mid x)$ or $l(\theta \mid x)$, note the reverse order of θ and x) is the likelihood function and is the probability mass function or probability density function associated with the probability of the observing data x given the parameter θ. We also use the prior density $\pi(\theta)$, which is the prior information or analyst's judgment of the possibilities of the various values of the parameter θ. The above expression for $p(\theta \mid x)$ is then constructed using Bayes' theorem (or Bayes' law) and is the probability of parameter θ given that we observe data x.

Typical steps for Bayesian analysis are as follows:

1. Determine the objectives of the analysis.
2. Determine the prior density $\pi(\theta)$ for the experiment.
3. Collect sample data $x = \{x_1, x_2, \ldots, x_n\}$.
4. Form the likelihood function $p(x \mid \theta)$.
5. Find the posteriori density $p(\theta \mid x)$.
6. Use the posteriori density to develop inferences and therefore parameter estimations, predictions, hypothesis evaluations, or other results. For example, for a point estimate of a parameter θ given a posteriori density $p(\theta \mid x)$, we may choose a deterministic $\hat{\theta} = \hat{\theta}(x)$ that minimizes the posterior expectation of some loss function $l(\theta, \hat{\theta})$, i.e.

$$\min_{\hat{\theta}} E[l(\theta, \hat{\theta})] = \min_{\hat{\theta}} \left[\int_{\theta \in \Omega} l(\theta, \hat{\theta}) \, p(\theta \mid x) \, d\theta \right]$$

where θ ranges over values in Ω. A typical loss function is of the form

$$l(\theta, \hat{\theta}) = \left(\theta - \hat{\theta} \right)^{\mathsf{T}} \left(\theta - \hat{\theta} \right)$$

(where θ and hence $\hat{\theta}$ are vectors and the superscript T denotes taking the transpose of a vector or matrix), although a different form may be appropriate for a given problem.

Important considerations when employing Bayesian techniques are as follows:

1. Choose the prior density carefully. It can have a significant impact on the results, particularly if there is not a lot of empirical data yet. Most criticisms of a Bayesian analysis will be criticisms of the prior density. Early in the analysis, plan how the prior density will be chosen and be sure to choose it before incorporating sample data into the calculations. Doing this will reduce the risk of choosing a particular prior density in order to get a particular result. Also consider performing sensitivity analysis to determine how sensitive the results are to the choice of prior density.

   Prior densities are often considered to be either uninformative or informative. Bayesian statistics always use prior information, but sometimes, there is no information to use. In these situations, we try to use an uninformative prior. An uninformative prior density, in theory, adds no information to the calculation, although all priors add some information. Proposed ways of constructing uninformative priors include maximum entropy, Jeffreys prior, principle of indifference, frequentist matching, and reference priors.

2. Whereas classical statistical analysis provides a single value for a statistical parameter, Bayesian analysis treats the parameter as a random variable and so provides a probability distribution or probability density for the parameter. Although more work is required to obtain such an answer, more can be done with this answer. For example, rather than only finding the expected value for a parameter, with the probability distribution, we can also find the probability that the parameter is greater than a specific value, or the probability that it is in a particular interval of values.

When the sample size is large, the difference between Bayesian results and those from classical statistics will most likely be small. Also, if an uninformative prior is used, the results from the two different approaches may be similar. If the sample size is small and an informative prior is used, differences in the two approaches may be significant. For example, the confidence interval associated with the Bayesian approach may be smaller than that of the classical approach. For more information on Bayesian techniques, see [12, 17, 72, 73], and [26].

Example 6.5.2 For a practical example of a Bayesian technique, consider a situation in which code inspections of a sample of 23 modules of software code determine that only 17 of them are following the project-mandated style guide. The software manager is interested in the probability that a randomly chosen software module will be found to follow the style guide. Using a classical statistics approach, we take the ratio of the number of sampled modules that use the style guide to the total number of sampled modules and get $17/23 \approx 0.73913$. How do we approach the problem from a Bayesian point of view? Let q be the probability of following the style guide and let $X = (x_1, x_2, \ldots, x_{23})$ be the "use–not use" record for style guide use, *i.e.* $x_1 = 1$ if the style guide is used for the first software module and $x_1 = 0$ if it is not used and so on up to and including x_{23}. We assume that the use/not use results from module to module are independent, and because of this independence, we summarize X by $x =$ "17 uses in 23 modules." We make the following assumptions:

1. The use of the style guide on modules is statistically modeled as a collection of Bernoulli trials. As a result, $p(x \mid q)$ has the binomial distribution:

$$p(x \mid q) = \binom{23}{x} q^x (1-q)^{23-x}$$

2. The prior distribution $\pi(q)$ has a beta distribution with parameters α and β:

$$\pi(q) = \frac{\Gamma(\alpha + \beta)}{\Gamma(\alpha)\,\Gamma(\beta)}\, q^{\alpha-1}\,(1-q)^{\beta-1}$$

Therefore,

$$
\begin{aligned}
p(q \mid 17) &= \frac{p(17 \mid q)\,\pi(q)}{p(17)} \\
&\propto \left[\binom{23}{17} q^{17}(1-q)^{23-17}\right] \left[\frac{\Gamma(\alpha+\beta)}{\Gamma(\alpha)\,\Gamma(\beta)}\, q^{\alpha-1}(1-q)^{\beta-1}\right] \\
&\propto q^{17+\alpha-1}(1-q)^{23-17+\beta-1}.
\end{aligned}
$$

The right-hand side must be a probability density function, and we recognize the form of it as a beta density function with parameters $\alpha_0 - 1 = 17 + \alpha - 1$ and $\beta_0 - 1 = 6 + \beta - 1$ (so that $\alpha_0 = 17 + \alpha$ and $\beta_0 = 6 + \beta$). Note that $p(q| x)$ is a "fusion" of both the prior information $\pi(q)$ and the information from the current data, $p(x \mid q)/p(x)$. The mean value μ for a beta distribution with parameters α_0 and β_0 is

$$\mu = \frac{\alpha_0}{\alpha_0 + \beta_0}$$

and the mode m for $\alpha_0, \ \beta_0 > 1$ is

$$m = \frac{\alpha_0 - 1}{\alpha_0 + \beta_0 - 2}$$

If we set $\alpha = \beta = 1$, the beta density is constant. More specifically, it is the uniform probability density function on $(0, 1)$, meaning that it has a value of one throughout this interval. Using this prior density, the expected value of the probability of a style guide use is

$$E[\text{use}] = \frac{17 + \alpha}{23 + \alpha + \beta} = \frac{18}{25} = 0.72$$

To obtain the value obtained from the "classical" approach, $17/23 \approx 0.73913$, we must have $6 \ \alpha = 17 \ \beta$. For example, if $\alpha = 17$ and $\beta = 6$, the mean value is $17/23$. If instead of the mean value, we choose the mode of $p(q \mid 17)$, we have

$$\frac{17 + \alpha - 1}{23 + \alpha + \beta - 2} = \frac{17 + 17 - 1}{23 + 17 + 6 - 2} = \frac{3}{4} = 0.75$$

which is greater than the mean value and the value we obtain using only the data (17 uses out of 23 modules) and using the classical statistical approach.

If we use a different beta density for the prior probability, we get a different expected value. For example, if we use $\alpha = 5$ and $\beta = 1$, we have $E[\text{use}] \approx 0.758621 > 0.72$. Note that the expected value of a random variable with a beta distribution with $\alpha = \beta = 1$ is 0.5, and for one with $\alpha = 5$ and $\beta = 1$, it is about 0.83333, so if our prior estimate is too high, our posteriori probability is likely to be high also. The same conclusion holds for prior estimates that are too low. This example demonstrates why it is important to develop a rigorous and defensible approach for determining the prior probability.

Another important point from this example is that if we solve this problem using the classical approach to statistics, our result is a single value for the parameter of interest. With the Bayesian approach, we find $p(q \mid x)$ to solve our problem, and we can use this function to answer many different questions about q. As a result, we have more information (if we need it). For example, we can find quantities such as $P(q > 0.5)$, *i.e.* find the probability that the value of q is greater than 0.5.

Benchmarking

Benchmarking is a technique for identifying world-class standards for a product, service, or process and then modifying the team's product, service, or process to meet or exceed those standards. It can also identify opportunities for improvement.

A benchmark is a measure that has already been achieved by some other company or organization. With the process of benchmarking, we look at competitors who are recognized leaders for a particular product, service, or process and use information about them to determine what is possible for our project. We may also be able to borrow ideas from these companies. Typical steps for this technique are as follows:

1. Identify the product, service, or process to be benchmarked
2. Identify the aspect or aspects of interest, such as quality, cost, or timeliness
3. Identify comparative companies or organizations
4. Determine the data collection methods and collect the data
5. Estimate current performance levels
6. Predict future performance levels
7. Generate ideas for improvements and determine which should be implemented
8. Communicate the results and obtain buy-in
9. Implement changes and monitor results
10. Update as needed

Sources for benchmarking data include the following:

1. Surveys and interviews with industry experts
2. Published articles
3. Trade or profession organizations
4. Prior experience of company personnel

More details are found in [26] and [71].

Brainstorming

Brainstorming is an idea generation technique that uses a group of people rather than a single individual. Each brainstorming session addresses a single question. The two main principles behind the technique are as follows:

1. Produce a large quantity of ideas: This principle assumes that generating more ideas increases the likelihood of generating good ideas.
2. Do not judge the ideas: Keep generating ideas separate from criticizing ideas. Criticizing ideas during idea generation may reduce the quantity and diversity of ideas. Also, people should avoid approving ideas as expressing approval of an idea may encourage people to generate similar ideas and potentially reduce the diversity of the set of ideas.

Unconventional ideas are encouraged and combinations of ideas should be explored. There are many variants of brainstorming, but typical steps include the following:

1. Decide what the problem or question is and determine if brainstorming is an appropriate technique for it.
2. Choose the group of people to participate in the brainstorming session.
3. Explain the situation, problem, or question that the session is devoted to.
4. Explain the rules to be followed.
5. Do not allow discussions until after the ideas are generated but do allow questions to provide clarifications.
6. Give everyone a few minutes to think about the problem or question individually and write down their thoughts. This step is an individual effort and is not collaborative.
7. Initiate the session. Usually, people take turns adding an idea.

8. Record the ideas as they are generated. This process is usually done so that everyone can see all of the ideas as they are generated.
9. The session ends when new ideas are no longer being generated or a preset time period has been reached.

Often there is a second session that starts with adding more ideas and then progresses to consolidating and narrowing down the list of ideas. See [71] and [74] for more details on this technique.

Checklists

A checklist is a useful tool for reminding and tracking small steps in a task or process. It is often used when all of the tasks are performed by the same person, although this is not required. For some checklists, the order of the steps is important and for others it is not. A checklist can reduce human errors by reminding the user of the required steps and their required order. It also helps the user keep track of progress by "checking off" steps as they are completed. Projects may decide to construct their own checklists based on experience; however, there are plenty of sources of checklists for various activities. For example, [11] provides a number of checklists for various aspects of software development.

Data Collection for Metrics

Metrics can provide important information that enables a project to anticipate or recognize problems so that preventative or corrective actions can be taken. Metrics, however, require data collection. Data for metrics may be continuous or discrete, quantitative (a number), or non-quantitative (such as a condition or a name). The data may be from the input or the output of a process or obtained from the process itself. Several techniques that can aid data collection are listed below:

1. Data collection plan: Meaningful metrics require timely, high-quality data, and obtaining such data requires planning. Such a plan should as a minimum:
   (a) Decide what data to collect. Decide on the metrics for products and processes and based on these decisions, determine the data required. A measurement selection matrix (see below) can be a useful tool for this step.
   (b) Identify the stratification factors that will be used for the data. Stratification factor identification is covered below.
   (c) Develop operational definitions to clearly define what to measure and how to do it. Operational definitions are explained in more detail below.
   (d) Define what adequate sampling constitutes. Some metrics require an adequate sample size with suitable sample characteristics. For example, reliability estimations require a sufficiently large sample size to meet the required confidence level, and these samples must be taken from runs closely matching the OP.
   (e) Consider the use of checksheets to standardize data collection. Checksheets are described below.
   (f) Assign collection responsibilities.
   (g) Train the data collectors.
2. Measurement selection matrix: The purpose of a measurement selection matrix is to find the measurements that best support the needs of the customer. Steps include the following:

(a) Collect the results of QFD or some other activity that elicits customer requirements. For a measurement selection matrix, these requirements are listed down the left side of a table.

(b) The columns of the table are the output measures or metrics. These are obtained through brainstorming, SIPOC diagrams (see the topic on process improvement techniques later in this section), or other techniques.

(c) For each place in the table (a row/column intersection), assign a "strong," "moderate," "weak," or "no relation" indicator. These indicators can be numbers or other symbols. The choices are usually determined by a knowledgeable team.

(d) Based on the results in the table, plan to collect data on the metrics that most strongly relate to the customer requirements.

See Example 6.5.3 for a simple example of this technique.

3. Stratification factors: Stratification factors are used to help identify patterns in collected data. For example, in the process of buying a house, we may stratify our options by neighborhood, size, price, and other factors that allow us to see patterns in the set of options and therefore make a better decision. One set of steps that can be used to identify these factors is as follows:

(a) Identify an output measurement.

(b) Next, list questions about this output. To start, consider questions related to "who," "what," "where," and "when" that are pertinent to the metric in question.

(c) List description characteristics that define different subgroups of the data. Choose subgroups that are expected to be relevant to the questions from the previous step. These description characteristics are the stratification factors.

(d) Create measurements for each stratification factor.

(e) Determine how to make the measurements.

See Example 6.5.4 below for more on this technique.

4. Operational definitions: Operational definitions are precise instructions on how to take a particular measurement. These instructions are important because useful metrics require accurate and consistent measurements. In developing operational definitions, consider who will make the measurements, what instruments are needed and if calibrations are required, how the measurements will be recorded and what other information may need to be recorded with the measurement, the schedule for measurements, and any other information needed to ensure precise and accurate measurements each time.

5. Checksheets, frequency plot checksheets: A common form for collecting measurement data is a checksheet. There are many ways to construct a checksheet and a project should tailor their checksheets to their needs. One example, sometimes called a basic checksheet and used often for defect recording, lists the type of defect in the left-most column and the time period (for example, the day) for the next columns. A tally mark is recorded on the appropriate defect type row under the time period that the defect is found. Often, a right-most column is included that consists of the total number of defects (tally marks) in a given row.

A second type of checksheet is called a frequency plot checksheet. Suppose that we want to measure the number of completed and tested software corrections on a weekly basis. With this type of checksheet, the left-most column is the week number, and a wide second column is where "X's" are placed each time that week that a software defect correction is completed and successfully tested. This type of checksheet gives a way to visually see a histogram of results.

See [71] for more on these techniques. Also, see [31] for software reliability data collection in general.

Example 6.5.3   As a simple example of the use of the measurement selection matrix technique, we consider a company that produces commercial software. After market research, they determine that the following characteristics are critical to the customer:

1. Software price,
2. Software MTBF,
3. Software installation size,
4. Required random access memory (RAM), and
5. Ease of software use.

At this point, they have completed the first step in creating a measurement selection matrix.

In the second step, output metrics are selected. The software product price should be low, preferably under one hundred dollars. A high MTBF is desirable, so a goal of at least 1000 hours is set. It is also determined that an installation size of no larger than 100 MB is desired. The software should not demand too much RAM. A RAM requirement of less than 1 GB is set as a goal. Finally, the software should be easy to use. The company decides to measure this characteristic by the number of complaints about usability that they receive from the beta testers.

In the third step, these required characteristics and metrics are put on a measurement selection matrix, and a team of subject matter experts rates each metric with each characteristic, giving it a value of 3 if the metric has a strong relationship to the corresponding characteristic, 2 if it is moderate, 1 if it is weak, and 0 if it is unrelated. Table 6.23 shows the results.

Finally, the table is used in creating a metrics plan. For example, an MTBF of at least 1000 hours has a significant impact on almost all of the required characteristics, whereas the installation size has a low impact on all of them but installation size and MTBF.

Table 6.23   Measurement Selection Matrix.

	≤ 100 Dollars Price	≥ 1000 Hours MTBF	≤ 100 MB Installation size	≤ 1 GB Required RAM	Number of Complaints
Price	2	3	1	2	2
MTBF	3	3	1	0	2
Installation size	0	2	2	1	0
Required RAM	0	0	0	3	0
Ease of use	2	2	1	0	3

Example 6.5.4   In this next example, we consider measurement stratification using house prices. A house buyer is looking at several houses in a town and price is a major criterion. Recall that the set of steps for measurement stratification outlined above are as follows:

1. Identify an output measurement.
2. Next, list questions about this output.
3. List description characteristics that define different subgroups of the data. Choose subgroups that are expected to be relevant to the questions from the previous step. These description characteristics are the stratification factors.
4. Create measurements for each stratification factor.
5. Determine how to make the measurements.

For the first step, the output measurement is house price. In the next step, the buyer asks the following questions:

1. "How big is the house?"
2. "How big is the yard?"
3. "How much crime is nearby?"
4. "How much traffic is there?"

In the third step, the buyer lists descriptive characteristics that define the subgroups. These characteristics are used to choose subgroups of the data that may show relevant patterns in the data. Using the list of questions from the previous step:

1. "How big is the house?": For this characteristic, the buyer considers the square footage of a given house.
2. "How big is the yard?": Here, she considers the acreage of the lot that the house is on.
3. "How much crime is nearby?": For this characteristic, the buyer considers burglary, robbery, assaults, and murders in the houses' neighborhoods and in surrounding neighborhoods.
4. "How much traffic is there?": The buyer considers when she expects to leave and return home for each day of the week and typical places that she may go to or return from and is interested in the time that the trip can be expected to take.

Combining the fourth and fifth steps, she creates measurements and measurement processes for each of the above stratification factors. Again using the list:

1. "How big is the house?": Use a contractually binding measure of square footage for each house.
2. "How big is the yard?": Use a contractually binding measure of acreage for the lot that the house is on.
3. "How much crime is nearby?": Use the publicly available police records for burglary, robbery, assaults, and murders in the houses' neighborhoods and in surrounding neighborhoods for the past three years. Add the instances of each type of crime to give a total number of serious crimes. Do this for each of the three years. Consider both the size of the crime problem and if there are trends.
4. "How much traffic is there?": For each house in the final list, drive from the house to work at the time that she will typically go to work, and from work to home at a typical time for that trip. Also, make a trip from the house to the main shopping area of town

and back. Record the time required for the trips. Weight the trip times by importance and add them, resulting in a single measurement per house.

Decision-Making Techniques

All projects require decision-making. Most of these decisions are made by one person on the spot, but some decisions are both important and complicated, and more care needs to be taken with them. There are many decision-making techniques. Five popular choices are described below:

1. Pugh decision matrix: A Pugh decision matrix is a device that can be used to compare a standard approach, called the baseline, with various alternatives. Multiple criteria are selected, and each alternative is compared to the baseline on each of the criteria. A group or a single person can use this technique. To prepare and use this matrix:
   (a) Determine what the baseline approach and the alternate approaches are. Define them clearly so that comparisons will be easier.
   (b) Come up with a list of selection criteria to be used in rating the various approaches.
   (c) Create a Pugh matrix:
       i. The first column of the matrix lists the selection criteria.
       ii. The header for the second column names the baseline approach and the remaining column headers name the various alternative approaches.
       iii. For each element of the matrix (at the intersection of a criterion and an approach), enter a score of −1, 0, or 1, meaning that the approach is worse than the baseline, equivalent to the baseline, or better than the baseline, respectively, for the criterion in the associated row. Note that the baseline column will have all 0 values as in this column, the baseline is compared with itself.
   (d) Sum each column to find the best approach.
   There are numerous variations on this approach. For example, each criterion could be given an importance weighting. Each assessment in the corresponding row is then multiplied by the weighting of the row, and the column sums are of the weighted scores. Another variation is to assign scores of −2, −1, 0, 1, and 2 to increase the distinctions in the scores.

2. Paired comparison matrix: A paired comparison matrix is used to compare each pair of a list of options. Pairwise comparisons tend to be easier than trying to compare several options all together, especially if the options are very different or not as well defined as one would like. To perform this analysis
   (a) Determine the list of options that will be compared. Assign each of the options a letter or number as a label.
   (b) Create the list of criteria that will be used to compare the options.
   (c) Create a square matrix with as many rows and columns as there are options. Label each column and each row with an option label. For example, if there are options A, B, C, and D, then the columns will be labeled from A to D and likewise for the rows. Shade or otherwise block in the diagonal matrix positions and all positions to the right of the diagonal.
   (d) Compare each pair of options. For the comparison of the first and second options, record an "A" in the matrix position that intersects the "A" column and the "B"

row if option "A" is judged superior. Otherwise, record a "B." Do this for each option pair.

(e) After all pairs have been compared and recorded, count the number of "wins" for each option. The option that wins the most pairwise comparisons is chosen as the best option. If there is a tie in the largest number of wins, look at the pairwise comparison of these two options and choose the winner as the overall winner.

3. Multivoting: Multivoting is a group decision-making technique. The steps are as follows:
   (a) Clearly define the decision to be made.
   (b) Decide who will be in the group that performs the voting.
   (c) Determine the list of options to compare.
   (d) Determine the rules for voting. Considerations include the following:
      i. Determine how many votes each member of the group will have. A "rule of thumb" is that each group member should be able to cast enough votes to cover one-third of the total number of options.
      ii. Determine if members can cast multiple votes for the same option or must cast each vote for a different option.
      iii. Decide if voting will be confidential or open for all to see who votes for which options.
   (e) Perform the voting.
   (f) Count the votes.

The results of the voting may determine a single winner or may be used to reduce the number of options.

4. Analytic hierarchical process: The analytic hierarchical process (AHP) decision-making technique is a rigorous process developed to address human strengths and weaknesses when making decisions. (See [75] for more details.) The process uses pairwise comparisons with a recommended numerical scale for quantifying these preferences. The technique then uses a mathematically rigorous method for assessing the best option and to determine if the pairwise comparisons are consistent. Steps for this approach are as follows:
   (a) Hierarchically define the problem. There are typically a minimum of three levels in an AHP problem definition. The first level is a high-level statement of the decision's goal. The next level is a list of decision criteria to be used in making the decision. The third level is a list of options to compare. There may be other levels. If any of these three levels is too abstract, sub-levels are called for.
   (b) Perform a pairwise comparison of the criteria and of the options. For many decision problems, it is difficult to quantify how much one prefers one option over another. Part of the difficulty is due to the trying to keep track of multiple options. Pairwise comparisons help with this difficulty. Another aspect of the difficulty is, even with pairwise comparisons, quantifying the amount of preference is subjective. To help with this issue, Saaty produced the following quantification, where values go from 1/9 to 9.
      i. 1 – Same (neither option is preferred over the other)
      ii. 3 – Weak (one option is preferred slightly over the other)
      iii. 5 – Clear (one option is preferred clearly over the other)
      iv. 7 – Strong (one option is preferred strongly over the other)

    v. 9 – Very strong (one option is preferred very strongly over the other)

   vi. 2, 4, 6, 8 – Compromise (values used to add graduation between the above evaluations)

  vii. Reciprocal values of the values above – If option A has one of the above values assigned to it when compared to option B, then option B will have the reciprocal of that value when compared with option A

For example, if there is a preference for option A over option B but the preference is slight, a value of 3 is appropriate. If later we see that there are other comparisons that are also weak, but the preference of A over B is the weakest of them, we may change from a value of 3 to a value of 2 for option A over option B.

In this step, create a matrix of preference values for the criteria and a separate matrix for the options for each of the criteria. For example, if there are five decision criteria, labeled 1 through 5, the criteria matrix will be a 5×5 matrix where the columns correspond to the criteria, 1 through 5 in order, and the rows are also ordered in this manner. The diagonal values always have a value of 1 for obvious reasons. If criterion i, when compared with criterion j, has a value of k, then the (i, j) matrix position has a value of k in it and the (j, i) position has a value of $1/k$.

In the same manner, create matrices for the collection of options. For example, if there are three options under consideration, each option matrix will be a 3×3 matrix. There will be one option matrix for each criterion. The preferences will be a pairwise comparison of options based on the single criterion chosen for that matrix.

(c) Determine the weights for the criteria matrix and for each option matrix. After creating the criterion matrix and the collection of option matrices, we determine the weights that each matrix assigns each criterion or option. There are two approaches that are frequently used.

   i. Eigenvector method: Consider a given criterion or option matrix. For this approach, the eigenvalues are found for the matrix, and the eigenvector for the largest of these eigenvalues is also found. This eigenvector is normalized by dividing each component of the vector by the sum of all of the components. These normalized eigenvector components are then the weights for the matrix.

  ii. Geometric mean method: The geometric mean method approximates the eigenvector method and is computationally easier. Suppose that the matrix is a $k \times k$ matrix. With this method, take the first row of the matrix and multiply all of the components in the row together, then take the kth root of this product. The result is the weight for the first row, which corresponds to the first option. Repeat for the other rows, resulting in a complete weighting of the matrix.

(d) Compute the contribution of each option to the overall goal. At this point, we have weights for the individual criteria and a weight for each option relative to each criterion. Now, we combine these to obtain a score for the contribution that each option contributes to the overall goal. Assume that there are n criteria and m options. Let w_i, $i = 1, 2, \ldots, n$, be the weight for the ith criterion and let $u_{i,j}$, $i = 1, 2, \ldots, n$, $j = 1, 2, \ldots, m$, be the weight for the ith criterion with the jth option. Then, the overall contribution W_j from the jth option is

$$W_j = \sum_{i=1}^{n} w_i \, u_{i,j}$$

(e) Check for consistency. To understand what is meant by "consistency" in this context, suppose that we have three options, labeled A, B, and C. If in our pairwise comparisons, we determine that option A scores twice as high as option B, and option B scores three times as high as option C, then if our scoring is perfectly consistent, option A will score six times as high as option C. In the real world, scoring is not expected to be perfectly consistent, but if it is too inconsistent, there is reason to be concerned and more investigation is needed.

To check the consistency of a decision matrix (whether a matrix for criteria or for options), find the largest eigenvalue $\lambda_{\max}$ of the matrix. Assuming that the matrix is an $n \times n$ matrix, find the consistency index (CI) using

$$CI = \frac{\lambda_{\max} - n}{n - 1}$$

This value needs to be compared with an average value of the CI taken from random matrices (that have been constructed like an AHP decision matrix). These values are called random index (RI) values and are dependent on the matrix size. In [75], Saaty provides the following RI values:

    i. $n = 3$: RI $= 0.52$
    ii. $n = 4$: RI $= 0.89$
    iii. $n = 5$: RI $= 1.11$
    iv. $n = 6$: RI $= 1.25$
    v. $n = 7$: RI $= 1.35$
    vi. $n = 8$: RI $= 1.40$
    vii. $n = 9$: RI $= 1.45$
    viii. $n = 10$: RI $= 1.49$
    ix. $n = 11$: RI $= 1.52$
    x. $n = 12$: RI $= 1.54$
    xi. $n = 13$: RI $= 1.56$
    xii. $n = 14$: RI $= 1.58$
    xiii. $n = 15$: RI $= 1.59$

To make our comparison, we divide the CI by the RI, resulting in the consistency ratio (CR). If the CR is less than 0.1, we consider the decision matrix to be consistent. If the value is above 0.1, we do three things:

    i) Find the most inconsistent judgment in the matrix.
    ii) Determine a range of values that the judgment can be changed to so that the consistency can be improved.
    iii) Consider if a change to this range is plausible.

(f) Perform an overall evaluation of the situation. At this point in the process, there is a tendency to use the results to choose an option. While this choice may be perfectly acceptable, it is also a time to explore the information that the analysis has provided to better understand the problem. Are the results consistent with intuition and if not, where do they differ? Do all of the assessments still seem correct? Did the approach capture all of the information that you think is relevant?

Example 6.5.5 provides more on this technique.

5. Simple multiattribute rating technique: Another popular decision-making technique is the simple multiattribute rating technique (SMART). It combines performance scores for alternatives and weights for criteria to obtain a total score. It can be used when there is a mixture of objective and subjective criteria, and it also enables the user to emphasize the more important criteria and put less emphasis on the less important ones. The steps for this technique are as follows:

(a) Identify the decision makers: A single person or a group can use the SMART approach. If a group is to be employed, determine who the participants will be.

(b) Identify the issue: Clearly define the issue to be addressed and the decision that needs to be made.

(c) Identify the alternatives: Create a list of alternatives to choose from. These alternatives may be obvious or may require research and creativity.

(d) Identify the criteria: Next, develop a list of criteria to base the decision on. Make sure that the list includes all of the important criteria but try not to make the list too long.

(e) Assign values to the alternatives for each criterion: We want normalized values. To obtain these, we can use the following steps:

   i. Assign raw values to the criteria: If a criterion is quantitative, use its quantitative value. For example, if the price of an item is a criterion, its price in dollars (or some other currency) can be used. If the criterion is subjective, such as appearance, rate it on a scale of 0 to 100 (where 100 is the best that you can hope for). At the end of this step, we have assigned a raw value to each alternative for each criterion.

   ii. Develop a scoring scale for each criterion: Determine a best value and a worst value for each criterion. For example, if the criterion is price, it may be that the best price that you could hope for is 100 dollars and the worst that you would consider is 500 dollars. Do this for each criterion. This pair of values will be used to create a scale to convert each criterion score to a value between 0 and 1.

   iii. Find normalized values for each criterion: Often, a linear scale is used. For a linear scale, let V_W and V_B be the worst value and the best value for the scoring scale for a criterion as found in the previous step, respectively. Let V_R be the raw value for the criterion. The normalized value for V_R may be found using

$$V_{norm} = \frac{V_R - V_W}{V_B - V_W}$$

where V_{norm} is the normalized value. This normalization puts the values for each criterion on the same scale. It also converts all criteria values to where larger values are better values. At the end of this step, we have assigned a normalized value to each alternative for each criterion.

(f) Apply a weight to each criterion: Usually, some criteria are more important than others. To account for this issue, we assign an importance weight to each criterion. We can use the following steps to develop these weights:

   i. Order the criteria in terms of their importance to the decision-making process.

ii. Assign the least important criterion a value of 10. Assign the next most important criterion a value that reflects how important it is relative to the least important criterion. Continue until all criteria have been assigned a weight.

iii. Finally, we normalize these values to obtain the weights to be used in the decision-making process. Take the sum of all of the values found in the previous step and divide each of the values by this sum, resulting in normalized weights.

(g) Determine the weighted average of values for each alternative: We are now ready to find scores for each alternative. Assume that there are m alternatives and n criteria. Let the weights assigned to the criteria in the previous step be $w_j, j = 1, 2, \ldots, n$ and let the normalized values assigned to the alternative be $V_{i,j}$ where $i = 1, 2, \ldots, m$ and $j = 1, 2, \ldots, n$. For example, $V_{1,1}$ is the normalized value for the first alternative with respect to the first criterion. $V_{1,2}$ is the normalized value for the first alternative with respect to the second criterion and so on.

To find the score for the ith alternative, we use

$$S_i = \sum_{j=1}^{n} w_j \, V_{i,j}$$

At the end of this step, we have a score for each alternative.

(h) Make an initial decision: For the initial decision, choose the alternative with the highest score. This decision may also be the final decision, but in reaching this decision, subjective judgments were used, and it is a good idea to add confidence to the decision with a sensitivity analysis.

(i) Perform a sensitivity analysis: Usually, when performing this decision-making technique, there are times when judgment calls are used. With a sensitivity analysis, we take our initial decision and change values that we are less sure of and repeat the analysis to see if still get the same answer. If we do, we have more confidence in the answer. If we do not, we consider our choices in more detail.

See [76] for more detail.

See [71, 76], and [75] for more on decision-making techniques.

Example 6.5.5 In this example, we redo Example 6.5.4 by using the AHP as an aid to choosing a house to buy. For this example, the buyer considers three houses that cost roughly the same. She therefore decides to base her decision on the four characteristics used as stratification factors in Example 6.5.4, namely, house size, yard size, crime, and traffic. Therefore, in hierarchically defining the problem (step a of the AHP approach)

1. The goal of the decision is to choose which of three houses to buy.
2. The list of criteria are house size, yard size, amount of crime, and traffic times.
3. The list of options to compare consists of the list of three houses.

We next make a pairwise comparison of the criteria using the pairwise comparison values stated above, resulting in the following matrix:

$$C = \begin{bmatrix} 1 & 3 & 1 & 1/5 \\ 1/3 & 1 & 1/3 & 1/6 \\ 1 & 3 & 1 & 1/5 \\ 5 & 6 & 5 & 1 \end{bmatrix}$$

The columns are, in order, house size, yard size, crime, and traffic. The rows are also these characteristics (in the same order). We then compare each house, H1 through H3, with each of the characteristics, resulting in four matrices:

$$C1 = \begin{bmatrix} 1 & 7 & 3 \\ 1/7 & 1 & 1/2 \\ 1/3 & 2 & 1 \end{bmatrix}$$

$$C2 = \begin{bmatrix} 1 & 3 & 1 \\ 1/3 & 1 & 1/3 \\ 1 & 3 & 1 \end{bmatrix}$$

$$C3 = \begin{bmatrix} 1 & 1/7 & 1/3 \\ 7 & 1 & 1 \\ 3 & 1 & 1 \end{bmatrix}$$

$$C4 = \begin{bmatrix} 1 & 2 & 3 \\ 1/2 & 1 & 1 \\ 1/3 & 1 & 1 \end{bmatrix}$$

Matrix C1 is the pairwise comparison of the three houses, H1 through H3 (in that order), using the first characteristic, house size. Matrices C2 through C4 are similar but using yard size (C2), crime (C3), and traffic (C4). These results complete step b of the AHP approach.

Next, we determine the weights (step c of the AHP approach). Using the eigenvector approach, the largest eigenvalue of matrix C is 4.11. The corresponding (normalized) eigenvector is [0.15 0.07 0.15 0.62] (in row form). If we had used the geometric mean approach, we would have [0.16 0.07 0.16 0.62]. Using the eigenvector approach for matrices C1 through C4, we have

1. C1: largest eigenvalue = 3.0026, eigenvector = [0.68 0.10 0.22]
2. C2: largest eigenvalue = 3.0000, eigenvector = [0.43 0.14 0.43]
3. C3: largest eigenvalue = 3.0803, eigenvector = [0.10 0.51 0.39]
4. C4: largest eigenvalue = 3.0183, eigenvector = [0.55 0.24 0.21]

Now, we perform step d, compute the contribution of each option to the overall goal. Starting with H1, the house size characteristic has a weight of 0.15. The corresponding house comparison matrix is C1. The weight for H1 for this matrix is 0.68, so these two numbers are multiplied together. We continue with the second characteristic and the C2 matrix and multiply 0.07 and 0.43 together. Continuing through all of the characteristics and the corresponding weights for H1, we have $(0.15 \times 0.68) + (0.07 \times 0.43) + (0.15 \times 0.10) + (0.62 \times 0.55) = 0.49$. For H2, we have $(0.15 \times 0.10) + (0.07 \times 0.14) + (0.15 \times 0.51) + (0.62 \times 0.24) = 0.25$, and for H3, $(0.15 \times 0.22) + (0.07 \times 0.43) + (0.15 \times 0.39) + (0.62 \times 0.21) = 0.25$. At the end of this step, H1 appears to be the best choice, but we still have two more steps.

In step e, we check for consistency. We start with the matrix C. It has a largest eigenvalue of 4.11 and is a 4×4 matrix, so $n = 4$. Therefore, its CI is

$$CI = \frac{\lambda_{max} - n}{n - 1} = \frac{4.11 - 4}{4 - 1} = 0.03667$$

Since $n = 4$, the RI used as a comparison value is 0.89, so the CR is $0.03667/0.89 = 0.0412$. This value is well below 0.1, and so we can consider decision matrix C to be consistent. For the decision matrices C1 through C4, we have

1. C1: largest eigenvalue = 3.0026, n = 3, CI = 0.00130, CR = 0.0025,
2. C2: largest eigenvalue = 3.0000, n = 3, CI = 0.00000, CR =0.0000,
3. C3: largest eigenvalue = 3.0803, n = 3, CI = 0.04015, CR =0.07721,
4. C4: largest eigenvalue = 3.0183, n = 3, CI = 0.00915, CR =0.01760.

These results indicate that each of C1 through C4 is consistent.

Finally, step f is to perform an overall evaluation. This step is important, but whether the results are consistent with intuition depend on the evaluator. As a result, we leave the evaluation to the reader.

Defect Education

Project personnel who are aware of typical defects in the products that they work on are likely to put fewer defects in their products than personnel who are unaware of them. One way that a project can reduce defects is to introduce task performers to the types of defects likely to occur from their tasks. Each type of job has its own set of "typical" errors and associated defects. Part of on-the-job training can be to familiarize employees with the defects that they may encounter or produce. Do this early in the project to train the employees and make them sensitive to these errors and defects and to teach them how to recognize them and prevent them. Even experienced task performers can benefit from being reminded of these typical errors and defects.

The list and description of defects may initially come from one or more task experts, supplemented by brainstorming or a task premortem. Defect information from previous similar projects can also be valuable. As the project progresses and defects are found and go through DRACAS, these newly found defects can be added to the list. The training material can be in the form of a document, a presentation, on-line material, or other forms.

Design of Experiments (DOE)

We often need to collect data in order to assess the software or to analyze and understand a situation, but the process of collecting these data may be expensive or time-consuming. We want to collect data as efficiently as possible by carefully designing the collection activities or "experiments" that we perform so that we obtain the needed results from the experiments while minimizing the amount of data and effort required for the experiments. In other words, we want to reduce the data collection effort while still ensuring that the collected data are appropriate for the required statistical inferences. This careful planning and designing is often referred to as DOE.

With DOE, we choose independent variables, often called "factors," that we can control and that we are interested in with respect to how they affect the dependent variables (if at all). These dependent variables are typically the measured results from our experiment. Frequently, we analyze the resultant data using ANOVA. There are several principles used to control an experiment in order to ensure that the statistical results will be valid and that the experiment will be performed efficiently:

1. Randomization: The first principle is randomization, which is the process of introducing certain experimental treatments at random so as to turn some of the systematic errors into random errors. Systematic errors, also called systematic biases, are consistent and repeatable errors because of the nature of the experiment. Examples include bias in an

instrument, wear of a measurement device, or human bias in making an assessment. Random errors, also called system noise, are generally unavoidable errors that have no discernable pattern.

2. Replication: The second principle is replication. A replication is a complete repetition of an experimental treatment or observation. It allows us to measure and account for statistical variance because of experimental errors. It also helps us reduce this variance. Note that replication as used here is not simply repeating a measurement of a treatment or observation several times. By replication, we mean repeating the experimental setup that went into making the measurement as well.

3. Blocking: The third principle is blocking. This principle minimizes systematic errors by making the variation within each experimental block as small as possible. Blocking is useful in reducing the impact of nuisance factors. Nuisance factors are factors that are not of prime interest for the experiment but could impact the results of the experiment. For example, if we are interested in whether the concentration of cement affects the density of concrete, we are not primarily interested in who mixes the concrete, but if we have several people mixing it, we should consider the person mixing because different people may mix it differently. In this experiment, the variance due to the person mixing is a nuisance factor.

   To perform blocking, we create two or more relatively homogeneous groupings of cases, called "blocks," in which the nuisance factors are constant and the factor of interest varies. Continuing with our concrete experiment example, let us suppose that there are three people, each making two different mixtures of concrete. We denote these people as "A," "B," and "C." Let us suppose that each person creates a batch that is 10% cement and a batch that is 15% cement. Unfortunately, we cannot assume that all three people mix the concrete in the same way, and this nuisance factor should be accounted for. To do this, we create three blocks, one for each person doing the mixing. The first block would only have samples from person "A." The second would only have samples from person "B," and the third, only from person "C." However, each block would have both types of mixtures from the person mixing for that block, *i.e.* the first block would have samples from the 10% mixture and also from the 15% mixture from person "A," and so on. In general, we want to remove (or minimize) nuisance factors by blocking, if we can, and if we cannot, we use randomization. We can usually use blocking if each level of the factor that we are interested in occurs the same number of times for each level of the blocking factor.

4. Orthogonality: A fourth principle is orthogonality, which is grouping the data into uncorrelated sets so that conclusions can be drawn more easily. As each grouping is uncorrelated to any other grouping, each grouping can be considered to convey different information.

5. Factorial experiments: A factorial experiment is an experiment consisting of two or more factors, each with two or more discrete values or levels. These experiments consider all combinations of values and factors. Most factorial experiments use factors with two values. If we have n factors, each with two values, there are a total of 2^n combinations of factors. When this value is too large, a fractional factorial design can be chosen. For these designs, a subset of the total number of combinations is used.

If three values are used for each factor, there are 3^n combinations with n factors. As should be obvious, the number of combinations increases rapidly as the number of factors increases, or as the number of values allowed for the factors increases.

With these principles in mind, we can list some basic steps for the process of DOE:

1. Determine the objective or objectives of the experiment. There are several types of objectives or goals:
   (a) Comparative: Our goal may be *comparative*, in which we are interested in whether a change in a factor has changed the observed results. Here, we are interested in whether this factor is significant.
   (b) Screening: We may be interested in *screening*, by which we mean using experiments to understand a system by ranking factors in order of their influence on the behavior of the experimental results. If screening is our objective, we typically want to "screen out" the less important factors so that we can concentrate on the important ones.
   (c) Modeling: A third type of objective is *modeling*. Here, we want to develop a mathematical function that fits the behavior of the system under investigation.
   (d) Optimization: Finally, we may want to *optimize* some aspect of the behavior of the system.
2. Determine the dependent and independent variables. The dependent variables are the quantities that are being measured. The independent variables are those that we control in the experiment. The independent variables are also called factors. We also need to determine the number of levels for each factor. More levels give us greater granularity but at the cost of more experimental runs.
3. Select an experimental design. The design chosen should be guided by the objective of the experiment.
   (a) If the objective is comparative, consider a completely randomized design, a randomized block design, or a version of a Latin square design.
   (b) If the goal is selective, consider a full factorial, fractional factorial, blocked full factorial, or Plackett–Burman design.
   (c) If the goal is to predict and possibly optimize responses, consider a response surface experimental design.
4. Perform the experiment. Record all aspects of the experiment that may be useful. Potential data to log include time of day, data, temperature, and person performing each run. Ultimately, the nature of the experiment should guide the choices of what to record.
5. Analyze the results, interpreting them in accordance with the way that the experiment is performed. It is also wise to perform two or three confirmatory runs at the same settings as used for the original results to see if we get the same results. Some analysis suggestions include the following:
   (a) Look at the data. The answer to the original question may be obvious. If not, try plotting the data in various ways. Produce histograms. Plot the results as a function of run sequences, *i.e.* the result from the first run followed by that of the second and so on. Look to see if there is a temporal pattern.
   (b) If the objective is to determine which treatment is best, ANOVA may be the best analysis approach.

(c) Create a theoretical model for the data. For example, when using ANOVA, we assume a model of the form $X_{jk} = \mu + \alpha_j + W_{jk}$ for a single factor and similarly for multiple factors. Here, $j = 1, 2, \ldots m$ where there are m independent groups of samples, and $k = 1, 2, \ldots, n$, where each group has n members. (For example, we could have m different brands of fertilizer and each brand is used on n different one-acre plots of land.) For this model form, μ is the part of the expected value that applies to all m groups of the data and α_j is the part of the expected value specific to the jth group. The collection of all of the α_j values sum to zero. W_{jk} is a normally distributed random variable with zero mean and variance σ^2. What model is needed depends on the effects that need to be estimated.

(d) Test the model with the data.

    i. Start by checking the residuals. Residuals are the values of r_i,

$$r_i = Y_i - y_i$$

where Y_i is the ith response from the experiment and y_i is the result that the theoretic model predicts should occur. These residuals should be random and distributed in accordance with the noise distribution assumed in the model. The vast majority of theoretic models assume normally distributed errors or "noise" with a mean of zero. Remember that we expect there to be "noise" in the experimental results. If there is not, then the model is in error.

    ii. Next, check the R^2- and R^2-adjusted values. The R^2 value is found from

$$R^2 = 1 - \frac{\sum_{i=1}^{n} (Y_i - y_i)^2}{\sum_{i=1}^{n} \left(Y_i - \overline{Y}\right)^2}$$

where

$$\overline{Y} = \frac{1}{n} \sum_{i=1}^{n} Y_i$$

The closer R^2 is to a value of 1, the better. The adjusted R^2 value is found using

$$\text{Adjusted } R^2 = 1 - \frac{(1 - R^2)(n - 1)}{n - k - 1}$$

where R^2 is the value of R^2 described above and k is the number of predictors (independent variables or coefficients, excluding the constant) that we fit to for the model.

(e) Use the results to answer the question or questions that we want answered.

One of the most common analysis techniques used when employing DOE is ANOVA. ANOVA is a statistical technique used when we have several collections of data and we want to determine if the mean values for these collections are the same or if at least one of the samples has a different mean value. There are many versions of ANOVA covering various different situations; however, a simple version is a one-factor analysis with each collection of data having an equal number of observations as described below.

A factor is a variable, such as how the samples are processed, that may affect the outcome of the experiment. With one-factor analysis, we have a single factor that is changed and we wish to determine if this factor is significant. We will assume that we have collected a independent groups of samples, each grouping corresponding to a specific treatment. Each group has b members. As an example, we may have a different brands of fertilizer, and each brand is used on b different one-acre plots of land.

To provide more details, let us assume that the data can be viewed as random variables $X_{11}, X_{12}, \ldots, X_{1b}, X_{21}, \ldots, X_{jk}, \ldots, X_{ab}$. The first index refers to the treatment, i.e. there are a different treatments, and the second index refers to the instance of that treatment. In this case, there are b instances of each of the a treatments. The instances for a given treatment, say X_{j1} through X_{jb}, are assumed to be independent and identically distributed and take on values x_{j1} through x_{jb}. Each X_{jk} is assumed to be of the form

$$X_{jk} = \mu_j + W_{jk}$$

where μ_j is the expected value and W_{jk} is a normally distributed random variable with zero mean and variance σ^2. The W_{jk} are further assumed to be independent in both j and k. We can also consider

$$X_{jk} = \mu + \alpha_j + W_{jk}$$

where

$$\mu = \frac{1}{a} \sum_{j=1}^{a} \mu_j$$

and

$$\sum_{j=1}^{a} \alpha_j = 0$$

In this interpretation, μ is the population mean and α_j is the effect from treatment j. The null hypothesis H_0 is that $\alpha_j = 0$ for all $j = 1, 2, \ldots, a$. If this is true, the data from each of the treatments have the same mean value and can be treated as statistically the same. To use the ANOVA to test for hypothesis H_0, we perform the following steps:

1. Find the means of the samples from each treatment. The sample mean from the jth treatment is

$$\bar{x}_{j\,.} = \frac{1}{b} \sum_{k=1}^{b} x_{jk}$$

where this value is sometimes called the jth *treatment mean*. The subscript dot denotes that the average is over the second subscript.

2. Find the overall sample mean. The mean of all of the samples is

$$\bar{x} = \frac{1}{a} \sum_{j=1}^{a} \bar{x}_{j\,.} = \frac{1}{ab} \sum_{j=1}^{a} \sum_{k=1}^{b} x_{jk}$$

3. Find the variation between treatments, v_b:

$$v_b = b \sum_{j=1}^{a} \left(\bar{x}_{j\,.} - \bar{x} \right)^2$$

The degree of freedom associated with this variation is $a - 1$.

4. Find the total variation, v:

$$v = \sum_{j=1}^{a} \sum_{k=1}^{b} \left(x_{jk} - \bar{x} \right)^2$$

The degree of freedom associated with this variation is $ab - 1$.

5. Find the within treatment variation, $v_w = v - v_b$. This variation has degree of freedom $a(b-1)$.

6. Find the mean square for between treatments:

$$\hat{s}_b^2 = \frac{v_b}{a-1}$$

7. Find the mean square for within treatments:

$$\hat{s}_w^2 = \frac{v_w}{a(b-1)}$$

8. Compare the ratio $\hat{s}_b^2/\hat{s}_w^2$ using the F-distribution with degrees of freedom $v_1 = a - 1$ and $v_2 = a(b-1)$ at the desired confidence level of $1 - \alpha$. If

$$\frac{\hat{s}_b^2}{\hat{s}_w^2} > F_{1-\alpha}(v_1, v_2)$$

then we can reject the null hypothesis with a confidence of $1 - \alpha$. Here, $F_{1-\alpha}(v_1, v_2)$ is the value that covers the fraction $1 - \alpha$ of an F-distribution random variable with degrees of freedom v_1 and v_2, i.e. $P(Y \leq F_{1-\alpha}(v_1, v_2)) = 1 - \alpha$, where Y is an F-distributed random variable with degrees of freedom v_1 and v_2. As a result, we can have a confidence of $1 - \alpha$ that the means differ. If

$$\frac{\hat{s}_b^2}{\hat{s}_w^2} \leq F_{1-\alpha}(v_1, v_2)$$

then we cannot reject the null hypothesis with a confidence of $1 - \alpha$.

This test works because if the null hypothesis is correct and $\alpha_j = 0$ for all j, then treating v_b and v_w as random variables V_b and V_w, respectively,

$$E\left[\frac{V_b}{a-1}\right] = Var\,(X_{ij}) = Var\,(W_{ij}) = \sigma^2$$

and

$$E\left[\frac{V_w}{a(b-1)}\right] = \sigma^2$$

If, however, the null hypothesis is not true,

$$E\left[\frac{V_b}{a-1}\right] = \sigma^2 + \frac{b}{a-1}\sum_{j=1}^{a}\alpha_j^2$$

Since this value is larger if the null hypothesis is not true, $\hat{s}_b^2/\hat{s}_w^2$ can be used as a test statistic, where the larger this test statistic is, the more likely that the null hypothesis is not true. It can be shown that this test statistic has an F distribution with $a - 1$ and $a(b-1)$ degrees of freedom.

Example 6.5.6   As an example of the use of DOE, suppose that a software project has a software product that is used on several operating systems and the project has concerns that this software is more prone to failures on a particular such system. To determine if this is the case, they design the following test. There are four different operating systems of interest to them, so $a = 4$. There are also eight primary scenarios for the software, so $b = 8$. The software is run on each of the four operating systems, and for each operating system, each of the eight scenarios is run. All operating systems run the scenarios for the same amount of time and the number of failures for each operating system–scenario pair is recorded. We wish to determine if the operating system choice affects the number of failures. The null hypothesis is that the operating system choice does not affect the number of failures. Table 6.24 shows the number of failures per operating system–scenario pair.

Following the above steps:

1. Find the means of the samples from each operating system. For operating system 1:

$$\bar{x}_{1.} = \frac{45 + 43 + 32 + 34 + 50 + 51 + 46 + 37}{8} = 42.250$$

Taking the other three averages, we have $\bar{x}_{2.} = 41.000, \bar{x}_{3.} = 51.375$, and $\bar{x}_{4.} = 38.500$.

2. Find the overall sample mean. This mean is

$$\bar{x} = \frac{42.250 + 41.000 + 51.375 + 38.500}{4} = 43.281$$

3. Find the variation between operating systems, v_b:

$$v_b = b \sum_{j=1}^{a} (\bar{x}_j. - \bar{x})^2 = 757.094$$

The degree of freedom associated with this variation is $a - 1 = 3$.

4. Find the total variation, v:

$$v = \sum_{j=1}^{a} \sum_{k=1}^{b} (x_{jk} - \bar{x})^2 = 1792.469$$

The degree of freedom associated with this variation is $ab - 1 = 31$.

5. Find the within operating system variation, $v_w = v - v_b = 1035.375$. This variation has degree of freedom $a(b - 1) = 28$.

Table 6.24   Software Failures.

	Scenario 1	Scenario 2	Scenario 3	Scenario 4	Scenario 5	Scenario 6	Scenario 7	Scenario 8
Operating system 1	45	43	32	34	50	51	46	37
Operating system 2	32	43	45	54	34	43	39	38
Operating system 3	54	45	56	43	58	57	49	49
Operating system 4	33	45	43	36	38	40	38	35

6. Find the mean square for between operating systems:

$$\hat{s}_b^2 = \frac{v_b}{a-1} = 252.365$$

7. Find the mean square for within operating systems:

$$\hat{s}_w^2 = \frac{v_w}{a(b-1)} = 36.978$$

8. Compare the ratio $\hat{s}_b^2/\hat{s}_w^2$ with the F-distribution with degrees of freedom $v_1 = a - 1 = 4 - 1 = 3$ and $v_2 = a(b-1) = 4(8-1) = 28$ at the desired confidence level. We have $\hat{s}_b^2/\hat{s}_w^2 = 6.825$ and $F_{0.95}(3, 28) = 2.947$. Therefore,

$$\frac{\hat{s}_b^2}{\hat{s}_w^2} = 6.8248 > 2.9467 = F_{0.95}(3, 28)$$

so we can reject the null hypothesis with a confidence of 0.95. If we require a confidence of 0.999, then $F_{0.999}(3, 28) = 7.1931$, and we cannot reject the null hypothesis at this level of confidence.

More on DOE may be found in [71, 77], and [78].

Inspections and Reviews

A proven approach for improving the reliability in hardware is to provide a parallel hardware system. For a software developer, inspections and reviews can serve as such a parallel mechanism for use during development. Inspections and reviews are potentially highly effective ways of detecting defects in nearly any kind of document, such as requirements specifications, plans, processes and procedures, narrative documents, software code, test cases, and various other document products. There are many steps that can be used, but for purposes of this work, we are referring to a distinct activity in which humans read parts of the document for the purpose of detecting defects. All of the versions of this technique involve one or more humans reading a document. There are various reading techniques that can be employed. In [4], the author considers requirements specifications and the following four reading techniques:

1. Checklist-based reading techniques: For this type of reading, the reader uses a checklist of yes/no questions based on previous knowledge of typical defects.
2. Defect-based reading techniques: For this technique, defects are classified, questions are posed for each defect class, scenarios are built, and each scenario is assigned to a reader to detect a particular type of defect.
3. Perspective-based reading techniques: With this technique, perspectives based on the different roles of people in the organization are developed. For each perspective, scenarios are defined and questions are developed.
4. Combined reading technique: Based on an analysis of the preceding three techniques, the authors of [4] created the combined reading technique. For this, each part of the requirements document is analyzed separately. For each part, a search for defects is performed based on the previous knowledge of defects in the search part of the document and the purpose of that part. Readers read the document and answer yes/no questions from a list.

In [79], software reviews are classified into five types of systematic reviews:

1. Management reviews: Management reviews are reviews under the leadership of management and are to monitor progress, assess status, and determine the effectiveness of a process.
2. Technical reviews: Technical reviews are under the leadership of an engineer and are typically used for a software product or service to determine if it conforms to requirements, adheres to standards, guidelines, and regulations and to verify that there is proper documentation for changes. Role responsibilities include a decision maker, a review lead, a recorder, and technical staff. Technical reviews have entrance and exit criteria. Entrance criteria include authorization for the review and preconditions. Preconditions include a statement of the objectives of the review and availability of the review material. The review material includes the item to be reviewed as well as review procedures and potentially other items. Exit criteria include completion of all steps of the review and availability of all planned output material.
3. Inspections: An inspection is an examination of a software product or document performed by peers with the purpose of detecting defects. The most well-known type of inspection is the Fagan inspection that is outlined below.

   A Fagan inspection is a multistep and detailed process involving four roles:

   (a) Moderator: The moderator manages the inspection and preserves objectivity. The moderator finds a suitable meeting location, ensures that inspection results are reported in one day, and that there is follow-up. The moderator also acts as a coach. In [80], it is recommended that the moderator be a competent programmer but not a technical expert on the program being inspected. Fagan [80] also states that it is advantageous to choose a moderator from a different project to preserve objectivity.

   (b) Author: The author is the person responsible for producing the work that is to be inspected.

   (c) Reader: This position is for the person who describes the work to the rest of the group. The reader should be a subject matter expert on the work.

   (d) Inspector: While each member of the group is an inspector, the group usually has a member chosen for a particular perspective. For example, if the inspection is of software code, someone who writes test cases may be chosen as an inspector.

   Different sources sometimes specify different role positions. In [80], the four positions for a software code inspection are listed as Moderator, Designer (the programmer responsible for producing the program design), Coder/Implementor (the programmer responsible for translating the design into code), and Tester (the programmer responsible for writing or executing test cases).

   With these four positions, the steps in a Fagan inspection are as follows:

   (a) Planning: The planning step is where people are chosen for the inspection, meeting locations are found and scheduled, and other preparatory tasks are handled.

   (b) Overview: During the overview, the group is educated on the nature of the material to be inspected. Roles are also assigned in this meeting.

   (c) Preparation: During preparation, participants review the material to be inspected to understood it. Each participant usually works alone in this step. Defects may be found in this step, but most are expected to be found in the inspection step. Checklists may be provided to aid the preparation.

(d) Inspection: This step is the meeting where most of the defects are found, including their type and severity. It also includes the reader's description of the work. Typically, most of the defects are found during the reader's discourse. To keep the defect detection efficiency high, keep the inspection meeting to around two hours, although longer or shorter meetings may be appropriate in some cases.

The purpose of this step is to find defects, not to correct them. Even if there is a question as to whether some aspect of the material is defective or not, note that it as a potential defect and move on. The moderator should keep the discussion on track and not let it deviate to solutions, blame, or other discussions not related to finding defects.

Within a day of the inspection meeting, the moderator should produce an inspection report that lists each defect found, its type, and its severity. (Note that the moderator may use a recorder to record the results.) This report may be formally entered into the DRACAS database (to track rework progress and to aid tracking inspection effectiveness). It may also be in the form of an e-mail.

(e) Rework: In this next step, defects are removed. This step is usually the responsibility of the author of the work.

(f) Follow-up: Finally, we see that all defects, problems, and issues have been resolved. It is important that all of these are resolved. In [80], it is recommended that if more than 5% of the original material has been reworked, the moderator should verify the quality of the rework or reconvene the team to reinspect either the rework or the entire material.

It is important to schedule these inspections and rework with the same level of importance as other project activities to prevent them from being pushed to the background. Fagan inspections have been very successful in finding defects and improving the quality of products. See [80] and [11] for more details. If done properly, they are cost-effective. However, they can be expensive and time-consuming, so various "light" versions have been explored.

4. Walk-throughs: A walk-through is a peer review usually with the objective of finding product defects, to consider alternate implementations, and to assess conformance to standards and specifications. Role responsibilities include a review leader, a recorder, the author, and team members. The material to be reviewed is distributed sufficiently ahead of the walk-through meeting that participants can review the material before the meeting and participants are usually expected to have questions on the material at the time of the meeting.

5. Audits: An audit of a software product or process provides an independent evaluation of conformance to regulations, standards, and guidelines applicable to the item audited. There are often actions based on audit findings and recommendations that are tracked to completion.

It is also useful for the author of a software product or a document to conduct frequent self-audits. Although a self-audit is not as similar to "hardware parallel" as the other versions listed here, it is "parallel" in the sense of a different and "parallel" action. A self-audit of a document is the activity of checking a document and is performed by the creator of the document. This activity may be very informal and sporadic or it may be more structured and planned. One example of an effective occasion for and type of self-audit is to use checklists for a self-audit of software code after the code compiles but before unit tests. The checklists ensure a level of completeness in the self-audit. Document producers often choose to perform a self-audit before walk-throughs, peer reviews, or formal inspections of their document to reduce defects. The result is usually a better product.

Both code testing and reviews and inspections detect software defects, although different defects are more readily caught by one technique than the other. Also, while testing has been shown to have an average defect detection rate of from 30% to 35%, design and code inspection defect detection rates are from 55% to 60% (see [11]). Finally, peer reviews give the program a greater depth of personnel in that if the software developer leaves the project, there are others who know that code. A few general guidelines for reviews include the following:

1. Make sure that the product is ready for the review. For example, before a code peer review, the code should at least compile or be runnable. An example checklist of conditions that are required before the code is ready for peer review may include the following items:
   (a) The code complies.
   (b) The code is neat and readable.
   (c) The code has readable, insightful comments.
   (d) The code has adequate logging to enable debugging.
   (e) The writer of the code has performed a self-audit of the code.
   (f) The code has successfully been through the developer's unit tests.
   (g) The unit tests have met code coverage criteria. Code coverage tools can be used to ensure adequate coverage.
2. Limit the amount of material reviewed at one time. The review process requires concentration and humans start making significantly more errors after 60 to 90 minutes. For code reviews, this corresponds to a review of no more than 200 to 400 lines of code at a time.
3. Limit the rate of material coverage. Reviewing faster is not necessarily better. For code reviews, a rate of no more than 200 to 400 lines of code per hour is recommended.
4. It is helpful for the author of the material to prepare an explanation of the material for the reviews before the review. Although helpful for the reviewers, the author may also find defects during this preparation.
5. Establish metrics for the review and set metric goals for it. Some examples of metrics for a review are review rate (the amount of material reviewed per hour), defect density (the number of defects per unit of the material reviewed), and defect rate (the number of defects found per hour of review). Later, these metrics are compared with other defect metrics, such as field failures. These metrics can be used to assess how well the review process is working.

   Additionally, consider the use of automated review tools to help collect these metrics. For example, measuring the defect rate is likely to be very approximate if a human reviewer has to stop and remember to record a time. An automated tool can also reduce the human workload so that the person can focus on the review and not the metrics process.
6. The effectiveness of a peer review can be increased by incorporating checklists ideally based on historical data. Multiple checklists can be used, such as a checklist to

ensure that the product is ready for the review and one to ensure that the right aspects of the product are being reviewed. The effectiveness of the review can also be improved by assigning different perspectives to different reviewers. As a result, there may be different checklists for different reviewers. Different types of peer reviews should be performed. Representative examples of things to look for in several different types of reviews are as follows:

Code peer reviews: Examples of different aspects to consider for a code review include the following:

(a) Standards compliance

(b) Style guide compliance

(c) Compliance with program best practices

(d) Readability and clarity:

    i. Does the code have adequate comments to explain what it is doing?

    ii. Is the code aligned so that the blocks of code are meaningful and easily identifiable? Is the code laid out so that it is easy to read and follow and so that different functions are clearly distinguished?

    iii. Is a uniform, meaningful, and easily understood naming convention used?

    iv. Does all of the code fit easily on the computer screen? Having to scroll left or right to see the rest of a line increases the workload of a software developer and a reviewer and increases the likelihood of errors.

    v. Does the code have old "commented out" code in it? Removing non-used code enables the reviewer to focus on what is important.

(e) Error handling

(f) A detailed review of each line of code for correctness

(g) A review for testability:

    i. Is the code easy to unit test?

    ii. Can root causes be assigned to the code easily if an error occurs?

(h) A review for code maintainability

(i) A review for requirements compliance

(j) A review for interface compliance

(k) A review for the types of defects that historically the program or similar programs have encountered

(l) Structure of the code:

    i. Is the code following the defined architecture?

    ii. Are the functions or classes too big?

    iii. Is the same code repeated more than twice?

Requirements peer reviews: Suggested considerations for a requirements peer review include the following:

(a) Are the requirements organized in a well-thought-out hierarchical manner?

(b) Are the requirements as a whole clear and understandable so that there is little chance of confusion?

(c) Are there conflicting requirements?

(d) Do the set of requirements cover all that is needed to specify the system, *i.e.* if the product meets all of these requirements, will the product be acceptable?

(e) Do any of the requirements over specify, telling the designer what the design should be rather than what the design should do?

(f) Are all inputs and all outputs specified?

(g) Check each requirement for ambiguity.

(h) Is each requirement verifiable and, if so, how?

(i) Are external hardware and software interfaces specified?

(j) Do the requirements specify sufficient execution times and response times to ensure that suitably designed software has adequate execution time characteristics?

(k) Are non-functional requirements such as software reliability, availability, maintainability, security, safety, and data protection specified?

(l) Do the requirements sufficiently specify the host platform?

Software architecture peer reviews: Examples for consideration for a software architecture peer review include the following:

(a) Is there a software architecture specification covering the constraints on the overall software design?

(b) Does the specification covers the fault tolerance approaches?

(c) Is error handling described?

7. Consider using some version of a DRACAS and DRB when these processes find defects. The DRACAS and DRB used at this phase may be less formal than for later phases, and it may be internal to the organization that produces the product, but it is still important to verify that the defects have been fixed and to assess the results for trends and possible problems with the project processes. It also provides a way to log the results, follow-through, and metrics, keeping them in a permanent record.

8. Make sure that the project emphasizes the importance of the reviews. For example:

(a) Use qualified people for the reviews. If reviews are considered to be important, highly qualified people will see the "prestige" of being a reviewer and may value the position.

(b) Make the review an important and scheduled activity for the reviewers. Too often, reviews are performed in which review team members are pulled away from their "main task" and so the review is considered to be a disruption to be completed as quickly as possible.

9. Develop and keep a good review environment. Finding defects should be considered good. Nothing should be done that reduces the likelihood of finding defects. For example, the results of a review should not be used to make negative comments in the author's performance review. Also, management should view the review as a team-building activity, and review team members should take pride in the quality of the resulting document. Consider having the name of each review team member permanently attached to the document to aid in developing pride of ownership of the results of the review.

10. Reviews should be performed systematically. As a minimum, they should be performed before considering the product under review to be validated.

These techniques have been shown to be one of the most effective ways of detecting defects, and as an added advantage, they can be used to detect defects early, before a potentially defective product (such as a specification) is passed to others, and these defects are then embedded in the next product (such as a software design document). Also, the types of defects found with these techniques tend to be different from those found with software testing or with formal methods, so these techniques can be complementary.

Determining which variation of the above approaches to take should be determined by project needs, balancing time, and expense with thoroughness and evidence of thoroughness. For more on inspections and reviews, see [79, 80], and [11]

Lessons Learned

Throughout a project, unexpected events occur that we can learn from. These events may have a negative impact on the project and we want to learn how to anticipate and avoid them, or they may have a positive impact, and we want to be able to take advantage of them and attempt to encourage more such events. In either case, documenting these events for future reference can be very useful. Such a list can be especially important to future projects, helping them to avoid negative consequence (or encourage and take advantage of positive events) experienced by other projects.

Typically, there is a procedure for how a project or a company creates and maintains a lessons learned list. Often, projects use a lessons learned template to ensure the consistency, uniformity, and completeness of the list. Many times, a database or spreadsheet is used and the columns are similar to those for an FMEA. Example columns include a unique identifier, the date and time of the event, a category such as the project area that the event and its lesson apply to, a description of the event, the results of the event, the lesson or lessons learned, and recommended actions to mitigate or take advantage of such events in the future. There is frequently a column for comments and remarks that can be used to clarify the situation.

Items may be added to the list as they occur or at pre-determined times. A combination of these two may be appropriate given that the details of the event are most readily available soon after the event, but without a scheduled reminder to add to the list, we tend to forget that we are building such a list. There may also be scheduled meetings to create a more formal "official" list. These meetings can also be used to determine what actions the project should take based on the items in the list. Furthermore, each project should be encouraged to make such a list and the company should make lists available across projects.

Finally, it is the well-thought-out applications of these lessons learned that make a positive difference, not just their collection. Often projects focus their lessons learned effort on producing a list of lessons. While important, a concerted effort should be made toward their application. For example, in the concept development and planning phase of a project, take lessons learned from other projects and determine which might be applicable to the current project and how to apply them.

Models and Simulations

Models are abstractions of the real world. As abstractions, they allow us to work in a less complex world. If the model is adequate for its intended use, working in this less complex world will give useful results that apply to the original real world that we have modeled. The purpose of a model usually falls into at least one of the three categories: describe a system, explain some result, or predict what will happen. Various uses for models include the following:

1. Particularly early in a project, models are used to represent the aspects of the project to aid communication and understanding.
2. Models are created for design concepts to validate their usefulness. Other phases of the project also use models for similar purposes. For example, prototypes may be developed to validate some aspects of the design or of certain processes.
3. Models may be made of the system environment for design validation activities.
4. We sometimes create operational effectiveness models to determine how well the system is expected to meet the overall objectives of the system. These usually use a set of scenarios. Note that it is often important to model non-operational modes also.
5. Models are also used to represent the aspects of the system that are not as far along in their development but are needed to test or further develop a specific part of the overall system. For example, two software modules may share an interface, but one module is developed before the other, so we model the interface from the undeveloped module so that we can test the developed module.
6. Sometimes, a model is made specifically to address a decision that must be made, such as with trade-off analysis.
7. SRE uses models for a variety of purposes, such as reliability and availability allocations, predictions, and estimations. These types of models are covered in more detail in the associated topic in this chapter.

A model may have a "one-time use" or may evolve as the system matures. In this latter situation, the model is likely to start as a simple "low-fidelity" model and become increasingly realistic as development progresses.

Models should be chosen based on the needs of the project. One consideration in model choice is how accurate the results need to be. Always remember that the model is not identical with what it models. A model is an approximation of a few aspects of the real system. Because of this, it can be much quicker and less expensive to develop, but we must also know its limits. Other considerations in model choice are when results are needed (which affects how much time can be devoted to model development and validation), resources available, and priority of the results.

Another important aspect of the choice of model is whether or not it can be validated adequately. "Adequate" varies depending on the purpose and criticality of the model, but all models need some validation to ensure that their representation of the "real world" meets their intended purpose. Some models, such as simulations, may also need to be verified.

There are many types of models. Some are graphical, spreadsheets, or schematic, such as diagrams or charts. A model may be physical, such as a mock up or a prototype. Some models are mathematical, using one or more mathematical expressions to calculate a result. These models may be based on test or operational data (such as using statistical analysis of collected data), based on rules (such as modeling a spring based on Hooke's law), or a combination of both. A good mathematical model should be, among other things, fruitful (meaning that it meets the needs required of it), accurate to the degree needed for its purpose, precise (giving the same or close to the same answer each time), robust (meaning that small changes in a parameter do not make large changes in the results unless this is the behavior of the "real system"), easy to use, understandable, simple (to the degree practical), and validatable.

The quality of the results found with a mathematical model depends on the data used, the accuracy of the model with respect to the desired output, and the accuracy of the method used to find the results. The results found with a mathematical model depend on the solution method used. A solution method may be exact (to the accuracy of the calculations), approximate (such as truncating an infinite series or linearizing an expression to make finding a solution easier), or numerical (such as using numerical methods to solve a system of partial differential equations or running a simulation). Note that there may be a trade-off in the accuracy of the model and the accuracy of the method used to find the results. Also, the accuracy required of the model depends on the accuracy of the input data.

A simulation is an important type of numerical mathematical model. Simulations allow us to gain an understanding of how the system will behave before the system has been made. Unlike some models, they let us see behavior as a function of time. They are particularly important for modeling complex systems. Simulations always add simulation error to the answer. Increasing the number of runs in a simulation can reduce this error but at the expensive of taking more time to produce the final output.

Another type of model is the use of a system modeling language, such as UML or SysML. UML is software oriented and can produce models to represent both structural and behavioral aspects of a system. SysML is an extension of UML that better includes hardware aspects of the system. These modeling languages provide a standard way to visualize the system as it is developed. These languages often employee use cases, sequence diagrams, and other techniques. Model-based systems engineering (MBSE) expands on these system modeling languages to where they are used as the primary representation of the design. These models are developed early in the project life cycle, and as the design matures, the models mature to be the build-to-baseline.

See [2, 81, 82], and [83] for more details on models and modeling.

Example 6.5.7　There are many types of models for many types of applications. In this example, we consider an operational effectiveness model for a communications system. The system communicates using radios, and there are six communication nodes. The network of communication links is as shown in Figure 6.2, and links are labeled from A to H. The objective is for node N1 to communicate with node N2.

Each communication path from N1 to N2 is capable of handling the required communications throughput by itself. The network routes messages based on the number of links required for the communications path. If all links are operational, the network will route the communication from Node N1 to Node N2 through link B and then link F. If either link B or link F is down, it goes from link C and then link G. If neither of these paths are operational, it chooses BEH, followed by ADH, and finally, if all else fails, it routes it through ADEF.

Recall that this model is an operational effectiveness model. The chosen measure of operational effectiveness is the probability of successful communications between nodes N1 and N2. For a given link to be operational, it must be available, reliable, and survivable. Here, survivable means that the weather is not too extreme for the radio to transmit to the other node of the link. There are infrequent but very local and severe rain cells that, if they occur between two nodes, can prevent adequate communications. The probability of link A being operational is then $P_A = A_A R_A S_A$, where A_A is the availability of the link, R_A is its reliability, and S_A is its survivability. We assume that these three quantities are independent and

Figure 6.2 Example Communication
Network.

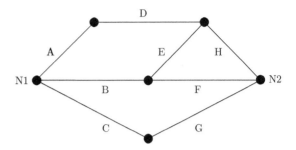

that the probabilities for the different links are also independent. (For example, the transmitter and the receiver at a given node are independent.) Using these link probabilities, we can calculate the probability of successful communications between nodes N1 and N2 and use this to make design trades. For example, if we increase the reliability of just one node, which node should we choose to obtain the most improvement in the probability of overall successful communications between nodes N1 and N2?

There are multiple ways to calculate the overall probability of communications from the individual link probabilities. One common approach is to use a Boolean table. With this method, we consider each link, A through H, individually. Each can either be "UP" and assigned a value of 1, or "DOWN" and assigned a value of 0. Each row of the Boolean table has one combination of possible up/down values for the eight links. The first row of the table has values 0 0 0 0 0 0 0 0 corresponding each of the nodes A through H being down. The next row has 0 0 0 0 0 0 0 1 corresponding to all links being down except H, which is up. The last row has 1 1 1 1 1 1 1 1 corresponding to all links being up.

We then assign a probability to each row. Each link is assigned a value of P or $1 - P$ depending on whether the corresponding link is up or down. The value of P is specific to the particular link in question. For example, in the second row (corresponding to the values 0 0 0 0 0 0 0 1), link H, which is up, is assigned a value of $P_H = A_H\,R_H\,S_H$, whereas link G, which is down, is assigned a value of $1 - P_G$. After assigning these probabilities to a row, the probability that this exact combination of up and down links will occur is the product of all of these link probabilities in the row. We can multiply the probabilities because the links are all assumed to be independent.

The final step is to determine which rows correspond to successful communications between nodes N1 and N2 and which do not. For example, the first row is not successful because no links are operational; however, the last row is successful as all links are operational. We then add the probabilities of the successful rows to find the probability of communications success. We can add the probabilities because the row probabilities are mutually exclusive.

The Boolean table approach is straightforward but can become very large. This example has $2^8 = 256$ rows and so is relatively small, but large networks can have huge tables. A second approach is less mechanical but is more insightful and results in a more succinct model. For it, we use the following steps:

1. Start with the main path and calculate its probability of success. For this example, it is the BF path, which has a probability of success of $P_B\,P_F$.

2. Next, find the probability of success through the second path and multiply it by the probability that the first path is unsuccessful. This probability is $P_C P_G (1 - P_B P_F)$. It is important to multiply by the probability of the first path being unsuccessful because we want to add these path probabilities, and to do this, they need to be mutually exclusive. The factor $(1 - P_B P_F)$ may be interpreted as the probability of not having both link B and link F.

3. We continue with the probability of success with the next path, multiplying this probability of success by the probability that both of the first two paths are unsuccessful.

4. Continue until we have covered all of the paths, and then add the mutually exclusive probabilities to find the probability of communications success.

For the five potential paths, we have the following mutually exclusive probability expressions:

1. $P_B P_F$
2. $P_C P_G [1 - (P_B P_F)]$
3. $P_B P_E P_H (1 - P_F) [1 - (P_C P_G)]$
4. $P_A P_D P_H [1 - (P_C P_G)] \{(1 - P_B) + (1 - P_E)(1 - P_F) - (1 - P_B)(1 - P_E)(1 - P_F)\}$
5. $P_A P_D P_E P_F (1 - P_B) [1 - (P_C P_G)] (1 - P_H)$

The path through links A, D, and H deserves mention. To make it mutually exclusive with the previous paths, we need to exclude going through the BF, CG, or BEH paths. Note that link B is in both the BF path and the BEH path. There are two ways to prevent these paths. If link B is down or if both links E and F are down, then both of these paths are down. We use the inclusion–exclusion law for expressing this. The probability of B being down, plus the probability of both E and F being down, then subtracting the probability that all three are down. Both the $(1 - P_B)$ term and the $(1 - P_E)(1 - P_F)$ term cover the probability that all the three links, B, E, and F, are down, and as we only want to count it once, we subtract $(1 - P_B)(1 - P_E)(1 - P_F)$.

As a final comment, we have considered two approaches for finding the probability of successful communications through the network. Unless a model is simple enough that it is obviously constructed correctly, it is a good idea to produce more than one model, each using a different modeling technique. In this case, we model the network probabilities using a Boolean table and also using mutually exclusive paths. By doing this, one model can serve as a check of the other. If both models agree, we have more confidence that whichever model we choose as the main model is correct. To use one model as a check of the other, vary the inputs to the models, but put the same inputs into both models during the checking. Other model checks include the following:

1. If we assume that all links are perfect and cannot fail, does the model give a network probability of one?

2. If all of the links have a probability of success of zero, is the network probability of success equal to zero?

3. If all of the links have a success probability of 0.9 and we increase the link probability of one of the links to 0.95, does the network probability increase?

Knowing how to check a model is an important part of creating the models.

Pareto Analysis

Pareto analysis is a technique for prioritizing actions when it is not feasible to address all of the actions immediately. The basic idea behind the technique is that when there are repeated problems, such as defects in a process, there are usually a few causes that produce most of the defect instances, with the remaining causes producing relatively few instances. This observation is simplified into the famous "80/20 rule": 80% of the problems are from 20% of the causes.

Steps for performing a Pareto analysis on problem occurrences are as follows:

1. Collect data on the problem occurrences and their causes.
2. Calculate the percentages of occurrences per cause.
3. Arrange the causes in decreasing order with the most frequently occurring cause first.
4. Calculate the cumulative frequency. If the percentage for the most common cause is $X_1\%$ and for the second most common is $X_2\%$, then the cumulative for the most common is $X_1\%$ and for the second is $X_1\% + X_2\%$, and so on.
5. Plot the causes and their cumulative frequencies on a vertical bar chart, with the causes listed along the horizontal axis and the associated frequencies on the vertical axis.
6. Plot a line graph of the cumulative percentages on the same chart.
7. Draw a horizontal line at the 80% mark.
8. Draw a vertical line where the 80% horizontal line intersects the cumulative frequency plot.
9. The causes to the left of this vertical line represent the most important causes.

The analysis is used to prioritize effort to ensure that the top priorities are handled first. See [71] for more details.

Process FMEA/FMECA

A process failure modes and effects analysis (PFMEA) is a bottom-up analysis of a process and is used to detect potential process failure modes and their effects at higher levels. An FMECA is an FMEA plus a CIL, which is a prioritized list of the most critical items. In this narrative, we will use the acronym PFMEA whether a CIL is created or not. A PFMEA is often associated with manufacturing processes, although it can be used for other types of processes as well. A software FMEA is covered in a separate topic in Section 6.4.

A PFMEA starts with defining the process to be analyzed. It is useful to produce a flowchart or some other representation of the process to ensure that it is well understood before analysis and that each step of the process is identified. A PFMEA is typically constructed in a tabular form with a row for each failure mode. Typical columns for a row include the following:

1. Row number: The row number uniquely identifies the row.
2. Step name: The step name is a descriptive name of the process step analyzed in the row.
3. Failure mode: The failure mode is an example of what can go wrong with this step. Each step in a process typically has multiple failure modes.
4. Effects: This column considers what could happen if the failure mode occurs. Often, there are multiple columns for effects, such as one for the effects on the current step, another for the effects on the process as a whole, and a third for the effects to the project or the final product.

5. Targets: This column states who will be impacted if the failure mode occurs.
6. Root causes: This column lists things or events that could cause the failure mode. There may be several potential root causes for a given failure mode.
7. Detection method: The detection method column lists a method or methods to detect the failure mode.
8. Mitigation: This column lists one or more activities that can prevent or reduce the negative impacts of the failure mode.
9. Risk priority number (RPN): The RPN is a numerical prioritization of the risks involved with the failure mode. It is usually found by multiplying the severity of the failure mode (1 for insignificant to 5 or 10 for catastrophic) by the failure mode likelihood of occurrence (1 for nearly impossible to 5 or 10 for almost certain) by the detectability of the failure mode (1 for nearly guaranteed to be detected to 5 or 10 for almost undetectable).
10. Comments: This column contains any comments that provide additional useful information.

After performing a PFMEA on a process, the higher risk failure modes should be collected for potential actions. See [71] for more information on PFMEAs.

Example 6.5.8   For this process FMEA example, we consider the process of performing a code inspection. One of the steps in performing a code inspection is the inspection meeting. During the inspection meeting, defects are found, and their types and severities are determined. Early in this meeting, a reader describes the work that is to be inspected. This reading is an important part of the overall process and often a significant number of the defects are found during this reading. This process FMEA example considers this part of the inspection. Note that the example has similarities with Example 6.4.4.

This example considers one aspect of the inspection meeting, namely, the reading of the description of the work to be inspected. It considers failure modes that can occur during this reading, such as the reading having too much detail, not enough detail, being unclear, or loss of control of the meeting during the reading. For example, the reader may get too involved with the reading and provide too much detail or unnecessary details. The reader may provide too little detail or not explain the material very well. Finally, the reader may lose control of the meeting during the reading, with side conversations and other activities going on rather than participants paying attention the reader's descriptions. These failure modes can result in loss of interest, misunderstanding the material, or an overall confusion as to what is taking place. Table 6.25 shows how the process FMEA can document these, along with who is impacted by them and what causes them. For this example, we assign row numbers 130–134 to these failure modes.

Table 6.26 continues the example by considering the detection methods, criticality, likelihoods, and RPN values, along with mitigations and comments. For this process FMEA, the following failure criticality levels are used:

1. Level 1: Negligible, no noticeable impact to the process
2. Level 2: Minor, noticeable but does not impact the main objectives of the process
3. Level 3: Moderate, does impact the process but there is a work-around that prevents process degradation

Table 6.25   Process FMEA.

Row Number	Step Name	Failure Mode	Effects	Targets	Root Causes
130	Reading of code description	Too much detail	Loss of interest	Code inspectors	Inexperienced reader
131		Too little detail	Misunderstanding of material	Code inspectors	Reader does not understand the code
132		Unclear description	Misunderstanding of material	Code inspectors	Reader does not understand the code
133		Loss of control of meeting	Confusion	Code inspectors	Insufficient training
134			Misunderstanding of material	Code inspectors	Insufficient training

4. Level 4: Serious, it impacts the process and at least some of the process objectives are negatively impacted
5. Level 5: Critical, process failure

Similarly, the following failure likelihoods are used:

1. Level 1: Less than 1% likelihood of the failure mode occurrence
2. Level 2: Between 1% and 5% likelihood of the failure mode occurrence
3. Level 3: Between 5% and 25% likelihood of the failure mode occurrence
4. Level 4: Between 25% and 50% likelihood of the failure mode occurrence
5. Level 5: Greater than 50% likelihood of the failure mode occurrence

Using these values, an RPN of 1 is likely to be acceptable, whereas an RPN of 25 is a serious matter that should not be allowed.

Process Improvement Techniques

Different projects rely on processes to differing degrees. Some projects are relatively disorganized and do not have many well-defined processes, while others are heavily dependent on rigid processes. Some projects use an agile philosophy and emphasize people and their interactions more than processes. However, for any project, processes are important but must be properly managed. A few process development and improvement techniques from [71] are listed below:

1. Process observation: Before improving a process, it is a good idea to know what process is actually used. Note that the process that is actually used may not be the same as the documented process, and the results that a project obtains are from the process as followed, not as documented. Process observation involves choosing a process to observe

Table 6.26 Process FMEA (cont.).

Detection Method	Criticality	Likelihood	RPN	Mitigation	Comments
Human detection	2	3	6	Stop processing when detected. Request more concise explanation.	Make sure that the reader is trained and sufficiently experienced
Human detection	3	2	6	Stop processing when detected. Request more details or reconvene when reader is ready.	Make sure that the reader is trained and sufficiently experienced
Human detection	4	2	8	Stop processing when detected. Request more clarity or reconvene when reader is ready.	Make sure that the reader is trained and sufficiently experienced
Human detection	5	1	5	Stop processing when detected. Request that participation rules be followed.	Make sure that the entire inspection team is sufficiently trained
Human detection	4	2	8	Stop processing when detected. Request that participation rules be followed.	Make sure that the entire inspection team is sufficiently trained

and identifying why the process is to be observed. Next, plan the observation activities and train the observers. It is usually a good idea to create an observation form to capture pertinent observations.

2. SIPOC diagrams: SIPOC stands for Suppliers, Inputs, Process boundaries, Outputs, Customers. A SIPOC diagram is a high-level map of a process that helps identify the basic elements of the process and match process inputs and outputs with sources and needs outside the process. The diagram has five headings: Suppliers, Inputs, Process, Outputs, and Customers. Usually, construction of the diagram starts with the process under consideration.

   (a) Process: Under the process column, briefly list the process steps. Keep it at a high level with no more than five or six steps. Sometimes, a simple vertical flowchart is drawn.

   (b) Outputs: Next, identify the key outputs of the process and write them in the outputs column.

   (c) Customers: After this, list the customers of the process under the customers column. These customers may be internal customers rather than an external customer of the project. These are the people or organizations that benefit from the process.

   (d) Inputs: Next, list the inputs to the process under the inputs column. Focus on the main inputs with a goal of listing no more than five or six inputs.

(e) Suppliers: After inputs, list the suppliers of the inputs. Sometimes, a short list of "critical-to-quality" requirements are included. Requirements for the inputs are listed with the inputs and likewise for requirements for the outputs.

3. Process flowcharts: Processes are often documented using a flowchart. A flowchart is a collection of boxes, diamonds, ovals, and sometimes other shapes, connected by arrows. An arrow goes from one box or shape to another, indicating that the output of one box is an input to the other. A common symbology uses ovals for inputs and outputs to the process, rectangular boxes for steps or basic sub-processes, and diamonds for decisions. Text is used to label each symbol and each output from a decision diamond. Many flowcharts are also color coded for easier use.

4. Swim-lane flowcharts: A swim-lane flowchart is a type of flowchart that more clearly shows who is responsible for a given input, output, or task. For this type of flowchart, the responsible people or job functions are listed down the left side of the page. The "swim-lanes" are designated by horizontal lines drawn so that each responsible person or job function has a horizontal strip of the page. The flowchart symbols are then placed in the strip corresponding to the responsible party and the arrows of the flowchart are drawn connecting them.

5. Value-added versus non-value-added analysis: The purpose of value-added (VA) versus non-value added (NVA) analysis, or VA/NVA analysis, is to reduce process complexity and to find hidden costs that do not add value for the customer. VA/NVA analysis classifies each process step into one of the following three categories:

(a) Value added: Value-added activities are those that are essential for the product or service delivered to the customer. These activities progress the item toward completion, are done right the first time (and so are not rework), and are activities that the customer would pay for. If an activity is such that if it were not performed, the customer would likely complain, then the activity is probably value added.

(b) Business non-value added: A business non-value-added activity is an activity that does not add real value from the standpoint of the customer but is required by the business to execute one or more value-added activities. These activities include legal requirements, activities that reduce risk, and activities that support reporting.

(c) Non-value-added: Non-value-added activities are activities that do not add value from the customer's standpoint and are not required from a business standpoint. Examples of these types of activities include overproduction, delays and idle time, and producing more features than the customer needs.

To perform VA/NVA analysis, first classify each step in the process as value added, business NVA, or NVA. Next, add up the time spent in each category. Finally, remove non-value-added activities, then try to reduce the time spent on business non-value-added tasks, and finally, try to optimize the value-added activities.

Project Best Practices

A project should determine and list what constitutes best practices for the given project. There are typically multiple lists with different but compatible items for different job areas. A few examples potentially applicable for someone writing software code include the following:

1. Keep it short. More code is not necessarily better code. Also try to keep functions short. If the code for a function goes past 30 or 40 lines, see if it can be broken up into two or more easy-to-understand and easy-to-test functions.

2. Keep it simple. The ability to write complex code is not an indication of a good software developer. An indication of a good software developer is the ability to make potentially complex code simple.

3. Try to avoid complicated logic, such as multiple nested if/then or other control statement blocks. As with the "keep it simple" best practice, complicated logic makes code more difficult to write, refactor, review, and maintain.

4. Use the highest possible language for the code. Use lower-level languages only when necessary. Per [11], programmers working in high-level languages are more productive and produce higher-quality code than those working is a low-level language.

5. Comment the code adequately. Good code also means well-commented code. Although the code may seem self-explanatory to the software developer right after it is written, it may be far less obvious to another person reviewing or maintaining the code. Also, what is obvious to the software developer today may be confusing to the same software developer in a year. Good comments should explain the code but not distract from the readability of the code.

6. Keep the code comments current. There is a tendency to update the code and wait until later to update the comments to the code. This tendency can result in comments that are incorrect or confusing and makes errors and defects more likely.

7. Have regularly scheduled code refactoring. Code complexity tends to grow the more the code is worked on, resulting in an increased likelihood of errors. By refactoring often, we make the refactoring easier and decrease the likelihood of difficult-to-find defects.

8. Make the code testable. Create well-defined interfaces and keep functions simple so that their tests can be reasonably simple. Design the code with testability in mind.

9. Test often. Software developers who test their code often tend to have fewer defects than those who test more infrequently.

10. Use automated tests, such as for unit testing, integration testing, and regression testing. Test automation reduces human errors and is generally more efficient and schedule friendly.

11. Use proper error-handling and cleanup techniques. This practice reduces software aging issues and reduces the likelihood of certain types of defects that can be very difficult to find and remove.

12. Design the code to be extensible. Make components easily replaceable with newer components. This practice makes code maintenance much easier and improves code reliability. It also simplifies code updates.

13. Programmers generally work best by making small incremental changes with frequent feedback. The sooner that errors are found, the less likely it is that they will be propagated to later products and phases where it is likely to be harder and more time-consuming to find and remove them.

14. Optimize code only after it works correctly. Correct performance is more important than optimization, and optimizing an incorrect code may be work that has to be redone after the code is corrected to work correctly.

A few comments on best practices are as follows:

1. Consider having a best practices plan for the company. The plan would establish where best practices are stored, a history of each (to help decide when it is applicable), and a procedure for establishing, using, and updating best practices on individual projects.
2. Also, consider scheduling periodic best practice reviews to assess if updates are needed.
3. Although best practices can be useful, they should also be understood and questioned. Yesterday's best practice may be today's weak link, but without understanding why it became a best practice, we do not understand the implications of changing it.
4. Each best practice should have a rationale for its inclusion but continually look for even better practices, even if it is just a better practice for one project.
5. There are many sources for best practices. One is from the experiences of project personnel and personnel of other projects in the company. Lessons learned, a techniques covered earlier in this section, can also provide best practices. Finally, there numerous other sources of potential best practices from books and articles, such as [84] and [11], but always make sure that the best practices are understood and apply to the project at hand.

Project Premortem

A project premortem is a useful technique for anticipating potential issues with a project so that actions can be taken to prevent them. There are different versions of the technique, but they all use as an analogy the concept of a postmortem. A postmortem starts after we recognize that there is a failure, such as the death of a patient at a hospital. We then perform a postmortem to determine why the patient died and what we might have done differently to prevent the death. A project premortem generally starts at the beginning of the project, although the technique can be used preceding other major events or project phases. For this technique, we start with the assumption that the project has failed and a meeting is held to determine why. All major project stakeholders are represented at the meeting. During the meeting, stakeholders list and record reasons for the failure. Everyone should feel free to list any reason, but do not propose solutions at this point. After it is agreed that a list covering each stakeholder's reasons has been created, pick the top reasons and find solutions. After creating solutions, decide which to implement and assign these to responsible parties.

Example 6.5.9   For a simple example of a project premortem, we consider a three-person premortem occurring at the start of a project with the intention of reducing the number of defects in the software. The three people are knowledgeable representatives from the requirements, software, and testing organizations. For this example, they are named Requirements, Design, and Testing, respectively. The meeting also has a recorder. Recall that at this time, the project has not developed any software, but the purpose of a premortem is to anticipate potential sources of defects before these sources cause problems.

After introductory remarks, Requirements, Design, and Testing start stating why the software has so many defects:

(1) Requirements: "The customer keeps changing what he wants."
(2) Design: "The requirements keep changing so much that the designers and coders cannot keep up."

(3) Testing: "By the time the testers have a product to test, schedule and cost overruns have shortened the test schedule and budget to where only a small amount of the originally-planned testing can get done."

After this first round, each discipline has contributed one suggestion. Often, in these early rounds, suggestions will be "typical" complaints heard from the discipline, as is the case here. However, it usually does not take long for participants to expand outside of their disciplines:

(4) Requirements: "The coding styles and practices of the various coders are inconsistent, resulting in confusing code."

(5) Design: "There are too many requirements, resulting in bloated code."

(6) Testing: "Nobody knows what is going on. Everybody finds out at the last minute that somebody wants something from them."

This process continues until either no one has anything else to contribute or a pre-agreed time limit has been reached. At this point, there is a list of potential reasons for defective software. The next step is to pick the top reasons and find solutions. Suppose that items 2 and 6 are among the chosen reasons. Item two is that there is excessive requirements churn. Potential solutions include creating project metrics for requirements churn to monitor for this problem. Also, requirement changes could be made progressively more difficult to make as the project progresses. For example, after the design organization has a set of approved requirements to work from, a higher level of approval would be necessary to add, remove, or change a requirement and the stated reason for the change would have to be more detailed and compelling. These and other potential solutions may be considered and adopted.

As for item six, "Nobody knows what is going on," this reason for defects amounts to issues with planning and communicating these plans. Suggested solutions may include developing a project schedule that lists internal products, who needs them, and when they are needed. There may need to be descriptions of these products so that the producer of the product knows what it takes to adequately support the product user. There may need to be a collection of internal "contracts" stating who needs what and when, and agreements to provide the items as stated. These contracts will need to be considered when there are project changes. Another part of the solution may be for each organization to have representatives from other organizations attend their status meetings or for status reports to be produced and distributed to all organizations.

This project premortem technique, like brainstorming, is very free-flowing and can be very effective in getting expertise and experience out to the rest of the project.

Prototypes

Prototypes are "shortcuts" to demonstrate certain aspects of the product. A prototype is a mock-up or partial implementation of the modeled system. Prototyping can be used for a variety of aspects of the product. Two examples are as follows:

1. Prototyping requirements: Prototyping for requirements elicitation improves user involvement and communication between users. Examples include storyboarding, screen mock-ups, and pilot systems. Requirements prototyping can result in a clearer

understanding of requirements and potentially reduce the number of requirements added after the requirements phase.

2. Prototyping software: One of the main applications of prototypes is prototyping software code. Prototyping code is usually a risk reduction technique for some higher risk software functionality. It demonstrates software functions and capabilities to assess if the proposed design is feasible and meets the intended needs. Prototypes are typically "throw-away" code and are not developed for usability, supportability, or other aspects that the final code should have.

Software prototypes are not intended to have the same standards applied to them as the actual product, so avoid trying to expand them into the product. Prototypes are valuable for what is learned from them, not for the code they produce. Also, prototypes should be planned to address one or more identified risks. Be sure to document what was done, the results and their implications, and any conclusions. See [31] for more on software prototyping. Also, prototypes are a type of model, so the topic in this section on models and simulations may be helpful when considering prototypes.

Quality Circles

A quality circle is a group of employees who perform similar work and who meet regularly to identify and solve work-related problems. The group may be self-organized by the employees or organized by management. The circles should meet and perform their work during regular work hours and not be a "burdensome" after work or unpaid task. One of the principles of quality circles is to break up the usual work hierarchy so that each member can contribute equally. Training in how to use quality circles is often needed. For more information on quality circles, see [85].

Rationale Documentation

Documenting the rationale for decisions is a very useful and general technique that serves multiple purposes. One purpose is to record why a decision was made so that later in the project, or on future projects, the decision intent can be found. Projects usually exist in dynamic environments and knowing the original rationale for a decision can help determine if the decision is still appropriate. Also, the process of determining the rationale for a decision and clearly documenting it forces the decision makers to make decisions consciously and systematically. Finally, documenting the rationale makes sure that everyone has a common understanding of why the decision was made, reducing the likelihood of different people having different understandings of the meaning of the decision situation.

Rationale documentation can be used for many products. When used for conceptual designs, it helps to clarify the strengths and weaknesses of each potential design and after one has been chosen, aids in its validation. When used for requirements, it clarifies each requirement, helps reduce requirement misunderstandings, and aids verification. It also tends to reduce the number of requirements by helping to identify redundant or overlapping requirements and unnecessary requirements.

It can be used to document plans, such as the reliability program plan, detailing why certain techniques and processes are used. It can also be used with design decisions, clarifying why certain decisions are made, improving cross-discipline communications and understanding, and reducing potential verification issues.

Root Cause Analysis

RCA is the process of finding the primary, or root, cause or causes of a defect, failure, or problem. It is an important part of reliability growth. Without knowing the root cause of a defect, we are working with symptoms, not the main causes, and the likelihood of this type of defect reoccurring is much greater. A root cause investigation typically uses the following steps:

1. Define what happened: RCA is initiated based on some unsatisfactory event. In this step, we precisely define this event. Do this as soon after the event as possible to preserve as much data as possible.
2. Data collection: Collect all data that may have bearing on a problem or undesired event. Again, do this as soon as possible to prevent loss of data.
3. Assessment: In this step, we identify the problem and its significance, identify the immediately preceding causes, and work back to find the root cause or causes. Various techniques may be used in this process. Some are listed below. The RCA for a project defect is usually reviewed and approved by the DRB.
4. Corrective actions: After determining the root cause or causes, the next step is to determine if corrective actions are appropriate. If they are, put measures in place to prevent a reoccurrence of the event by determining and implementing corrective actions and setting up monitoring to determine if the corrective actions are effective. The corrective actions for project defects are usually reviewed and approved by the DRB.
5. Record: Enter the problem, analysis, analysis results, corrective actions implemented, and expected results from the monitoring in the DRACAS database. Recording is performed incrementally as data become available. For example, a description of the problem and other associated data are entered first. As other data are collected, they are entered. After analysis has been performed, these results are entered, and so on. If it is determined that corrective actions are not to be performed, document the reasons for this decision.
6. Monitor and follow-up: Monitor the corrective actions to ensure that they are effective. If not, reassess the situation and update the corrective actions.

There are many techniques that can be used to find the root cause or causes of a failure or defect. Several are summarized below:

1. Five whys: Five whys is a technique for digging deeper in root cause analysis to prevent acceptance of superficial causes. To perform the technique, start by writing the problem outcome. Ask why this outcome occurred and list the answers. Select one of these answers and ask why it occurred. Continue until a root cause is found.
2. Appreciation: The appreciation technique is similar to the five whys technique. With the appreciation technique, we take a fact related to the event and ask "So what?" We are seeking to understand what this information implies and if it is important. Generally, we repeat this process several times to drill down as far as is needed to fully understand the information.
3. Fault tree analysis (FTA): FTA is a systematic top–down fault or failure analysis technique for taking a single failure and identifying all (or nearly all) potential reasons for

the failure. FTA can be used to support FMECA development and updates, design studies, failure investigations, and other activities. See [17, 25, 39], and [26] for more details. Also, SFTA is covered as a separate topic in Section 6.4.

4. Fishbone diagrams (cause/effect diagrams, Ishikawa diagrams): Fishbone diagrams are used to relate potential causes to effects. The technique typically involves listing the main problem at the "head of the fish," which is typically on the right side of the diagram. Draw a horizontal line from the "head" to the left. Next specifying major cause categories and labeling these above and below the horizontal line, connecting them to the horizontal line with angled lines that tend to make the resulting diagram looks similar to the skeleton of a fish (hence the name of the technique). After this, brainstorming or other "idea generation" techniques are used, and the results are organized hierarchically on the diagram associated with a given major cause. See [71] and [26] for more details.

5. Scatter plots and correlation analysis: When searching for a root cause, there may be several factors involved. It may be that two or more independent events are required to cause the observed failure, or it may be that one of the factors causes the others or that a hidden factor causes both of them. Part of the process of determining if any of these situations apply is to determine if one factor correlates with another. A popular measure for statistical correlation is the correlation coefficient, also known as the Pearson correlation coefficient or Pearson product–moment correlation coefficient. Suppose that we have collected two sets of numerical data, X and Y, and want to determine if they are correlated. The correlation coefficient $\rho(X, Y)$ for these data sets is defined by

$$\rho(X, Y) = \frac{E\left[\left(X - \mu_X\right)\left(Y - \mu_Y\right)\right]}{\sigma_X\,\sigma_Y}$$

where μ_X and μ_Y are the mean values for data sets X and Y, respectively, σ_X and σ_Y are the standard deviation values for the data sets, and $E[\,\cdot\,]$ is the expected value operator, which means that we take the mean value of what is in the brackets. If $\rho(X, Y)$ is close to zero, the data sets are not correlated. If it is close to 1, there is a positive correlation, and if it is close to -1, there is a negative correlation. When $\rho(X, Y) > 0.65$, it is usually a sign that there is significant positive correlation and $\rho(X, Y) < -0.65$ provides evidence of a significant negative correlation.

If there is correlation, is there a causal relation? Correlation analysis does not give us that information. It does however point to where we need to look. See [71] for more details on correlation.

6. FMEA/FMECA: After a software defect has been discovered, the software FMECA can be consulted to see what defects have the same characteristics as those observed with the subject defect. If no such defect is found, the software FMECA should be updated to include the new defect information. Software FMECAs are covered as a standalone technique in Section 6.4.

7. Brainstorming: Sometimes, the root cause is not obvious and we need to create a list of possible root causes to investigate. Brainstorming is a well-known technique for generating ideas. This technique can be used for idea generation in general and so is covered as a standalone technique above in this section.

8. Event and causal factor analysis: Event and causal factor analysis is used when a problem has a long or complex causal chain. It creates a cause-and-effect diagram describing the time sequence of tasks along with the conditions surrounding them, providing a visual display of the process and helps to identify where deviations from the correct method occur. See [86] for more details.

9. Barrier analysis: In barrier analysis, a barrier is something that separates an affected component or process from an unintended situation. Barrier analysis is a systematic approach for identifying barrier failures resulting in hardware, software, or process failures. See [86] for more details.

10. Change analysis: Change analysis is used when the cause is obscure and this type of analysis is generally used in support of a larger RCA. It compares the situation in which the failure occurred with other similar situations in which a failure did not occur and identifies differences. These differences are analyzed to determine a cause for the failure. See [86] for more details.

11. Human performance evaluation: Human performance evaluation identifies factors that influence task performance and is used when the situation has significant human involvement. Typically, the analysis covers input detection, input understanding, action selection, and action execution. For each, issues with operability, procedures, documents, and management attitude are considered. Human performance evaluation can be an effective analysis approach for human-related defects but requires training to apply correctly. See [86] for more details.

12. Orthogonal defect classification: ODC is a technique for categorizing defects for easier use. This categorizing can significantly reduce the time to find root causes and can also help identify issues earlier. ODC is covered in more detail in Section 6.4.

For more details on RCA, see [25, 86] and [71].

Example 6.5.10 As an example of the use of the five whys technique, suppose that we run some gaming software that our project has written and the software crashes. The problem is obviously that the software crashed. A potential sequence of questions and answers is as follows:

1. Why did it crash? It crashed because when the user tried to enter the quick graphics mode, the quick graphics mode software was not installed.

2. Why was the quick graphics mode software not installed? The quick graphics mode software was not installed because there is no requirement for it.

3. Why is there no requirement for the quick graphics mode software? There is no requirement for it because the requirements team thought that the quick graphics mode would be added in a later build.

4. Why did the requirements team think that the quick graphics mode would be added in a later build? The requirements team thought that the quick graphics mode would be added in a later build because they were using an outdated build schedule.

5. Why was the requirements team using an outdated build schedule? They were using an outdated build schedule because the build schedule was updated without their notification or input. (At this point, there are two obvious paths to take to continue the questions. The first is to consider why the requirements team was not notified of the build schedule

change and the other is to question why they had no input to the build schedule change. For this example, we consider the first of these questions.)

6. Why was the requirements team not notified of the build schedule change? The requirements team was not notified of the build schedule change because Alice sends those notices out and she was on vacation and no one thought to send the notification out for several days.

We could continue with this line of questioning, but one obvious problem is that this software organization seems to be lacking a process for making sure that processes are followed without regard to who is at work. One way of improving their processes is to ensure that each team, including the team responsible for the build schedule, follows process checklists so that a step similar to "send build change notifications to project personnel" must be performed before the build schedule change is complete. It should also be noted that the second line of questioning, why the requirements team had no input to the build schedule change, needs to be pursued as well.

Statistical Process Control

Widely used in hardware manufacturing, statistical process control (SPC) or statistical quality control can be an effective way to determine if processes are not producing the results that are expected of them. To efficiently and effectively manage a project, metrics should be taken for each significant product and process of each project phase and SPC can help monitor and improve them.

SPC makes extensive use of control charts. These charts provide a visual representation of variations and trends in the monitored quantities, helping the user distinguish between common or natural variation that is expected with a process and an abnormal deviation that requires project action to bring the process back in line. For most control charts, the abscissa ("x-axis") for a control chart is time. The ordinate ("y-axis") of the control chart is the value of the metric. A control chart has a horizontal line drawn at the average value for the samples and two other horizontal lines known as control lines. The purpose of these control lines is to indicate whether the plotted values, in the order that they were obtained, behave like a set of randomly drawn samples from a single population. If not, then we need to investigate why this deviation has occurred and possibly make changes to the process. These control lines are often drawn at three standard deviations above and below the average value, although other values are sometimes used. Usually, historical data are used to provide a "baseline" for these lines.

A different version of this type of control chart uses "rational subgroup" number for the abscissa ("x-axis") of a control chart. A rational subgroup is a collection of typically three to six measurements such that within this subgroup, variability is due to chance only but between these subgroups, variations due to a process issue might be detected. Data in a rational subgroup are usually taken over a short time period and under essentially the same conditions. The ordinate ("y-axis") of the control chart is the mean value for the members of the subgroup. This type of control chart also has a horizontal line drawn at the average value for the samples and two other horizontal lines used as control lines.

When rational subgroups are used, a second-type control chart is often created. For each subgroup, we calculate the range (maximum value minus minimum value) of the subgroup samples. As with the control chart for averages, there are three horizontal lines correspond-ing to the range: the average and the two control lines. There are a variety of other control charts, and their applicability depends on project objectives and the nature of the data to be analyzed. Control charts are used for data associated with what is assumed to be a stable process and the metrics used for the charts need to be collected "in real time."

It is important to choose metrics applicable to SPC. As an example, [87] recommends the following metrics for peer reviews:

1. Review speed: The size of the reviewed product divided by the number of hours spent reviewing it
2. Defect density: The number of defects found in the review divided by the size of the reviewed product
3. Early bug detection rate: The number of defects found in the peer review divided by the total number of defects found (evaluated at a set point in the project)
4. Review efficiency: The number of defects found in the peer review divided by the time spent on the review (or some other measure of review effort)

See [71, 77, 78], and [88] for more details on control charts.

It should be noted that SPC may be less effective when applied to software develop-ment than when used for its traditional manufacturing application. For example, [89] states that the assumptions underlying control charts are often violated when applied to software development. Some of these violated assumptions include using too little data, using a sin-gle control chart for data from multiple common-cause systems, and using control limits that are too wide to be useful.

Whether control charts are used or not, statistical analysis can still be applied to software metrics. Techniques such as regression analysis, various techniques for multivariate sta-tistical analysis, tests for trends, Taguchi methods, and tests of randomness, among other techniques, may provide insights and advanced warnings of problems. However, any time that statistical analysis is used, care must be taken to ensure that the data meet the assump-tions of the technique used or that the technique is sufficiently robust to the assumptions that the analysis results are correct. See [90–92], and [78] for more details.

Most control charts monitor a single variable. If a process has several metrics measuring its performance, we can construct a control chart for each metric and see if any of them indicate a problem. However, such an approach may miss problems that a multivariate technique can discover from the same data. In these cases, multivariate statistical analysis techniques are called for.

Suppose that we are taking metrics on the performance of a process, such as peer reviews of code. We take p metrics for each peer review and we perform a peer review on each of m software LRUs. More generally, we let m be the number of samples and let p be the number of variables obtained from each sample. We make a column p-vector of the peer review results for each LRU and find the sample mean vector μ for the m such vectors. We also find the associated sample covariance matrix S for these vectors. The statistics that we calculate is called the Hotelling T^2 statistics and is given by

$$T_i^2 = (x_i - \mu)^\mathsf{T} S^{-1}(x_i - \mu)$$

where x_i is the p-vector of the peer review results for the ith LRU and i varies from $i = 1$ to $i = m$.

We also need a way of knowing if a given value of T_i^2 is suspicious. Let $\beta_{1-\alpha,\, a,\, b}$ denote the value of a beta distribution with parameters a and b that gives a confidence level of $1 - \alpha$. To find the upper control limit (UCL) and lower control limit (LCL) for this statistic, use

$$\text{UCL} = \frac{(m-1)^2}{m}\, \beta_{1-\alpha/2,\, p/2,\, (q-p-1)/2}$$

and

$$\text{LCL} = \frac{(m-1)^2}{m}\, \beta_{\alpha/2,\, p/2,\, (q-p-1)/2}$$

where

$$q = \frac{2(m-1)^2}{3m-4}$$

When a one-sided test with only the UCL is needed (and the lower bound is zero), use

$$\text{UCL} = \frac{(m-1)^2}{m}\, \beta_{1-\alpha,\, p/2,\, (q-p-1)/2}$$

Expressed this way, α is the desired in-control false alarm probability (FAP) for each observation. If we want a desired overall in-control FAP of α_0 for all m observations, we find α for the UCL and LCL using

$$\alpha = 1 - (1 - \alpha_0)^{\frac{1}{m}}$$

Note that $1 - \alpha_0$ is the probability that there are no false alarms in m observations, whereas $1 - \alpha$ is the probability that a given observation is not a false alarm. This calculation assumes that the observations are independent.

Example 6.5.11 As an example of this use of multivariate statistics, we consider two metrics for code peer reviews. The first is the preparation time for the peer review measured in minutes. The second metric is review time for the peer review. This metric is also measured in minutes. Both metrics are on a per-reviewer basis. Table 6.27 shows the measured metrics for the 20 LRUs.

If we were to plot the preparation time or the review time versus the LRU number, there is no obvious outliers. Using upper and lower control lines provides a more rigorous approach for this assessment. The sample mean for preparation review time is 112.00 minutes and 194.95 minutes for review time. Their respective standard deviation values are 5.12 and 38.61, giving 2 σ upper bounds of 122.24 and 272.17, respectively, and the associated lower bounds are 101.76 and 117.73. The maximum and minimum values for preparation time are 120 and 105, respectively, and for review time are 238 and 118, so none of the metrics are outside of 2 σ bounds.

The list of T^2 values versus LRU numbers show that T_{13}^2 appears to be a significant outlier. Again, using a UCL is a more rigorous approach. A 2 σ bound is close to a 95% UCL for the T^2 test, so for this test, the UCL for all of the observations is found using

$$\alpha = 1 - (1 - 0.05)^{1/20} = 0.002561$$

Table 6.27 Code Review Metrics.

LRU Number	Preparation Time	Review Time	T^2 Value
1	109	218	1.76
2	119	238	2.00
3	105	118	4.01
4	111	212	0.53
5	106	175	1.43
6	106	170	1.38
7	110	180	0.19
8	114	206	0.16
9	115	222	0.53
10	108	166	0.73
11	116	223	0.71
12	115	217	0.42
13	118	118	12.83
14	107	173	0.95
15	105	160	1.88
16	116	237	1.21
17	115	233	0.97
18	120	238	2.49
19	107	160	1.11
20	118	235	1.54

With this value of α and

$$q = \frac{2(m-1)^2}{3m-4} = \frac{2\,(20-1)^2}{60-4} = 12.89$$

$$\begin{aligned} \text{UCL} &= \frac{(m-1)^2}{m}\, \beta_{1-\alpha,\ p/2,\ (q-p-1)/2} \\ &= \frac{(20-1)^2}{20}\, \beta_{1-0.002561,\ 2/2,\ (12.89-2-1)/2} \\ &= 12.65 \end{aligned}$$

Notice that $T_{13}^2 = 12.83 > 12.65$, so the T^2 test highlights this potentially anomalous sample. With further investigation, it appears that the review time for LRU 13 may have been too short. However, the review time for LRU 3 is also 118 minutes but $T_3^2 = 4.01$ rather than 12.83. The difference is in the preparation time. For LRU 3, it is 105, and for LRU 13, it is 118. This may not seem like a large difference, but when we subtract the mean of the preparation review time, the first component (preparation time) of $x_3 - \mu$ is -7, and for $x_{13} - \mu$, it is 6. The value of -7 for the first component for LRU 3 is one of the lower such values, whereas the value for LRU 13, 6, is one of the higher values. It is the combination of the two components that makes LRU 13 stand out differently. This result demonstrates that as a multivariate test, the T^2 test considers not just the individual variables but also the relation between the variables.

Style Guides

Style guides are used to promote uniformity in products. For example, they have been used successfully to promote the production of a readable code and a code that is consistent across a project. They also have been used to aid in the production of a consistent set of requirements. Style guides are something of a "double-edged sword." A software developer may be very comfortable and successful with certain style conventions, and a project's style guide may prevent some of those conventions, resulting in added hardship for that particular software developer. Although a software developer may object to the loss of a certain amount of freedom because of the style guide, it also reduces the unnecessary distractions that making style choices can cause and as a result, a style guide enables the software developer to focus on code functionality. In addition, it creates uniformity across the project, making tasks such as peer reviews, fault detections, and software maintenance easier.

Trend Tests

There are several statistical tests for data trends. These tests can be useful for assessing if the reliability of the software is improving, if the number of detected defects in a process is decreasing, or if there is a trend in some other quantitative metric. We briefly cover three such tests below. See Chapter 10 of [17] for more details.

1. Laplace test statistic: The Laplace test statistic is useful for detecting and quantifying trends in time series data of events, particularly if the underlying statistics of the data series follows a Poisson process. The Poisson process may be homogeneous or non-homogeneous. For this test, we generally use a sequence of arrival times of events. The occurrences of events are discrete, such as the occurrence of software failures in some system, and all arrival times are measured from some fixed starting time.

   Assume that we have collected failure data from software runs and want to know if we can have confidence that the software is trending toward fewer failures. Given that all of the data represents essentially the same OP and that failures occur based on a Poisson process $\{N(t) : t \geq 0\}$, the probability that by time t there have been exactly k failures be given by

$$P(N(t) = k) = \frac{\left(\lambda\, t^\beta\right)^k e^{-\lambda\, t^\beta}}{k!}$$

   with parameters $\lambda > 0$ and $\beta > 0$. The variable $t \geq 0$ is time and $N(t)$ is the number of failures that have occurred by time t. If $\beta = 1$, then the process is a homogeneous Poisson process. Otherwise, the distribution is non-homogeneous.

   Given data from such a process, say $t_1, t_2, \ldots, t_n$, where each t_i is an event or failure arrival time (not an inter-arrival time), and also given an end of observation time of $T > t_n$, we calculate

$$z = \frac{\sqrt{12}}{\sqrt{n}\, T} \sum_{i=1}^{n} \left(t_i - \frac{T}{2}\right) = \frac{\sqrt{12\, n}}{T} \left(\frac{\sum_{i=1}^{n} t_i}{n} - \frac{T}{2}\right) \tag{6.7}$$

If $T = t_n$, then use $n - 1$ rather than n in all five places for these expressions. If z is approximately equal to zero, then the sequence of inter-arrival times (the durations between the events) is unlikely to be either increasing or decreasing. If $z > 0$, then the inter-arrival times are tending to decrease, and if $z < 0$, the inter-arrival times are tending to increase. The statistic z approximately follows a standard normal distribution. So, for example, let our null hypothesis be that the process that we have obtained the data from is a homogeneous Poisson process and the alternate hypothesis be that it is an NHPP. If we choose a significance level of 0.1, then our bounds of interest are -1.645 and 1.645 because

$$0.05 = \frac{1}{\sqrt{2\pi}} \int_{-\infty}^{-1.645} \exp\left(-\frac{x^2}{2}\right) dx$$

and

$$0.05 = \frac{1}{\sqrt{2\pi}} \int_{1.645}^{\infty} \exp\left(-\frac{x^2}{2}\right) dx$$

or to put it another way,

$$0.9 = \frac{1}{\sqrt{2\pi}} \int_{-1.645}^{1.645} \exp\left(-\frac{x^2}{2}\right) dx$$

Therefore, if our test statistic z is such that $-1.645 < z < 1.645$, we cannot reject the null hypothesis that we have a homogeneous process at a significant level of $\alpha = 0.05 + 0.05 = 0.1$. This means that we cannot reject the hypothesis that there is no trend and that the inter-arrival times are neither increasing nor decreasing. If $z < -1.645$, we can assume (at our 0.1 significance level) that there is a trend of improving failures, meaning that the inter-arrival times are increasing. If, however, $z > 1.645$, then there is a deteriorating trend and the inter-arrival times are decreasing. Note that if the process is not a Poisson process, the Laplace test statistic may still work, but the further the process is from a Poisson process, the less confidence we can have in the test results. More on the Laplace trend test may be found in Chapter 18 of [19] and Chapter 10 of [17].

Example 6.5.12   As an example of the use of the Laplace test, suppose that we run tests on three software modules and record how long each test runs until there is a failure. We then collect the failure data and restart the test, resulting in a sequence of failure inter-arrival times. As we are simultaneously running tests on three modules, we have three such sequences as shown in the first three columns of Table 6.28. In this example, the first failure event for the first software module occurs after 12 000 time units, the second failure occurs 8000 time units later, and so forth. The second and third columns are two other sets of inter-arrival times and are for the second and third software modules, respectively. Notice that in this example, each of the three columns has the same inter-arrival numbers but each is with a different ordering. The next three columns of the table, columns 4–6, are the arrival times for the failures associated with the first three columns. The fourth column is the collection of arrival times for the first data set. The first value in this column is the time of the first failure arrival time for the first software module, the second value is 8000 time units after the first arrival, *i.e.* $12\,000 + 8000 = 20\,000$ units after the starting time, and so forth. The fifth and sixth

Table 6.28 Laplace Test Statistic Example.

Inter-Arrival Set 1	Inter-Arrival Set 2	Inter-Arrival Set 3	Arrival Time Set 1	Arrival Time Set 2	Arrival Time Set 3
12 000	2000	1000	12 000	2000	1000
8000	12 000	2000	20 000	14 000	3000
4000	1000	4000	24 000	15 000	7000
2000	4000	8000	26 000	19 000	15 000
1000	8000	12 000	27 000	27 000	27 000

columns give arrival times for the second and third data sets, respectively. The fourth, fifth, and sixth columns are the t_i values used in the Laplace test.

If we assume that we stop data collection at 30 800 time units and using Eq. 6.7, the z value for the first data set is 1.61, for the second is 0.00, and for the third is -1.21. We compare these values with critical values for a standard normal distribution. These critical values z_α are found from

$$\alpha = \frac{1}{\sqrt{2\pi}} \int_{-\infty}^{-z_\alpha} \exp\left(-\frac{x^2}{2}\right) dx$$

or

$$\alpha = \frac{1}{\sqrt{2\pi}} \int_{z_\alpha}^{\infty} \exp\left(-\frac{x^2}{2}\right) dx$$

For $\alpha = 0.1$ and a two-sided test, we cannot reject the null hypothesis of no trend if $-1.645 < z < 1.645$ because

$$0.05 = \frac{1}{\sqrt{2\pi}} \int_{-\infty}^{-1.645} \exp\left(-\frac{x^2}{2}\right) dx$$

and

$$0.05 = \frac{1}{\sqrt{2\pi}} \int_{1.645}^{\infty} \exp\left(-\frac{x^2}{2}\right) dx$$

so

$$1 - (0.05 + 0.05) = 0.9 = \frac{1}{\sqrt{2\pi}} \int_{-1.645}^{1.645} \exp\left(-\frac{x^2}{2}\right) dx$$

(Note that $-1.645 < z < 1.645$ corresponds to $-z_{\alpha/2} < z < z_{\alpha/2}$, where $\alpha = 0.1$.) Comparing these values with $z = 1.61$ for the first data set, we see that we cannot reject the null hypothesis (that we have a homogeneous Poisson process) with a 90% confidence. If we choose an 80% confidence, then we can reject the null hypothesis because $z > 1.282$ and the test indicates that at this level of significance, there is a decreasing inter-arrival time trend. Note that

$$1 - (0.1 + 0.1) = 0.8 = \frac{1}{\sqrt{2\pi}} \int_{-1.282}^{1.282} \exp\left(-\frac{x^2}{2}\right) dx$$

The second data set has a z value of 0.00, so there is no indication of a trend at any level of confidence. Finally, for the third data set, $z = -1.21$, potentially indicating an increasing inter-arrival time trend. For a 60% confidence, we have

$$1 - (0.2 + 0.2) = 0.6 = \frac{1}{\sqrt{2\pi}} \int_{-0.842}^{0.842} \exp\left(-\frac{x^2}{2}\right) dx$$

so we can have a 60% confidence that the inter-arrival times in the third data set are increasing because $z = -1.21 < -0.842$, but we cannot justify an 80% confidence as $z = -1.21 > -1.282$.

2. Mann–Kendall test: The Mann–Kendall test is a non-parametric test, meaning that the test does not assume a specific distribution for the data. The test looks for potentially nonlinear trends. Like the Laplace test, a sequence is used. The Mann–Kendall test assumes that if there is no trend, then the values in the sequence are independent and identically distributed.

For this test, the null hypothesis, H_0, is that there is no monotonic trend. The alternate hypothesis, H_1, is that a monotonic trend is present. Steps for performing the test are as follows:

(a) List the data in the order collected. Denote this sequence by $X_1, X_2, \ldots, X_n$.

(b) Calculate S, where

$$S = \sum_{k=1}^{n-1} \sum_{j=k+1}^{n} \text{sgn}\left(X_j - X_k\right) \tag{6.8}$$

and

$$\text{sgn}(x) = \begin{cases} 1 & \text{if } x > 0 \\ 0 & \text{if } x = 0 \text{ and} \\ -1 & \text{if } x < 0 \end{cases}$$

(c) Determine the number of tied values, i.e. values in the sequence that are repeated. Let p be the number of groups of tied values and let t_i be the number of tied values in the ith such group.

(d) Compute the variance of S:

$$\text{Var}(S) = \frac{1}{18}\left(n(n-1)(2n+5) - \sum_{i=1}^{p} t_i(t_i - 1)(2t_i + 5)\right) \tag{6.9}$$

(e) Compute the Mann–Kendall statistic Z:

$$Z = \begin{cases} (S-1)/\sqrt{\text{Var}(S)} & \text{if } S > 0 \\ 0 & \text{if } S = 0 \text{ and} \\ (S+1)/\sqrt{\text{Var}(S)} & \text{if } S < 0 \end{cases} \tag{6.10}$$

If $Z > 0$, then the data tend to increase as the sequence progresses. If $Z < 0$, they tend to decrease; however, in either case, we still need to determine if the trend is statistically significant.

(f) To test significance, we have three cases:

  i. Increasing trend: Let the null hypothesis be that there is no trend and the alternate hypothesis be that the data are increasing. Let the type I error rate be $0 < \alpha < 0.5$. This is the probability of falsely rejecting the null hypothesis. Let $z_{1-\alpha}$ be the $100(1-\alpha)$th percentile for the standard normal distribution. For example, if $\alpha = 0.05$, then $z_{1-\alpha}$ is the 95th percentile for the standard normal distribution, which is approximately 1.645. This is the value that is greater than 95% of the values taken from a standard normal distribution. We reject the null hypothesis H_0 if $Z \geq z_{1-\alpha}$.

  ii. Decreasing trend: Next, let the null hypothesis be that there is no trend and the alternate hypothesis be that the data are decreasing. Again let the type I error rate be $0 < \alpha < 0.5$ and let $z_{1-\alpha}$ be the $100(1-\alpha)$th percentile for the standard normal distribution. In this case, we reject the null hypothesis H_0 if $Z \leq -z_{1-\alpha}$.

  iii. Increasing or decreasing trend: Finally, let the null hypothesis be that there is no trend and the alternate hypothesis be that the data are trending either upward or downward. As before, $0 < \alpha < 0.5$ is the type I error rate. We reject the null hypothesis H_0 if $|Z| \geq z_{1-\alpha/2}$.

These steps assume that $n > 10$. If this is not the case, the standard normal distribution is not adequate, and there are tables with the appropriate values to use. Also, this test is less accurate if there are significant periodicities in the data. For such a situation, remove the periodicities before using the test.

Example 6.5.13 As an example of the use of the Mann–Kendall test, we again consider tests run on a software module and record how long each test runs until there is a failure. We collect the failure data and then restart the clock at zero and begin a new test run. The result is a sequence failure inter-arrival times. Table 6.29 lists the inter-arrival times for failures for this example. The first column of the table lists the run numbers and the second column lists the corresponding inter-arrival time for the run. There are a total of twelve runs in this example.

We first use Eq. 6.8 to find that $S = -4$. There are no tied values, so $p = 0$, and there is no need to find any t_i values. Using Eq. 6.9, we have that $Var(S) = 212.6667$. We then use Eq. 6.10 to determine that $Z = -0.2057$.

Suppose that we want to have a 95% confidence in our decision if we reject the null hypothesis that there is no trend. Since we have $n > 10$ runs, we can compare Z with critical values for a standard normal distribution. For a 95% confidence, $\alpha = 0.05$ and we compare $|Z| = 0.2057$ with $Z_{1-\alpha/2} = Z_{0.975} = 1.9600$. Since $Z_{0.975} = 1.9600 > 0.2057 = |Z|$, we cannot reject the null hypothesis that there is no trend with a 95% confidence.

3. Spearman's rank correlation coefficient: The Spearman's rank correlation coefficient ρ is a measure of the rank correlation between two data sets. Rank correlation is a measure of ordinal association. By comparing the observed values with the associated observation times, ρ can be used as a statistical test for a data trend. More specifically, we collect ordered pairs of data, $\{(\tau_1, y_1), (\tau_2, y_2), \ldots, (\tau_n, y_n)\}$, where data sample y_1 is collected at time τ_1, sample y_2 is collected at time τ_2, and so on. We only use the rank information about the data, so that if y_k is the smallest y value and y_i is the second smallest, then

Table 6.29 Mann–Kendall Test
Statistic Example.

Run Number	Inter-Arrival Times
1	14986
2	1883
3	10871
4	7705
5	3118
6	19438
7	9435
8	16521
9	1692
10	8170
11	7690
12	11155

$\text{rank}(y_k) = 1$, $\text{rank}(y_i) = 2$, and so on until the largest value. We want to determine if the y values are trending with the τ values.

To make this determination, we set the problem up as a hypothesis test where the null hypothesis is that there is no trend and the alternate hypothesis is either that there is a trend, or that there is an increasing trend, or that there is a decreasing trend. We then use the *rank-correlation coefficient* or *Spearman's rank correlation coefficient*. Let $\text{rank}(\tau_k)$ be the rank of τ_k and similarly for $\text{rank}(y_k)$. Define the rank correlation coefficient as

$$r' = 1 - \frac{6}{n(n^2 - 1)} \sum_{k=1}^{n} (\text{rank}(\tau_k) - \text{rank}(y_k))^2 \tag{6.11}$$

If the samples are statistically independent, this coefficient r' is asymptotically normal (*i.e.* as $n \to \infty$, the distribution of r' approaches a normal distribution) with zero mean and a variance of $1/(n-1)$. If $n > 20$ (some sources say $n > 10$), we can assume that the test statistic follows a Student's t-distribution with $n - 2$ degrees of freedom. Once r' has been found, find t by

$$t = \frac{r' \sqrt{n-2}}{\sqrt{1 - (r')^2}} \tag{6.12}$$

The random variable t follows a t-distribution with $n - 2$ degrees of freedom. The p-value for the test is the probability $P(t > u)$ that a random variable t that follows a t-distribution with $n - 2$ degrees of freedom will have a value of u or greater. If the p-value is close to one, any observed correlations are likely due to chance and we cannot reject the null hypothesis. If the p-value is close to zero, any observed correlations are not likely to be due to chance and we can reject the null hypothesis. Our null hypothesis is that there is no trend. The smaller the p-value, the less justification we have for assuming the null hypothesis. The p-value can be found from look-up table or with the help of computer programs. If we reject the null hypothesis, then $r' > 0$ provides evidence for an increasing trend and $r' < 0$ provides evidence for a decreasing trend.

For values of $n \leq 20$, tables can be used. For larger values of n (usually $n > 30$), we can create the zero mean, unit variance normal random variable

$$Z = \frac{r'}{1/\sqrt{n-1}} = r'\sqrt{n-1}$$

and test for significance using the standard normal distribution. Therefore, under the null hypothesis, r' should be near zero. If r' is close to 1, the y values are trending with the τ values (i.e. there is an increasing trend), and if r' is close to -1, there is a decreasing trend. More on Spearman's rank correlation coefficient may be found in [93, 94], and [95].

Table 6.30 Spearman's Rank Correlation Coefficient Example.

Run Number	Inter-Arrival Times y_k	Rank(y_k)	d_k^2
1	953	2	1
2	8367	5	9
3	121	1	4
4	9956	9	25
5	8956	6	1
6	9844	8	4
7	11766	12	25
8	10149	10	4
9	15262	18	81
10	2921	3	49
11	17234	20	81
12	9791	7	25
13	12012	13	0
14	8248	4	100
15	19552	22	49
16	11544	11	25
17	12076	14	9
18	16153	19	1
19	14315	15	16
20	14572	16	16
21	19141	21	0
22	14990	17	25

Example 6.5.14   As an example of the use of the Spearman's rank correlation coefficient to test for trends, we again consider tests run on a software module and record how long each test runs until there is a failure. The result is a sequence of inter-arrival times for failures. Table 6.30 lists the inter-arrival times for failures for this example. The first column of the table lists the run numbers, which we treat as our τ values, and the second column lists the corresponding inter-arrival times for the runs. The third column lists the rank of the inter-arrival times where the smallest inter-arrival time has rank 1, the second smallest has rank 2, and so on until the largest inter-arrival time which has rank 22. For this example, there are a total of 22 test runs. Finally, the fourth column lists the values of $d_k^2 = (\text{rank}(\tau_k) - \text{rank}(y_k))^2$ used in Eq. 6.11 to find r'.

Using Eq. 6.11 with the data in Table 6.30, we find that $r' = 0.6894$, potentially indicating a positive correlation and an increasing trend for the inter-arrival times. We have 22 samples, so we cannot justify using a standard normal distribution but can use a Student's t-distribution of degree $n - 2 = 20$. Using Eq. 6.12, we have that $t = 4.2567$.

Next, we find the p-value for the test. This can be done with look-up tables or computer programs. We use a computer program to find a p-value of 0.000193 for a one-tailed analysis with $n - 2 = 20$ degrees of freedom. This value is quite small, giving us strong justification to reject the null hypothesis (that there is no trend) and accept the alternate hypothesis that there is an increasing trend in the inter-arrival times.

References

1 ISO 16355-1. *Application of statistical and related methods to new technology and product development process - Part 1: General principles and perspectives of quality function deployment (QFD)*, 2021. International organization of standards, Geneva, Switzerland.

2 Kossiakoff A, Sweet W, Seymour S, Biemer S. *System engineering principles and practice.* Wiley series in system engineering and management. John Wiley and Sons Publishing, Hoboken, NJ, 2011.

3 Hook I. Writing good requirements. *Proceedings of the 3rd international symposium of the NCOSE*, 2, 1993. Available via https://reqexperts.com/wp-content/uploads/2015/07/writing_good_requirements.htm. Accessed 22 Aug 2020.

4 Alshazly A, Elfatatry A, Abougabal M. Detecting defects in software requirements specifications. *Alexadria Eng. J.*, 53(3):513–527, 2014.

5 Nuseibeh B, Easterbrook S. Requirements engineering: A roadmap. *ICSE proceedings of the conference on the future of software engineering*, pages 35–46, 2000. Available via https://www.cs.toronto.edu/ sme/papers/2000/ICSE2000.pdf. Accessed 22 Aug 2020.

6 Cohn M. *User stories applied: For agile software development.* Addison-Wesley Professional, Boston, MA, 2004.

7 Data Item Description DI-EDRS-82219. *Algorithm description document*, 2018. US Department of Defense.

8 Bachmann, F., Bass, L., Clements, P., Garlan, D., Ivers, J., Little, M., Merson, P., Nord, R., Stafford, J. *Documenting software architectures: Views and beyond*, 2nd edition. Addison-Wesley Professional, 2010.

9 Kazman R, Klein M, Clements P. *ATAM^{SM}: Method for architecture evaluation*. Technical report CMU/SEI-2000-TR-004, ESC-TR-2000-004, Carnegie Mellon Software Engineering Institute, Pittsburgh PA, 2000. Available via https://resources.sei.cmu.edu/asset_files/TechnicalReport/2000_005_001_13706.pdf. Accessed 22 Aug 2020.

10 Shore J, Warden S. *The art of agile development*. O'Reilly Media Inc., Sebastopol, CA, 2008.

11 McConnell S. *Code complete: A practical handbook of code construction*. Microsoft Press, Washington, DC, 2004.

12 Military Handbook 338B. *Electronic reliability design handbook*, 1998. US Department of Defense.

13 Fitzgerald B, Stol K-J. Continuous software engineering and beyond: Trends and challenges. *Proceeding of the 1st international workshop on rapid continuous software engineering*, pages 1–9, 2014. Available via https://www.researchgate.net/publication/260338818_Continuous_Software_Engineering_and _Beyond_Trends_and_Challenges. Accessed 22 Aug 2020.

14 Laukkanen E, Itkonen J, Lassenius C. Problems, causes and solutions when adopting continuous delivery - a literature review. *Inf. Softw. Technol.* 82:55–79, 2017. Available via http://dx.doi:10.1016/j.infsof.2016.10.001, via Elsevier Science Direct. Available via https://www.sciencedirect.com/science/article/pii/S0950584916302324. Accessed 22 Aug 2020.

15 Scaled Agile Inc. Continuous delivery pipeline. Available via http://www.scaledagileframework.com /continuousdeliverypipeline/ Accessed 5 July 2021.

16 Kapur P.K., Pham H., Gupta A., Jha P.C. *Software reliability assessment with OP applications*. Springer-Verlag, London, 2011.

17 Lyu M, editor. *Handbook of software reliability engineering*. Computer Society Press and McGraw-Hill Book Company, New York, 1996.

18 Lyu M. Software reliability engineering: A roadmap. *Future of software engineering 2007*, pages 153–170, 2007. Available via https://www.researchgate.net/publication/4250863_Software_Reliability_Engineering _A_Roadmap. Accessed 22 Aug 2020.

19 Pham H, editor. *Handbook of reliability engineering*. Springer-Verlag, London, 2003.

20 Pham H, editor. *Handbook of software reliability engineering*. Springer-Verlag, London, 2006.

21 Pullum L. *Software fault tolerance: Techniques and implementations*. Artech House, Boston, MA, 2001.

22 Torres-Pomales W. *Software fault tolerance: A tutorial*. Technical report, Langley Research Center, Virginia, 2000. Available via https://ntrs.nasa.gov/citations/20000120144. Accessed 22 Aug 2020.

23 Etzhorn L, Delugach H. Towards a semantic metric suite for object-oriented design. *Proceedings on the 34th international conference on technology of object-oriented languages and systems*, pages 71–80, 2000. Available via https://ieeexplore.ieee.org/document/868960. Accessed 22 Aug 2020.

24 Helali R. Software sematic metrics: A survey. *Int. J. Adv. Res. Comput. Commun. Eng.*, 4(2), February 2015. Available via https://www.asee.org/documents/zones/zone1/2014/Student/PDFs/133.pdf. Accessed 22 Aug 2020.

25 IEEE Standard 1633. *Recommended practice on software reliability*, 2017. Software Engineering Technical Committee of the IEEE Computer Society.

26 US Department of Defense. DoD guide for achieving reliability, availability, and maintainability. 2005. Available via http://www.acq.osd.mil/sse/docs/RAM_Guide_080305 .pdf, OUSD(AT&L) DS/ED Accessed 22 Aug 2020.

27 Military Handbook 189C. *Reliability growth management*, 2011. US Department of Defense.

28 Military Handbook 781A. *Handbook of reliability test methods*, 1996. US Department of Defense.

29 Kuhn R, Chandramouli R. Cost effective use of formal methods in verification and validation. *Foundations verification and validation workshop*, 2002.

30 Bertolino A. Software testing research: Achievements, challenges, dreams. *Future of software engineering 2007, IEEE-CS Press*, 2007. Available via https://www.cs.drexel.edu/ spiros/teaching/CS576/papers/fose07-testing.pdf. Accessed 22 Aug 2020.

31 Neufelder A. *Ensuring software reliability*. Marcel Dekker, Inc., New York, 1993.

32 Anderson P. The use and limitations of static-analysis tools to improve software quality. *Crosstalk, J. Def. Softw. Eng.*, 21(6), June 2008. Available via https://www .semanticscholar.org/paper/The-Use-and-Limitations-of-Static-Analysis-Tools-to-Anderson/3b8ea53e62bfc753dabf1deb83be7572d069f6b1. Accessed 22 Aug 2020.

33 Vladu A. Software reliability prediction model using Rayleigh function. *U.P.B. Sci. Bull., Ser. C*, 73(4), 2011. Available via https://www.scientificbulletin.upb.ro/rev_docs_arhiva/ full18273.pdf. Accessed 22 Aug 2020.

34 Software Productivity Consortium Services Corporation. Software error estimation program (SWEEP) user manual, SPC-93017-CMC. Available via https://apps.dtic.mil/dtic/tr/ fulltext/u2/a274697.pdf/ Accessed 5 July 2021.

35 Military Standard 2155. *Failure reporting, analysis and corrective action system (FRACAS)*, 1985. US Department of Defense.

36 Koziolek H. Operational profiles for software reliability. *Seminar "Dependability Engineering", Summer term 2005, Carl von Ossiesky University of Oldenburg, DFG Graduate school "TrustSoft"*, 2005. Available via https://sdqweb.ipd.kit.edu/heiko/Koziolek2005c .pdf. Accessed 26 Mar 2021.

37 Chillarege R, Bhandari I, Chaar J, Halliday M, Moebus D, Ray B, Wong M-Y. Orthogonal defect classification - a concept for in-process measurements. *IEEE Trans. Softw. Eng.*, 18(11), November 1992. Available via http://www.chillarege.com/articles/odc-concept. Accessed 22 Aug 2020.

38 Kusiuk A, Larson N. *System reliability and risk assessment: A quantitative extension of IDEF methodologies*. AAAI technical report SS-94-04, 1994. Available via https://www .aaai.org/Papers/Symposia/Spring/1994/SS-94-04/SS94-04-018.pdf. Accessed 22 Aug 2020.

39 IEC Standard 61025. *Fault tree analysis*, 2006. International Electrotechnical Commission, Geneva, Switzerland.

40 Military Standard 1629A. *Procedure for performing a failure modes, effects, and criticality analysis*, 1980. US Department of Defense.

41 Houston D, Buettner D, Hecht M. Dynamic COQUALMO: Defect profiling over development cycles. *International conference on software process, Trustworthy development* *software processes*, pages 161–172, 2009. Available via https://link.springer.com/chapter/ 10.1007/978-3-642-01680-6_16. Accessed 22 Aug 2020.

42 Denson W, Keene S, Caroli J. New system reliability assessment methodology. *Reliability Analysis Center, RAC Project No. A06839*, 1997. Available via https://www.researchgate .net/scientific-contributions/2071053516-S-Keene. Accessed 22 Aug 2020.

43 Duane J. Learning curve approach to reliability monitoring. *IEEE Trans. Aerosp.*, 2(2):563–564, April 1964. Available via https://ieeexplore.ieee.org/document/4319640. Accessed 22 Aug 2020.

44 Jalote P, Murphy B. Reliability growth in software products. *Proceeding of 15th annual symposium on software reliability engineering, IEEE computer society*, 2004. Available via https://https://www.microsoft.com/en-us/research/wp-content/uploads/2016/02/tr-2004-144.pdf. Accessed 23 Dec 2020.

45 Ellner P, Hall J. *Planning model based on projection methodology (PM2)*. AMSAA technical report No. TR-2006-9, U.S. Army material system analysis activity, Aberdeen proving grounds, Maryland, 2006. Available via https://apps.dtic.mil/dtic/tr/fulltext/u2/a448130 .pdf. Accessed 22 Aug 2020.

46 Jalote P, Murphy B, Garzia M, Errez B. *Measuring reliability of software products*. MSR-TR-2004-145, Microsoft Corporation, Redmond Washington, 2004. Available via https://www.microsoft.com/en-us/research/wp-content/uploads/2016/02/tr-2004-145.pdf. Accessed 22 Aug 2020.

47 ANSI/GEIA-STD-0009. *Reliability program standard for systems design, development, and manufacturing*, 2008. Information Technology Association of America (ITAA).

48 Jelinski Z, Moranda P. *Software reliability research*. In: Freiberger Walter (ed.) *Statistical computer performance evaluation*. Academic Press, New York, 1972.

49 Goel A, Okumoto K. Time-dependent error-detection rate model for software reliability and other performance measures. *IEEE Trans. Reliab.*, R-28(3):206–211, 1979.

50 Musa J, Iannino B, Okumoto K. *Software reliability: Measurement, prediction, application*. McGraw-Hill, New York, 1987.

51 Gokhale S, Triveda K. Log-logistic software reliability growth model. *Proceedings IEEE high-assurance system engineering symposium*, pages 34–41, 1998. Available via https:// ieeexplore.ieee.org/document/73159.

52 Kenny G. Estimating defects in commercial software during operational use. *IEEE Trans. Reliab.*, 42(1):107–115, March 1993.

53 Musa J, Okumoto K. A logarithmic Poisson execution time model for software reliability measurement. *Proceedings of the 7th international conference on software engineering*, pages 230–238, March 1984. Available via https://citeseerx.ist.psu.edu/viewdoc/ summary?doi=10.1.1.111.2201 Accessed 21 May 2021.

54 Yamada S, Ohba M, Osaki S. S-shaped reliability growth modeling for software error detection. *IEEE Trans. Reliab.*, R-32(5):475–484, 1983.

55 Virene E. Reliability growth and its upper limit. *Proceedings of the 1968 annual symposium on reliability, IEEE catalog no. 68 C33-R*, pages 265–270, January 1968.

56 Ohba M. Inflection S-shaped software reliability growth model. *In*: Osaki S, Hatoyama Y (ed.) *Stochastic models in reliability theory, lecture notes in economics and mathematical systems*. Springer-Verlag, Berlin Heidelberg, vol. 235, pages 144–162, 1984.

57 Tsoularis A. Analysis of logistics growth models. *Res. Lett. Inf. Math. Sci.*, 2: 21–55, 2002.

58 AL-Saati N, Alabajee M. Selecting best software reliability growth models: A social spider algorithm based approach. *Int. J. Comput. Appl. (0975–8887)*, 181(8):16–24, August 2018. Available via https://arxiv.org/ftp/arxiv/papers/2001/2001.09924.pdf.

59 Littlewood B. Rationale for a modified Duane model. *IEEE Trans. Reliab.*, R-33(2):157–159, June 1984.

60 Yamada S, Tokuno K, Osaki S. Imperfect debugging model with fault introduction rate for software reliability assessment. *Int. J. Syst. Sci.*, 23(12): 2241–2252, 1992.

61 Pham H, Nordmann L, Zhang Z. A general imperfect-software-debugging model with s-shaped fault-detection rate. *IEEE Trans. Reliab.*, 48(2):169–175, July 1997.

62 Pham H, Zhang X. An NHPP software reliability model and its comparison. *Int. J. Qual. Saf. Eng.*, 14(3):269–282, 1997.

63 Zhang X, Teng X, Pham H. Considering fault removal efficiency in software reliability assessment. *IEEE Trans. Syst. Man Cybern. Part A Syst. Humans*, 33(1): 114–120, January 2003.

64 Pham H, Pham M. Software reliability models for critical applications, EGC-2883. *Idaho national engineering laboratory, EG and G Idaho*, December 1991. https://doi.org/10.2172/10105800 Accessed 21 Mar 2021.

65 Li Q, Pham H. A generalized software reliability growth model with consideration of the uncertainty of operating environment. *IEEE Access*, 7:84253–84267, June 2019. https://doi.org/10.1109/ACCESS2019.2924084 Available via https://ieeexplore.ieee.org/document/8742600. Accessed 12 March 2021.

66 Chow Y. *The Theory of Optimal Stopping*. Dover Publications, New York, 1991.

67 Huang C-Y, Kuo S-Y, Lyu M. An assessment of testing-effort dependent software reliabilty growth model. *IEEE Trans. Reliab.*, 56(2), June 2007. Available via https://ieeexplore.ieee.org/abstract/document/4220785. Accessed 3 April 2021.

68 Scott H, Wohlin C. Capture-recapture in software unit testing - a case study. *ESEM '08 proceedings of the 2nd ACM-IEEE international symposium on empirical software engineering and measurement*, pages 32–40, 2088. Available via https://dl.acm.org/doi/10.1145/1414004.1414012. Accessed 22 Aug 2020.

69 Briand L, Emam K, Freimut B, Laitenberger O. A comprehensive evaluation of capture-recapture models for evaluating software defect content. *IEEE Trans. Softw. Eng.*, 26(6):518–540, June 2000.

70 Military Handbook 189A. *Reliability growth management*, 2009. US Department of Defense.

71 George M, Rowlands D, Price M, Maxey J. *The lean six sigma pocket toolbox*. McGraw Hill, New York, 2005.

72 Hogg R, Craig A. *Introduction to mathematical statistics*, 4th edition. McMillan Publishing, New York, 1978.

73 Mittelhammer, RC, Judge, GG, Douglas JM. *Econometric foundations*. Cambridge University Press, USA, 2000.

74 Osborn A. *Applied imagination: Principles and procedures of creative thinking*. Scribners, New York, USA, 1957.

75 Saaty T. Relative measurement and its generalization in decision making why pairwise comparisons are central in mathematics for measurement of intangible factors. *Rev. R. Acad. Exact Phys. Nat. Sci., Ser. A Math.*, 102(2):251–318, 2008.

76 Taylor J, Love B. Simple multi-attribute rating technique for renewable energy deployment decisions. *J. Def. Model. Simul. Appl. Methodol. Technol.*, 11(3):227–232, 2014. Available via https://digitalcommons.unl.edu/pkifacpub/. Acessed 17 Oct 2021.

77 Mandel J. *The statistical analysis of experimental data*. Dover Publications, New York, 1984.

78 Ryan T. *Statistical methods for quality improvement*. John Wiley and Sons, New York, 1989.

79 IEEE Standard 1028. *IEEE standard for software reviews and audits*, 2008. Software and Systems Engineering Standards Committee of the IEEE Computer Society.

80 Fagan M. Design and code inspection to reduce errors in program development. *IBM J.*, 15(3), 1976. Available via https://link.springer.com/chapter/10.1007/978-3-642-59412-0_35. Accessed 22 Aug 2020.

81 Law A. *Simulation modeling and analysis*, 5th edition. McGraw-Hill Book Company, New York, 2015.

82 Bender E. *An introduction to mathematical modeling*. Dover Publications, New York, 2000.

83 Meyer W. *Concepts of mathematical modeling*. Dover publications, New York, 2004.

84 Capers J. *Software assessments, benchmarks, and best practices*. Addison-Wesley, Boston, MA, 2000.

85 Ishikawa K. *What is total quality control? The Japanese way*. Prentice Hall, Englewood Cliffs, NJ, 1985.

86 DOE. DOE-NE-STD-1004-92, root cause analysis. 1992.

87 Komuro M. Experiences of applying SPC techniques to software development processes. *International conference on software engineering: Proceedings of the 28th international conference on software engineering*, pages 577–584, 2006. Available via https://www.researchgate.net/publication/221556197_Experiences_of_applying_SPC_techniques_to_software_development_processes. Accessed 22 Aug 2020.

88 Shewhart W. *Statistical method from the viewpoint of quality control*. Dover Publications, New York, 1986.

89 Raczynski R, Curtis B. Point/Counterpoint: Software data violate SPC's underlying assumptions. *IEEE Softw.*, 25(3):49–51, May/June 2008. Available via https://ieeeexplore.ieee.org/document/4497763. Accessed 22 Aug 2020.

90 Rencher A. *Methods of multivariate analysis*. John Wiley and Sons, New York, 1995.

91 Manly B. *Multivariate statistical methods: A primer*. Chapman and Hall, New York, 1991.

92 Press S. *Applied multivariate analysis*. Dover Publications, New York, 2005.

93 Kokoska S, Zwillinger D. *Standard probability and statistics tables and formulae*. Chapman Hall, Boca Raton, FL, 2000.

94 Freund J. *Modern elementary statistic*, 3rd edition. Prentice-Hall, Englewood Cliffs, New Jersey, 1967.

95 Spiegel M. *Schaum's outline series: Theory and problems of probability and statistics*. McGraw-Hill, New York, 1975.

Index

Software Reliability Techniques for Real-World Applications, First Edition. Roger K. Youree.
© 2023 John Wiley & Sons Ltd. Published 2023 by John Wiley & Sons Ltd.